TEXTBOOKS in MATHEMATICS

ELEMENTARY NUMBER THEORY

James S. Kraft

Gilman School
Baltimore, Maryland, USA

Lawrence C. Washington

University of Maryland
College Park, Maryland, USA

CRC Press is an imprint of the
Taylor & Francis Group, an **informa** business

A CHAPMAN & HALL BOOK

TEXTBOOKS in MATHEMATICS

Series Editors: Al Boggess and Ken Rosen

PUBLISHED TITLES

ABSTRACT ALGEBRA: AN INQUIRY-BASED APPROACH
Jonathan K. Hodge, Steven Schlicker, and Ted Sundstrom

ABSTRACT ALGEBRA: AN INTERACTIVE APPROACH
William Paulsen

ADVANCED CALCULUS: THEORY AND PRACTICE
John Srdjan Petrovic

ADVANCED LINEAR ALGEBRA
Nicholas Loehr

ANALYSIS WITH ULTRASMALL NUMBERS
Karel Hrbacek, Olivier Lessmann, and Richard O'Donovan

APPLYING ANALYTICS: A PRACTICAL APPROACH
Evan S. Levine

COMPUTATIONS OF IMPROPER REIMANN INTEGRALS
Ioannis Roussos

CONVEX ANALYSIS
Steven G. Krantz

COUNTEREXAMPLES: FROM ELEMENTARY CALCULUS TO THE BEGINNINGS OF ANALYSIS
Andrei Bourchtein and Ludmila Bourchtein

DIFFERENTIAL EQUATIONS: THEORY, TECHNIQUE, AND PRACTICE, SECOND EDITION
Steven G. Krantz

DIFFERENTIAL EQUATIONS WITH MATLAB®: EXPLORATION, APPLICATIONS, AND THEORY
Mark A. McKibben and Micah D. Webster

ELEMENTARY NUMBER THEORY
James S. Kraft and Lawrence C. Washington

ELEMENTS OF ADVANCED MATHEMATICS, THIRD EDITION
Steven G. Krantz

EXPLORING LINEAR ALGEBRA: LABS AND PROJECTS WITH MATHEMATICA®
Crista Arangala

PUBLISHED TITLES CONTINUED

AN INTRODUCTION TO NUMBER THEORY WITH CRYPTOGRAPHY
James Kraft and Larry Washington

AN INTRODUCTION TO PARTIAL DIFFERENTIAL EQUATIONS WITH MATLAB®, SECOND EDITION
Mathew Coleman

INTRODUCTION TO THE CALCULUS OF VARIATIONS AND CONTROL WITH MODERN APPLICATIONS
John T. Burns

LINEAR ALGEBRA, GEOMETRY AND TRANSFORMATION
Bruce Solomon

THE MATHEMATICS OF GAMES: AN INTRODUCTION TO PROBABILITY
David G. Taylor

QUADRACTIC IRRATIONALS: AN INTRODUCTION TO CLASSICAL NUMBER THEORY
Franz Holter-Koch

REAL ANALYSIS AND FOUNDATIONS, THIRD EDITION
Steven G. Krantz

RISK ANALYSIS IN ENGINEERING AND ECONOMICS, SECOND EDITION
Bilal M. Ayyub

RISK MANAGEMENT AND SIMULATION
Aparna Gupta

TRANSFORMATIONAL PLANE GEOMETRY
Ronald N. Umble and Zhigang Han

CRC Press
Taylor & Francis Group
6000 Broken Sound Parkway NW, Suite 300
Boca Raton, FL 33487-2742

First issued in hardback 2019

© 2015 by Taylor & Francis Group, LLC
CRC Press is an imprint of Taylor & Francis Group, an Informa business

No claim to original U.S. Government works

ISBN-13: 978-1-4987-0268-3 (hbk)

This book contains information obtained from authentic and highly regarded sources. Reasonable efforts have been made to publish reliable data and information, but the author and publisher cannot assume responsibility for the validity of all materials or the consequences of their use. The authors and publishers have attempted to trace the copyright holders of all material reproduced in this publication and apologize to copyright holders if permission to publish in this form has not been obtained. If any copyright material has not been acknowledged please write and let us know so we may rectify in any future reprint.

Except as permitted under U.S. Copyright Law, no part of this book may be reprinted, reproduced, transmitted, or utilized in any form by any electronic, mechanical, or other means, now known or hereafter invented, including photocopying, microfilming, and recording, or in any information storage or retrieval system, without written permission from the publishers.

For permission to photocopy or use material electronically from this work, please access www.copyright.com (http://www.copyright.com/) or contact the Copyright Clearance Center, Inc. (CCC), 222 Rosewood Drive, Danvers, MA 01923, 978-750-8400. CCC is a not-for-profit organization that provides licenses and registration for a variety of users. For organizations that have been granted a photocopy license by the CCC, a separate system of payment has been arranged.

Trademark Notice: Product or corporate names may be trademarks or registered trademarks, and are used only for identification and explanation without intent to infringe.

Library of Congress Cataloging-in-Publication Data

Kraft, James S.
  Elementary number theory / James Kraft, Larry Washington.
    pages cm. -- (Textbooks in mathematics)
  "A CRC title."
  Includes bibliographical references and index.
  ISBN 978-1-4987-0268-3 (hardcover : alk. paper) 1. Number theory. I. Washington, Lawrence C. II. Title. III. Title: Number theory.

QA241.K727 2015
512.7'2--dc23   2014039473

Visit the Taylor & Francis Web site at
http://www.taylorandfrancis.com

and the CRC Press Web site at
http://www.crcpress.com

*To Kristi, Danny, and Aaron,
and to Miriam Kraft and the memory of Norman Kraft*

*To Susan and Patrick,
and to Ida H. Washington and Lawrence M. Washington*

Contents

Preface	xiii
Introduction	1
1 Divisibility	**9**
1.1 What Is a Proof?	9
1.1.1 Proof by Contradiction	15
1.2 Divisibility	17
1.3 Euclid's Theorem	19
1.4 Euclid's Original Proof	21
1.5 The Sieve of Eratosthenes	23
1.6 The Division Algorithm	25
1.6.1 A Cryptographic Application	26
1.7 The Greatest Common Divisor	27
1.8 The Euclidean Algorithm	30
1.8.1 The Extended Euclidean Algorithm	32
1.9 Other Bases	37
1.10 Fermat and Mersenne Numbers	40
1.11 Chapter Highlights	43
1.12 Problems	44
1.12.1 Exercises	44
1.12.2 Computer Explorations	50
1.12.3 Answers to "Check Your Understanding"	51
2 Linear Diophantine Equations	**55**
2.1 $ax + by = c$	55
2.2 The Postage Stamp Problem	60

	2.3	Chapter Highlights	64
	2.4	Problems	64
		2.4.1 Exercises	64
		2.4.2 Answers to "Check Your Understanding"	66

3 Unique Factorization — 67

- 3.1 Preliminary Results ... 67
- 3.2 The Fundamental Theorem of Arithmetic ... 69
- 3.3 Euclid and the Fundamental Theorem of Arithmetic ... 74
- 3.4 Chapter Highlights ... 75
- 3.5 Problems ... 75
  - 3.5.1 Exercises ... 75
  - 3.5.2 Answers to "Check Your Understanding" ... 76

4 Applications of Unique Factorization — 77

- 4.1 A Puzzle ... 77
- 4.2 Irrationality Proofs ... 79
- 4.3 The Rational Root Theorem ... 81
- 4.4 Pythagorean Triples ... 83
- 4.5 Differences of Squares ... 87
- 4.6 Chapter Highlights ... 89
- 4.7 Problems ... 90
  - 4.7.1 Exercises ... 90
  - 4.7.2 Computer Explorations ... 93
  - 4.7.3 Answers to "Check Your Understanding" ... 93

5 Congruences — 95

- 5.1 Definitions and Examples ... 95
- 5.2 Modular Exponentiation ... 102
- 5.3 Divisibility Tests ... 104
- 5.4 Linear Congruences ... 108
- 5.5 The Chinese Remainder Theorem ... 115
- 5.6 Fractions Mod m ... 119

	5.7	Queens on a Chessboard	121
	5.8	Chapter Highlights	123
	5.9	Problems	124
		5.9.1 Exercises	124
		5.9.2 Computer Explorations	130
		5.9.3 Answers to "Check Your Understanding"	130

6 Fermat, Euler, Wilson   133

	6.1	Fermat's Theorem	133
	6.2	Euler's Theorem	138
	6.3	Wilson's Theorem	145
	6.4	Chapter Highlights	147
	6.5	Problems	147
		6.5.1 Exercises	147
		6.5.2 Computer Explorations	150
		6.5.3 Answers to "Check Your Understanding"	151

7 Cryptographic Applications   153

	7.1	Introduction	153
	7.2	Shift and Affine Ciphers	156
	7.3	Vigenère Ciphers	160
	7.4	Transposition Ciphers	166
	7.5	RSA	169
	7.6	Stream Ciphers	176
	7.7	Block Ciphers	181
	7.8	Secret Sharing	184
	7.9	Chapter Highlights	187
	7.10	Problems	187
		7.10.1 Exercises	187
		7.10.2 Computer Explorations	195
		7.10.3 Answers to "Check Your Understanding"	196

8 Order and Primitive Roots   199

- 8.1 Orders of Elements 199
  - 8.1.1 Fermat Numbers 201
  - 8.1.2 Mersenne Numbers 203
- 8.2 Primitive Roots 203
- 8.3 Decimals 209
  - 8.3.1 Midy's Theorem 212
- 8.4 Card Shuffling 214
- 8.5 The Discrete Log Problem 216
- 8.6 Existence of Primitive Roots 218
- 8.7 Chapter Highlights 223
- 8.8 Problems 223
  - 8.8.1 Exercises 223
  - 8.8.2 Computer Explorations 227
  - 8.8.3 Answers to "Check Your Understanding" 227

9 More Cryptographic Applications   229

- 9.1 Diffie–Hellman Key Exchange 229
- 9.2 Coin Flipping over the Telephone 231
- 9.3 Mental Poker 233
- 9.4 Digital Signatures 238
- 9.5 Chapter Highlights 240
- 9.6 Problems 240
  - 9.6.1 Exercises 240
  - 9.6.2 Computer Explorations 243
  - 9.6.3 Answers to "Check Your Understanding" 243

10 Quadratic Reciprocity   245

- 10.1 Squares and Square Roots Mod Primes 245
- 10.2 Computing Square Roots Mod p 253
- 10.3 Quadratic Equations 255
- 10.4 The Jacobi Symbol 256
- 10.5 Proof of Quadratic Reciprocity 261

10.6	Chapter Highlights	268
10.7	Problems	268
	10.7.1 Exercises	268
	10.7.2 Answers to "Check Your Understanding"	273

11 Primality and Factorization   275

- 11.1 Trial Division and Fermat Factorization 275
- 11.2 Primality Testing 279
    - 11.2.1 Pseudoprimes 279
    - 11.2.2 Mersenne Numbers 284
- 11.3 Factorization 285
    - 11.3.1 $x^2 \equiv y^2$ 285
    - 11.3.2 Factoring Pseudoprimes and Factoring Using RSA Exponents 289
- 11.4 Coin Flipping over the Telephone 291
- 11.5 Chapter Highlights 293
- 11.6 Problems 293
    - 11.6.1 Exercises 293
    - 11.6.2 Computer Explorations 295
    - 11.6.3 Answers to "Check Your Understanding" 296

12 Sums of Squares   297

- 12.1 Sums of Two Squares 297
    - 12.1.1 Algorithm for Writing $p \equiv 1 \pmod 4$ as a Sum of Two Squares 300
- 12.2 Sums of Four Squares 301
- 12.3 Other Sums of Powers 306
- 12.4 Chapter Highlights 307
- 12.5 Problems 307
    - 12.5.1 Exercises 307

13 Arithmetic Functions   311

- 13.1 Perfect Numbers 311
- 13.2 Multiplicative Functions 315

13.3	Chapter Highlights	321
13.4	Problems	321
	13.4.1 Exercises	321
	13.4.2 Computer Explorations	324
	13.4.3 Answers to "Check Your Understanding"	325

14 Continued Fractions   327

14.1	Rational Approximations	328
14.2	Evaluating Continued Fractions	332
14.3	Pell's Equation	334
14.4	Chapter Highlights	336
14.5	Problems	337
	14.5.1 Exercises	337
	14.5.2 Computer Explorations	339
	14.5.3 Answers to "Check Your Understanding"	339

15 Recent Developments   341

15.1	Goldbach's Conjecture and the Twin Prime Problem	341
15.2	Fermat's Last Theorem	342
15.3	The Riemann Hypothesis	345

A Supplementary Topics   351

A.1	Geometric Series	351
A.2	Mathematical Induction	353
A.3	Pascal's Triangle and the Binomial Theorem	358
A.4	Fibonacci Numbers	364
A.5	Matrices	367
A.6	Problems	371
	A.6.1 Exercises	371
	A.6.2 Answers to "Check Your Understanding"	374

B Answers and Hints for Odd-Numbered Exercises   377

Index   389

Preface

Number theory has a rich history. For many years, it was one of the purest areas of pure mathematics, studied because of the intellectual fascination with properties of integers. More recently, it has been an area that also has important applications to subjects such as cryptography. The goal of this book is to present both sides of the picture, giving a selection of topics that we find exciting.

The main thing to remember is number theory is supposed to be fun. We hope you enjoy the book.

The book is designed for use in a basic undergraduate course in number theory, but the book has also been used in a course for advanced high school students and could work well for independent study.

How does this book differ from our book *An Introduction to Number Theory with Cryptography* (also published by CRC)? Since number theory is often the first theoretical course a student takes, we have added an introductory section on how to do proofs, including a brief discussion of what lemmas, propositions, theorems, and corollaries are. We have added a few more cryptographic topics such as Hill ciphers and transposition ciphers. To fulfill our goal of a more basic course, we have removed topics that are often not covered in a one-semester course: geometry of numbers, algebraic integers, analytic techniques, and some of the more advanced material on primality and factorization. The chapter on continued fractions contains the highlights, in particular the solution of Pell's equation, which is too beautiful to omit, but leaves out the rather technical proofs. We have also adjusted the exercises to fit with the present version of the book.

The Chapters. The flowchart (following this preface) gives the dependencies of the chapters. The section number 7.5 that occurs with an arrow means that only that section from Chapter 7 is needed for Chapter 9 at the end of the arrow.

The core material is Chapters 1, 2, 3, 5, 6, and 8, plus Section 10.1.

These should be covered if at all possible. At this point, there are several possibilities. It is highly recommended that some sections of Chapters 4, 7, and 9 be covered. These present some of the exciting applications of number theory to various problems, especially in cryptography. If time permits, additional topics from Chapters 11 through 14 can be covered. These chapters are mostly independent of one another, so the choices depend on the interests of the audience.

We have tried to keep the prerequisites to a minimum. Besides the introduction to proofs in Chapter 1, there is Appendix A, which treats some topics such as induction, the binomial theorem, and 2×2 matrices (for use in Hill ciphers). Our experience is that many students have seen these topics but that a review is worthwhile. The appendix also treats Fibonacci numbers, since they occur as examples in a few places throughout the book and are interesting in their own right.

Notes to the reader. At the end of each chapter, we have a short list of Chapter Highlights. We were tempted to use the label "If you don't know these, no one will believe you read the chapter." In other words, when you finish a chapter, make sure you thoroughly know the highlights. (Of course, there is more that is worth knowing.) At the end of several sections, there are problems labeled "CHECK YOUR UNDERSTANDING." These are problems that check whether you have learned some basic ideas. The solutions to these are given at the ends of the chapters. You should not leave a topic until you can do these problems easily.

Problems. At the end of every chapter, there are problems to solve. The *Exercises* are intended to give practice with the concepts and sometimes to introduce interesting ideas related to the chapter's topics. Computations have had a great influence on number theory, and the *Computer Explorations* introduce this type of experimentation. Sometimes they ask for specific data, sometimes they are more open-ended. But they represent the type of exploration that number theorists often do in their research.

Appendix B contains answers or hints for the odd-numbered problems. For the problems where the answer is a number, the answer is given. When the exercise asks for a proof, usually a sketch or a key step is given.

Computers. Many students are familiar with computers these days and many have access to software packages such as Mathematica®,

Maple®, MATLAB®, Sage, or Pari that perform number theoretical calculations with ease. Some of the exercises (the ones that use numbers of five or more digits) are intended to be used in conjunction with a computer. Many can probably be done with an advanced calculator. The Computer Explorations definitely are designed for students with computer skills.

Acknowledgments. Jim Kraft thanks the Gilman School for its generous support during the writing of this book. He also thanks his students Rishi Bedi, John Chirikjian, Anthony Kim, and John Lee, whose comments helped make this a better book. Many thanks are also due to Manjit Bhatia, who made many very useful suggestions. We both thank our many students over the years who have taught us while we have taught them. This book would not have been possible without them.

We welcome comments, corrections, and suggestions. Corrections and related matter will be listed on the website for the book (www.math.umd.edu/~lcw/elementarynumbertheory.html).

James S. Kraft
Gilman School
jkraft@gilman.edu

Lawrence C. Washington
University of Maryland
lcw@math.umd.edu

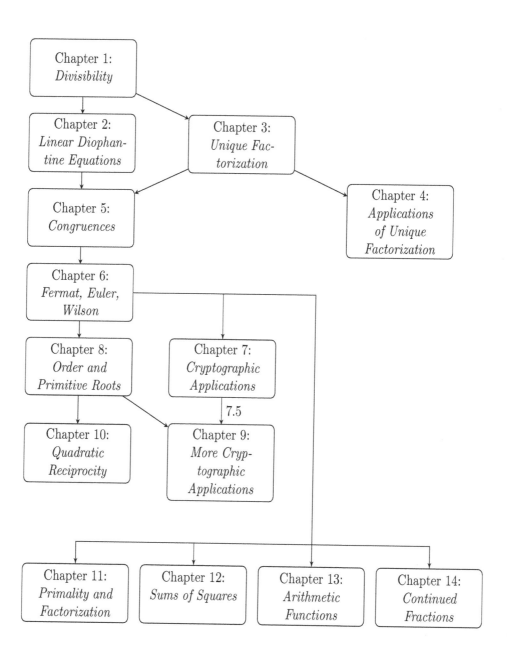

Introduction

At Columbia University there is a Babylonian clay tablet called Plimpton 322 that is more than 3800 years old and not much larger than a cell phone. Written in cuneiform script with 4 columns and 15 rows, it contains numbers written in base 60 (just as base 10 is standard today, base 60 was standard in Babylon). Each row gives a Pythagorean triple, that is, three whole numbers x, y, z satisfying

$$x^2 + y^2 = z^2$$

(for example, $3^2 + 4^2 = 5^2$ and $12709^2 + 13500^2 = 18541^2$ are triples from the tablet). This is one of the earliest examples where integers are studied for their interesting properties, not just for counting objects.

Throughout history, there has been a fascination with whole numbers. For example, the Pythagorean school (ca. 500 BCE) believed strongly that every quantity could be expressed in terms of integers or ratios of integers, and they successfully applied the idea to the theory of musical scales. However, this overriding belief received a sharp setback when one of the Pythagoreans, possibly Hippasus, proved that $\sqrt{2}$ is irrational. There is a story, which may be apocryphal, that he discovered this at sea and was promptly thrown overboard by his fellow Pythagoreans. Despite their attempt at suppressing the truth, the news of this discovery soon got out. Nevertheless, even though irrational numbers exist and are plentiful, properties of integers are still important.

Approximately 200 years after Pythagoras, Euclid's *Elements*, perhaps the most important mathematics book in history, was published. Although most people now think of the *Elements* as a book concerning geometry, a large portion of it is devoted to the theory of numbers. Euclid proves that there are infinitely many primes, demonstrates fundamental properties concerning divisibility of integers, and derives a formula that yields *all* possible Pythagorean triples, as well as many other seminal results. We will see and prove these results in the first four chapters of this book.

Number theory is a rich subject, with many aspects that are inextricably intertwined but which also retain their individual characters. In this introduction, we give a brief discussion of some of the ideas and some of the history of number theory as seen through the themes of Diophantine equations, modular arithmetic, the distribution of primes, and cryptography.

1 Diophantine Equations

Diophantus lived in Alexandria, Egypt, about 1800 years ago. His book *Arithmetica* gives methods for solving various algebraic equations and had a great influence on the development of algebra and number theory for many years. The part of number theory called **Diophantine equations**, which studies integer (and sometimes rational) solutions of equations, is named in his honor. However, the history of this subject goes back much before him. The Plimpton tablet shows that the Babylonians studied integer solutions of equations. Moreover, the Indian mathematician Baudhāyana ($\approx$ 800 BCE) looked at the equation $x^2 - 2y^2 = 1$ and found the solutions $(x, y) = (17, 12)$ and $(577, 408)$. The latter gives the approximation $577/408 \approx 1.4142157$ for $\sqrt{2}$, which is the diagonal of the unit square. This was a remarkable achievement, considering that, at the time, a standardized system of algebraic notation did not yet exist.

The equation
$$x^2 - ny^2 = 1,$$
where n is a positive integer not a square, was studied by Brahmagupta (598–668) and later mathematicians. In 1768, Joseph-Louis Lagrange (1736–1813) presented the first published proof that this equation always has a nontrivial solution (that is, with $y \neq 0$). Leonhard Euler (1707–1783) mistakenly attributed some work on this problem to the English mathematician John Pell (1611–1685), and ever since it has been known as **Pell's equation**, but there is little evidence that Pell did any work on it. In Chapter 14, we show how to solve Pell's equation.

Introduction 3

Perhaps $x^2 + y^2 = z^2$, the equation for Pythagorean triples, is the most well-known Diophantine equation. Since sums of two nonzero squares can be a square, people began to wonder if this could be generalized. For example, Abu Mohammed Al-Khodjandi, who lived in the late 900s, claimed to have a proof that a sum of nonzero cubes cannot be a cube (that is, the equation $x^3 + y^3 = z^3$ has no nonzero solutions). Unfortunately, our only knowledge of this comes from another manuscript, which mentions that Al-Khodjandi's proof was defective, but gives no evidence to support this claim. The real excitement began when the great French mathematician Pierre de Fermat (1601–1665) penned a note in the margin of his copy of Diophantus's *Arithmetica* saying that it is impossible to solve $x^n + y^n = z^n$ in positive integers when $n \geq 3$ and that he had found a truly marvelous proof that the margin was too small to contain. After Fermat's son, Samuel Fermat, published an edition of Diophantus's book that included his father's comments, the claim became known as **Fermat's Last Theorem**. Today, it is believed that he actually had proofs only in the cases $n = 4$ (the only surviving proof by Fermat of any of his results) and possibly $n = 3$. But the statement acquired a life of its own and led to many developments in mathematics. Euler is usually credited with the first complete proof that Fermat's Last Theorem (abbreviated as FLT) is true for $n = 3$. Progress proceeded exponent by exponent, with Adrien-Marie Legendre (1752–1833) and Johann Peter Gustav Lejeune Dirichlet (1805–1859) each doing the case $n = 5$ around 1825 and Gabriel Lamé (1795–1870) treating $n = 7$ in 1839. Important general results were obtained by Sophie Germain (1776–1831), who showed that if $p < 100$ is prime and xyz is not a multiple of p, then $x^p + y^p \neq z^p$.

The scene changed dramatically around 1850, when Ernst Eduard Kummer (1810–1893) developed his theory of *ideal numbers*, which are now known as *ideals* in ring theory. He used them to give general criteria that allowed him to prove FLT for all exponents up to 100, and many beyond that. His approach was a major step in the development of both algebraic number theory and abstract algebra, and it dominated the research on FLT until the 1980s. In the 1980s, new methods, based on work by Taniyama, Shimura, Weil, Serre, Langlands, Tunnell, Mazur, Frey, Ribet, and others, were brought to the problem, resulting in the proof of Fermat's Last Theorem by Andrew Wiles (with the help of Richard Taylor) in 1994. The techniques developed during this period have opened up new areas of research and have also proved useful in solving many classical mathematical problems.

2 Modular Arithmetic

Suppose you divide 1234^{25147} by 25147. What is the remainder? Why should you care? A theorem of Fermat tells us that the remainder is 1234. Moreover, as we'll see, results of this type are surprisingly vital in cryptographic applications (see Chapters 7 and 9).

Questions about divisibility and remainders form the basis of modular arithmetic, which we introduce in Chapter 5. This is a very old topic, and its development is implicit in the work of several early mathematicians. For example, the Chinese Remainder Theorem is a fundamental and essential result in modular arithmetic and was discussed by Sun Tzu around 1600 years ago.

Although early mathematicians discovered number theoretical results, the true beginnings of modern number theory began with the work of Fermat, whose contributions were both numerous and profound. We will discuss several of them in this book. For example, he proved that if a is a whole number and p is a prime then $a^p - a$ is always a multiple of p. Results such as this are best understood in terms of modular arithmetic.

Euler and Karl Friedrich Gauss (1777–1855) greatly extended the work done by Fermat. Gauss's book *Disquisitiones Arithmeticae*, which was published in 1801, gives a treatment of modular arithmetic that is very close to the present-day version. Many of the original ideas in this book laid the groundwork for subsequent research in number theory.

One of Gauss's crowning achievements was the proof of Quadratic Reciprocity (see Chapter 10). Early progress toward this fundamental result, which gives a subtle relation between squares of integers and prime numbers, had been made by Euler and by Legendre. Efforts to generalize Quadratic Reciprocity to higher powers led to the development of algebraic number theory in the 1800s by Kummer, Richard Dedekind (1831–1916), David Hilbert (1862–1943), and others. In the first half of the 1900s, this culminated in the development of *class field theory* by

Introduction

many mathematicians, including Hilbert, Weber, Takagi, and Artin. In the second half of the 1900s up to the present, the *Langlands Program*, which can be directly traced back to Quadratic Reciprocity, has been a driving force behind much number-theoretic research. Aspects of it played a crucial role in Wiles's proof of Fermat's Last Theorem in 1994.

3 The Distribution of Primes

> *There are two facts about the distribution of prime numbers of which I hope to convince you so overwhelmingly that they will be permanently engraved in your hearts. The first is that, despite their simple definitions and role as the building blocks of the natural numbers, the prime numbers belong to the most arbitrary and ornery objects studied by mathematicians: they grow like weeds among the natural numbers, seeming to obey no other law than that of chance, and nobody can predict where the next one will sprout. The second fact is even more astonishing, for it states just the opposite: that the prime numbers exhibit stunning regularity, that there are laws governing their behavior, and that they obey these laws with almost military precision.* – Don Zagier

Euclid proved that there are infinitely many primes, but we can ask for more precise information. Let $\pi(x)$ be the number of primes less than or equal to x. Legendre and Gauss used experimental data to conjecture that

$$\frac{\pi(x)}{x/\ln x} \approx 1,$$

and this approximation gets closer to equality as x gets larger. For example,

$$\frac{\pi(10^4)}{10^4/\ln 10^4} = 1.132, \quad \text{and} \quad \frac{\pi(10^{10})}{10^{10}/\ln 10^{10}} = 1.048.$$

In 1852, Pafnuty Chebyshev (1821–1894) showed that the conjecture of Legendre and Gauss is at least approximately true by showing that, for sufficiently large values of x,

$$0.921 \leq \frac{\pi(x)}{x/\ln x} \leq 1.106.$$

A few years later, Bernhard Riemann (1826–1866) introduced techniques from the theory of complex variables and showed how they could lead to more precise estimates for $\pi(x)$. Finally, in 1896, using Riemann's ideas, Jacques Hadamard (1865–1963) and Charles de la Vallée-Poussin (1866–1962) independently proved that

$$\lim_{x \to \infty} \frac{\pi(x)}{x/\ln x} = 1,$$

a result known as the **Prime Number Theorem**.

If we look at the list of *all* integers, we know that within that list there is an infinite number of primes. Suppose we look at a list like this:

$$1, 6, 11, 16, 21, 26, \ldots,$$

or like this:

$$3, 13, 23, 33, 43, 53, \ldots,$$

or like this:

$$1, 101, 201, 301, 401, \ldots.$$

Does each of the three lists contain an infinite number of primes as well? The answer is yes, and we owe the proof of this remarkable fact to Dirichlet. In 1837, he proved that every arithmetic progression of the form $a, a+b, a+2b, a+3b, \ldots$ contains infinitely many primes if a and b are positive integers with no common factor greater than 1. We will not prove this result in this book.

There are many other questions that can be asked about primes. One of the most famous is **Goldbach's Conjecture**. In 1742, Christian Goldbach (1690–1764) conjectured that every even integer greater than 2 is a sum of two primes (for example, $100 = 83 + 17$). Much progress has been made on this conjecture over the last century. In 1937, I. M. Vinogradov (1891–1983) proved that every sufficiently large odd integer is a sum of three primes, and in 1966, Jingrun Chen (1933–1996) proved that every sufficiently large even integer is either a sum of two primes or the sum of a prime and a number that is the product of two primes (for example, $100 = 23 + 7 \cdot 11$). In 2013, Helfgott completed Vinogradov's work by showing that every odd integer greater than or equal to 7 is a sum of three primes. These results require very delicate analytic techniques. Work on Goldbach's Conjecture and related questions remains a very active area of modern research in number theory.

4 Cryptography

For centuries, people have sent secret messages by various means. But in the 1970s, there was a dramatic change when Fermat's theorem and Euler's theorem (a generalization of Fermat's theorem), along with other results in modular arithmetic, became fundamental ingredients in many cryptographic systems. In fact, whenever you buy something over the Internet, it is likely that you are using Euler's theorem.

In 1976, Whitfield Diffie and Martin Hellman introduced the concept of public key cryptography and also gave a key establishment protocol (see Chapter 9) that uses large primes. A year later, Ron Rivest, Adi Shamir, and Len Adleman introduced the RSA cryptosystem (see Chapter 7), an implementation of the public key concept. It uses large prime numbers and its security is closely tied to the difficulty of factoring large integers.

Topics such as factorization and finding primes became very popular, and soon there were several major advances in these subjects. For example, in the mid-1970s, factorization of 40-digit numbers was the limit of technology. As of 2014, the limit was 230 digits. Some factorization methods will be discussed in Chapter 11.

Cryptography brought about a fundamental change in how number theory is viewed. For many years, number theory was regarded as one of the purest areas of mathematics, with little or no application to real-world problems. In 1940, the famous British number theorist G. H. Hardy (1877–1947) declared, "No one has yet discovered any warlike purpose to be served by the theory of numbers or relativity, and it seems unlikely that anyone will do so for many years" (*A Mathematician's Apology*, section 28). Clearly this statement is no longer true.

Although the basic purpose of cryptography is to protect communications, its ideas have inspired many related applications. In Chapter 9, we'll explain how to sign digital documents, along with more lighthearted topics such as playing mental poker and flipping coins over the telephone.

Chapter 1

Divisibility

1.1 What Is a Proof?

If you are studying chemistry and read that the boiling point of salt water changes as the concentration of salt changes, you might try an experiment to see if this is really true. How do you proceed? You use a thermometer to measure the water temperature and you calculate the concentration of the salt in the water. After performing this experiment several times, you conclude that the statement is, in fact, correct.

If you are studying mathematics and read that the sum of two even integers is always an even integer, you might also try an experiment to see if this is really true. How do you proceed in this case? You can write down two lists of even numbers, take a number from each list, and add them up. If you do this correctly, can you conclude that the sum of two even numbers is always even? If you suspect that this does not completely justify the statement, you're on to something. This is not how mathematics works. Although mathematicians can do experiments (often with computers) to try to decide if something seems to be true, they do not use experiments to conclude that things are true. Instead, mathematicians *prove* things. What is a proof? Why do we prove things? How is a proof different from experimental verification? We'll answer these questions in the next few paragraphs.

Simply put, we prove things to convince others (and ourselves) that a statement we're asserting to be true really is true. We start off with an initial assumption or hypothesis and use a chain of logical deductions to arrive at what we want to prove. The validity of our proof rests on our correct use of logic, not on the consistency of experimental data. Let's see how this works as we prove that the sum of two even integers is an even integer.

To begin, coming up with several examples

$$4 + 12 = 16, \quad 36 + 20 = 56, \quad 128 + 416 = 544$$

does not constitute a proof because it's impossible to write down the sum of every pair of even integers. Even if you calculated the sum of 1000 different pairs of even integers, it's possible that the next pair you'd try would contradict what you believe to be true. Instead, you need to come up with some general method that handles all possibilities. In order to do this, you need to begin with a clear and precise definition of even integers. (You can't prove something about even integers until you have defined them.) This is the motivation behind the following definition.

Definition 1.1. *An integer n is an* **even** *integer if n is a multiple of 2. In other words, there is an integer k with $n = 2k$.*

For example, 26 is even because $26 = 2 \cdot 13$, and 100 is even because $100 = 2 \cdot 50$.

Now that we've defined even, let's take another look at what we want to prove, namely that the sum of two even integers is also an even integer. When trying to prove something, it's essential to recognize what your assumptions are and be absolutely certain what it is you want to prove. In this case, we're assuming that we have two even integers and we want to prove that their sum is even. Sometimes it helps to link our assumption with what we want to prove in an "if, then" sentence called a conditional statement. Here's what a conditional statement looks like in this example.

If you have two even integers, **then** their sum is even.

The underlined clause that follows the "if" is called the **hypothesis**, and the underlined clause that follows the "then" is called the **conclusion**. The hypothesis is what you assume; the conclusion is what you want to prove.

Now, let's begin the proof. We start off with our even integers: we'll

Section 1.1   What Is a Proof?   11

call them m and n. (Giving the numbers names such as m and n makes it much easier to work with them. Don't skip this step.) We want to prove that $m + n$ is also even. The best way to begin is to make use of the hypotheses. Since both m and n are even, we'll write down the only thing we know about even integers, namely their definition. Since m is even, we know that $m = 2k_1$ for some integer k_1, and since n is even, $n = 2k_2$ for some integer k_2. What do we do now? Since we want to prove something about the sum of m and n, we'll add m to n and see what happens. We see that $m + n = 2k_1 + 2k_2$. Our goal is to show that $m + n$ is even, which means (by the definition) that we want to express $m + n$ as a multiple of 2. If fact, this is what we've done:

$$m + n = 2k_1 + 2k_2 = 2(k_1 + k_2).$$

This shows that $m + n$ is twice the integer $k_1 + k_2$, which tells us that $m + n$ is even.

This proof is in many ways a model for how you should think of any proof you try to do. First, make sure you have an unambiguous definition of all your terms. Next, write what you want to prove as a conditional statement, making sure you understand the hypothesis and conclusion. Then, translate the hypothesis into a mathematical statement. Finally, develop a strategy that will allow you to go from the hypothesis to the conclusion. Although many of the proofs in this book are more complex than the present example, the method that we just outlined will always apply.

When you finish reading a proof (yes, you should always try to read the proofs; that's how to learn how concepts fit together and how to do proofs), it is very good practice to look back and see where the hypotheses were used. In our example, the hypothesis was that m and n are even. How was this used? It allowed us to write $m = 2k_1$ and $n = 2k_2$, which is what got us started. If a proof is not using one of the hypotheses, it is likely that something is wrong (or that the hypothesis is not needed).

No discussion of how to prove things is complete without saying how to disprove things. Suppose you see the statement that all integers are less than 100. That's absurd, you say. For example, 123 is an integer and it is not less than 100. You have disproved the statement. Whenever you want to disprove a statement, all you need to do is find a counterexample. One counterexample suffices. This is in complete contrast to most proofs, where an example is not the same as a proof.[1]

For practice, let's try to prove or disprove the following statements. For each you should first try a few examples ("experiments") to get a feel for what is being said. If you find a counterexample, you are done, since you have disproved the statement. If your examples seem to indicate that the statement is true, try to give a proof. We'll give answers after the statements, but you should try each before looking at the solutions.

Prove or Disprove:

1. The sum of two odd integers is always even. (*Note:* First, you need a definition of *odd*; we say that n is odd if there is an integer k such that $n = 2k + 1$.)

2. The product of two even integers is always a multiple of 4.

3. Every multiple of 3 is odd.

4. Every multiple of 6 is even.

5. Every odd number larger than 1 is a prime number.

Solutions:

1. Some examples show that this seems to be true: $5 + 9 = 14$, $21 + 111 = 132$, $1 + 3 = 4$. Now let's prove it. First, write the statement as a conditional statement: If m and n are odd integers, then $m + n$ is even. The hypothesis is that m and n are odd. To write this mathematically, use the definition: $m = 2k_1 + 1$ for some integer k_1, and $n = 2k_2 + 1$ for some integer k_2. Our goal is to say something about the sum $m+n$, so we compute

$$m + n = (2k_1 + 1) + (2k_2 + 1) = 2k_1 + 2k_2 + 2.$$

What do we do now? Look at the conclusion. It says that $m+n$

---

[1] An exception: If the statement you want to prove is an existence statement, then an example might suffice. For example, if the statement says that there exists a cube that is two more than a square, then the example $3^3 = 5^2 + 2$ shows that the statement is true.

Section 1.1   What Is a Proof?   13

is even, which means that we need to be able to express $m+n$ as a multiple of 2. Our expression for $m+n$ lets us do this:

$$m + n = 2k_1 + 2k_2 + 2 = 2(k_1 + k_2 + 1).$$

Therefore, $m+n$ is even, which is what we wanted to prove.

2. Try some examples: $2 \cdot 6 = 12$, $8 \cdot 6 = 48$, $10 \cdot 10 = 100$. The products are multiples of 4, so the statement seems to be true. Let's prove it. The hypothesis is that we have two even integers. Let's call them m and n. Since m and n are even, we can write $m = 2k_1$ and $n = 2k_2$ for some integers k_1 and k_2. The conclusion says something about the product mn, so we write

$$mn = (2k_1)(2k_2) = 4(k_1 k_2).$$

This says that mn is a multiple of 4, so we have completed the proof.

3. Let's try some examples: $3 \cdot 1 = 3$, $3 \cdot 2 = 6$, $3 \cdot 3 = 9$. Two out of three; not bad. In most sporting events, winning two out of three is awesome. But in math, one counterexample to a statement is enough to disprove the statement. Since $2 \cdot 3 = 6$ shows that there is a multiple of 3 that is even, the statement that all multiples of 3 are odd is false. So we have disproved the statement.

4. Try some examples: $6 \cdot 1 = 6$, $6 \cdot 2 = 12$, $6 \cdot 3 = 18$. All of these are even, so the statement looks correct. Let's try to prove it. A conditional form of the statement is "If n is an integer, then $6n$ is even." The hypothesis is that n is an integer. There's not much that we can say about n, so let's look at what we're trying to do. The conclusion says something about $6n$, so let's look at this number. We're trying to prove that $6n$ is even, which means that we need to express it as a multiple of 2. In fact, this is easy:

$$6n = 2(3n).$$

Therefore, $6n$ is a multiple of 2, so $6n$ is even. This proves the statement.

5. Let's look at some odd numbers: 3, 5, 7, 9, 11, 13. The numbers 3, 5, 11, and 13 are primes, but 9 is $3 \cdot 3$, so it is not prime. Since

the odd number 9 is not prime, we have a counterexample, which means the statement is false.

Throughout this book you will see proofs and wonder how in the world anyone could have thought of them. Don't be disheartened. They might represent hours, or years, of work by brilliant people. We don't see their mistakes and false starts — we see only the final successes. Andrew Wiles, who proved Fermat's Last Theorem (see the Introduction and Chapter 15), explained it as follows:

> "Perhaps I could best describe my experience of doing mathematics in terms of entering a dark mansion. One goes into the first room, and it's dark, completely dark. One stumbles around bumping into the furniture, and gradually, you learn where each piece of furniture is, and finally, after six months or so, you find the light switch. You turn it on, and suddenly, it's all illuminated. You can see exactly where you were." *(PBS NOVA Broadcast, October 28, 1997)*

Although none of the proofs you will be asked to do in this book should take six months, don't be afraid of stumbling around and making mistakes. Learning what doesn't work is sometimes as important as learning what does work.

We'll wrap up this introductory section with a brief discussion of some of the terminology concerning proofs you'll be seeing throughout this book. A **proposition** is a statement that we'll be able to prove. A **theorem** is an extremely important proposition, and is usually a highlight of the topic under consideration. If something is called a theorem, you should make a special effort to remember its statement and to understand what it says. A **lemma** is a statement that is used to prove a proposition or theorem. Often, a lemma is singled out because it is useful and interesting in its own right, but is not considered to be as important as a proposition or a theorem. (Admittedly, this can be somewhat subjective. One person's lemma can, at times, be someone else's proposition.) A **corollary** is a result that is an easy consequence of a proposition or theorem.

A statement that says "A is true if and only if B is true" is a combination of two statements: "If A is true then B is true" and "If B is true then A is true." To prove such a statement, you assume A is true and deduce that B is true. Then you assume B is true and deduce that A is true.

Section 1.1   What Is a Proof?   15

Often, "A if and only if B" is written as "$A \Leftrightarrow B$." Similarly, "$A \Rightarrow B$" means that "A implies B."

Sometimes you will see a statement saying that A, B, and C are equivalent. This is the same as saying that A is true if and only if B is true, B is true if and only if C is true, and A is true if and only if C is true. However, it is not necessary to prove all six implications. Instead, prove that if A is true then B is true, then prove that if B is true then C is true, and finally prove that if C is true then A is true. This suffices to prove all of the necessary implications and is much more efficient. For example, since A implies B and B implies C, we automatically conclude that A implies C without proving it directly.

1.1.1   Proof by Contradiction

Sometimes, the easiest way to prove a statement is to suppose the statement is false and deduce a false consequence. This is known as *proof by contradiction*. Classically, it was called *reductio ad absurdum* (Latin for "reduction to absurdity").

In the following, we give three examples of proof by contradiction. In each case, proof by contradiction arises naturally because the conclusion is a negative statement ("there is no such number" and "the number is not rational"). The proof proceeds by assuming the positive statement ("there is such a number" or "the number is rational") and showing that this leads to a contradiction, so the negative statement must be true.

Recall that a rational number is a number that can be expressed as the ratio of two integers; for example, $2/3$, $-15/2$, $71 = 71/1$, and $0 = 0/1$ are rational. An irrational number is a number that is not rational. We will see in Chapter 4 that $\sqrt{2}$ and $\sqrt{3}$ are examples of irrational numbers.

I. *Prove that there is no largest integer.*

Proof. If someone told you that there exists a largest integer, you would probably say something like this: "That's ridiculous. Adding one to any integer gives you a larger integer. So there can't be a largest one."

This reasoning is essentially the proof of the statement: Suppose that the statement to be proved is false. Then there exists a largest integer n. Let $m = n + 1$. Then m is an integer and $m > n$, which contradicts the assumption that n was the largest integer. So, no largest integer can exist. This completes the proof. □

II. *Prove that there is no smallest positive rational number. (Recall that "positive" means greater than 0, and does not include 0.)*

Proof. Let's suppose the statement is false. Then there is a smallest positive rational number. Call it r. Since r is rational, we can write $r = a/b$, where a and b are integers. Then $r/2 = a/(2b)$ is positive, rational, and smaller than r, which contradicts the assumption that r is the smallest positive rational number. Therefore, the assumption that there is a smallest positive rational number has led to a contradiction. The only possibility that remains is that there is no smallest positive rational number. □

III. *Prove that a rational number plus an irrational number is irrational.*

Proof. Let's translate the statement into symbols: It says that if x is rational and y is irrational, then $z = x+y$ is irrational. We need to show that z is irrational, which means that z cannot be written in the form $z = c/d$ with integers c and d. Because we're saying that something is *not* rational, it's very natural to use a proof by contradiction. So let's assume that z is rational, which means that $z = c/d$ for some integers c and d. Moreover, one of the hypotheses is that x is rational, so x can be written in the form $x = a/b$, where a and b are integers. Therefore, $x + y = z$ tells us that $y = z - x$, which we can write as

$$y = \frac{c}{d} - \frac{a}{b} = \frac{bc - ad}{bd}.$$

But $bc-ad$ and bc are integers, so y can be expressed as the ratio of two integers. This contradicts the assumption that y is irrational. Therefore, the assumption that z is rational led to the false consequence that y is rational. Therefore, we conclude that z must be irrational. □

A consequence of what we just proved is that $3 - \sqrt{2}$ is irrational. How do we show this? As mentioned earlier, we'll show in Chapter 4 that $\sqrt{2}$ is irrational. Suppose $3 - \sqrt{2}$ is rational. Then $3 = (3 - \sqrt{2}) + \sqrt{2}$ is a rational plus an irrational, which is irrational by what we just proved. But 3 is rational, so we have a contradiction. Therefore, our assumption that $3 - \sqrt{2}$ is rational must be false, so $3 - \sqrt{2}$ is irrational.

For the record, we now note that an irrational plus an irrational can be either rational or irrational. For example, $\sqrt{2} + \sqrt{2} = 2\sqrt{2}$ is an example where the sum of two irrational numbers is irrational. On the other hand, $\sqrt{2} + (3 - \sqrt{2}) = 3$ is an example where the sum of two irrationals is rational.

When we have an "if–then" statement such as "if A is true then B is true," its *contrapositive* is "if B is false then A is false." A statement is true if and only if its contrapositive is true. A proof by contradiction is really just proving the contrapositive of the statement in question.

Warning: Proof by contradiction is a very useful tool, but do not overuse it. It is not uncommon to see something like the following: You are asked to prove that $x^2 - x = 12$ has a solution that is an integer. You write, "Suppose it does not have such a solution. Let $x = 4$. Then $4^2 - 4 = 12$, so the equation has a solution. Contradiction. Therefore, the equation has an integral solution." What's wrong? Technically, nothing. But what you've really done is say, "Suppose this does not have a solution. But it does! Let $x = 4$. Since $4^2 - 4 = 12$, the equation has a solution that is an integer." It makes much more sense to state directly that $x = 4$ is a solution. The moral is, don't use proof by contradiction when a direct proof is more straightforward.

1.2 Divisibility

A large portion of this book will be spent studying and proving properties of the integers. You can add, subtract, and multiply integers, and doing so always gives you another integer. Division is a little trickier — sometimes when you divide one integer by another you get an integer (12 divided by 3) and sometimes you don't (12 divided by 5). Because of this, the first idea we have to make precise is that of divisibility.

Definition 1.2. *Given two integers a and d with d nonzero, we say that d **divides** a (written $d \mid a$) if there is an integer c with $a = cd$. If no such integer exists, so d does **not** divide a, we write $d \nmid a$. If d divides a, we say that d is a **divisor** of a and that a is a **multiple** of d.*

Examples. $5 \mid 30$ since $30 = 5 \cdot 6$, and $3 \mid 102$ since $102 = 3 \cdot 34$, but $6 \nmid 23$ and $4 \nmid -3$. Also, $-7 \mid 35$, $8 \mid 8$, $3 \mid 0$, $-2 \mid -10$, and $1 \mid 4$.

Remark. There are two technical points that need to be mentioned. First, we never consider 0 to be a divisor of anything. Of course, we could agree that $0 \mid 0$, but it's easiest to avoid this case completely since we never need it. Second, if d is a divisor of a, then $-d$ is a divisor of a. However, whenever we talk about the *set of divisors* of a positive integer, we follow the convention that we mean the *positive* divisors. So we say that the divisors of 6 are 1, 2, 3, and 6 (and ignore $-1, -2, -3, -6$).

There are several basic results concerning divisibility that we will be using throughout this book.

Proposition 1.3. *Assume that a, b, and c are integers. If $a \mid b$ and $b \mid c$, then $a \mid c$.*

Proof. We first need to change the hypotheses into something we can use. We are assuming that $a \mid b$ and $b \mid c$. What can we deduce from this? Maybe the best thing to do is use the definition of divisibility: Since $a \mid b$, we can write $b = ea$ for some integer e, and since $b \mid c$, we can write $c = fb$ for some integer f. Since we are trying to show $a \mid c$, we need to find an equation that relates a and c. Let's try putting together what we have so far. Substitute $b = ea$ into $c = fb$ to obtain

$$c = fb = f(ea) = (fe)a.$$

Therefore, c equals a times an integer so, by definition, $a \mid c$. □

Example. The proposition implies, for example, that a multiple of 6 is even: Let $a = 2$ and $b = 6$, and let c be an arbitrary integer. Then $a \mid b$. If $6 \mid c$, the proposition says that $2 \mid c$, which says that c is even.

Proposition 1.4. *Assume that $a, b, d, x,$ and y are integers. If $d \mid a$ and $d \mid b$, then $d \mid ax + by$.*

Proof. Write $a = md$ and $b = nd$. Then

$$ax + by = (md)x + (nd)y = d(mx + ny),$$

so $c \mid ax + by$ by definition. □

Often, $ax + by$ is called a *linear combination* of a and b, so Proposition 1.4 says that every divisor of both a and b is also a divisor of each linear combination of a and b.

Corollary 1.5. *Assume that $a, b,$ and d are integers. If $d \mid a$ and $d \mid b$, then $d \mid a+b$ and $d \mid a-b$.*

Proof. To show that $d \mid a+b$, set $x=1$ and $y=1$ in the proposition and to show that $d \mid a-b$, set $x=1$ and $y=-1$ in the proposition. □

Examples. Since $3 \mid 9$ and $3 \mid 21$, the proposition says that

$$3 \mid 5 \cdot 9 + 4 \cdot 21 = 129.$$

Since $5 \mid 20$ and $5 \mid 30$, we have $5 \mid 20+30 = 50$. Since $10 \mid 40$ and $10 \mid 60$, we have $10 \mid 40-60 = -20$.

CHECK YOUR UNDERSTANDING[2]

1. Does 7 divide 1001?
2. Show that $7 \nmid 1005$.

1.3 Euclid's Theorem

Fundamental to the study of the integers is the idea of a prime number.

Definition 1.6. *A* **prime number** *is an integer $p \geq 2$ whose only divisors are 1 and itself. A* **composite number** *is an integer $n \geq 2$ that is not prime.*

You may be wondering why 1 is not considered to be prime. After all, its only divisors are 1 and itself. Although there have been mathematicians in the past who have included 1 in the list of primes, nobody does so anymore. The reason for this is that mathematicians want to say there's *exactly one* way to factor an integer into a product of primes. If 1 were a prime, and we wanted to factor 6, for example, we'd have $6 = 2 \cdot 3 = 2 \cdot 3 \cdot 1 = 2 \cdot 3 \cdot 1 \cdot 1, \ldots$ and we would have an infinite number of ways to factor an integer into primes. So, to avoid this, we simply declare that 1 is not prime.

---

[2]Answers are at the end of the chapter.

The first ten prime numbers are

$$2, 3, 5, 7, 11, 13, 17, 19, 23, 29.$$

Notice that 2 is prime because its only divisors are 1 and 2, but no other even number can be prime because every other even number has 2 as a divisor.

It's natural to ask if the list of primes ever terminates. It turns out that it doesn't; that is, there are infinitely many primes. This fact is one of the most basic results on number theory. The first written record we have of it is in Euclid's *Elements*, which was written more than 2300 years ago. In the next section, we'll discuss Euclid's original proof. Before we do that, here's a proof that is a variation of his idea. We begin with a lemma.

Lemma 1.7. *Every integer greater than 1 is either prime or is divisible by a prime.*

Proof. If an integer n is not a prime, then it is divisible by some integer a_1, with $1 < a_1 < n$. If a_1 is prime, we've found a prime divisor of n. If a_1 is not prime, it must be divisible by some integer a_2 with $1 < a_2 < a_1$. If a_2 is prime, then since $a_2 \mid a_1$ and $a_1 \mid n$, we have $a_2 \mid n$, and a_2 is a prime divisor of n. If a_2 is not prime, we continue and get a decreasing sequence of positive integers $a_1 > a_2 > a_3 > a_4 > \cdots$, all of which are divisors of n. Since you can't have a sequence of *positive* integers that decreases forever, this sequence must stop at some a_m. The fact that the sequence stops means that a_m must be prime, which means that a_m is a prime divisor of n. □

Example. In the proof of the lemma, suppose $n = 72000 = 720 \times 100$. Take $a_1 = 720 = 10 \times 72$. Take $a_2 = 10 = 5 \times 2$. Finally, take $a_3 = 5$, which is prime. Working backwards, we see that $5 \mid 72000$.

Euclid's Theorem. *There are infinitely many primes.*

Proof. We assume that there is a finite number of primes and arrive at a contradiction. So, let

$$2, 3, 5, 7, 11, ..., p_n \tag{1.1}$$

be the list of all the prime numbers. Form the integer

$$N = 2 \cdot 3 \cdot 5 \cdot 7 \cdot 11 \cdots p_n + 1.$$

To begin, N can't be prime since it's larger than p_n and p_n is assumed to be the largest prime. So, we can use the previous lemma to choose a prime divisor p of N. Since equation (1.1) is a list of every prime, p is equal to one of the p_i and therefore must divide $2 \cdot 3 \cdot 5 \cdot 7 \cdot 11 \cdots p_n$. But p now divides both N and $N - 1 = 2 \cdot 3 \cdot 5 \cdot 7 \cdot 11 \cdots p_n$. By Corollary 1.5, p divides their difference, which is 1. This is a contradiction: $p \nmid 1$ because $p > 1$. This means that our initial assumption that there is a finite number of primes must be incorrect. □

Since mathematicians like to prove the same result using different methods, we give an alternative proof.

Another proof of Euclid's Theorem. We'll show that for each $n > 0$, there is a prime number larger than n. Let $N = n! + 1$ and let p be a prime divisor of N. Either $p > n$ or $p \leq n$. If $p > n$, we're done. If $p \leq n$, then p is a factor of $n!$, so $p \mid N - 1$. Recall that p was chosen so that $p \mid N$, so we now have $p \mid N$ and $p \mid N - 1$. Therefore, $p \mid N - (N - 1) = 1$, which is impossible. This means that $p \leq n$ is impossible, so we must have $p > n$.

In particular, if n is prime, there is a prime p larger than n, so there is no largest prime. This means that there are infinitely many primes. □

CHECK YOUR UNDERSTANDING
3. Explain why $5 \nmid 2 \cdot 3 \cdot 5 \cdot 7 + 1$.

1.4 Euclid's Original Proof

Here is Euclid's proof that there is an infinite number of primes, using the standard translation of Sir Thomas Heath. Euclid's statements are written in italics. Since his terminology and notation may be unfamiliar, we have added comments in plain text where appropriate. It will be helpful to know that when Euclid says "*A measures B*" or "*B is measured by A,"* he means that A divides B or, equivalently, that B is a multiple of A. Rather than saying that there are infinitely many primes, he says that, given a finite set of primes, there is at least one prime that is not in this set.

Euclid's Statements	Explanation
Let A, B, and C be the assigned prime numbers.	This is the finite set of primes. Euclid assumes that there are only three. You can think of this as representing some arbitrary, unknown number of primes.
I say that there are more prime numbers than A, B, and C.	I will show that there is a prime that is not in our finite set.
Take the least number DE measured by A, B, and C. Add the unit DF to DE.	DE is the least common multiple of A, B, and C, and EF = DE + 1.
Then EF is either prime or not. Let it be prime.	Either EF is prime or it's not. First, assume that it's prime.
Then the prime numbers A, B, C, and EF have been found which are more than A, B, and C.	Then we have found a set of primes that is larger than our original set of primes.
Next, let EF not be prime. Therefore it is measured by some prime number. Let it be measured by the prime number G.	Next, assume that EF is not prime. Then, it is a multiple of some prime. Let it be a multiple of the prime G.
I say that G is not the same with any of the numbers A, B, and C.	We will now show that G is not in our set of primes.
For if possible, let it be so. Now A, B, and C measure DE, therefore G also will measure DE.	Assume that G is in our set. Since DE is a multiple of all of the primes in our set, G divides DE.
But it also measures EF.	But, EF is also a multiple of G.

Therefore G, being a number, will measure the remainder, the unit DF, which is absurd.

Since EF is a multiple of G, and DE = EF + 1 is a multiple of G, their difference (EF + 1 - EF), which equals 1, is also a multiple of G. This is a contradiction.

Therefore G is not the same with any one of the numbers A, B, and C. And by hypothesis it is prime. Therefore, the prime numbers A, B, C, and G have been found which are more than the assigned multitude of A, B, and C. Therefore, prime numbers are more than any assigned multitude of prime numbers. Q.E.D.

So G is a prime number that is not in our set of primes. Therefore, no finite set can contain all of the primes. □

1.5 The Sieve of Eratosthenes

Eratosthenes was born in Cyrene (in modern-day Libya) and lived in Alexandria, Egypt, around 2300 years ago. He made important contributions to many subjects, especially geography. In number theory, he is famous for a method of producing a list of prime numbers up to a given bound without using division. To see how this works, we'll find all the prime numbers up to 50.

List the integers from 1 to 50. Ignore 1 and put a circle around 2. Now cross out every second number after 2. This yields (we give just the beginning of the list)

$$1 \quad ② \quad 3 \quad \cancel{4} \quad 5 \quad \cancel{6} \quad 7 \quad \cancel{8} \quad 9 \quad \cancel{10}$$
$$11 \quad \cancel{12} \quad 13 \quad \cancel{14} \quad 15 \quad \cancel{16} \quad 17 \quad \cancel{18} \quad 19 \quad \cancel{20}.$$

Now look at the next number after 2 that is not crossed out. It's 3. Put a circle around 3 and cross out every third number after 3. This yields

$$1 \quad ② \quad ③ \quad \cancel{4} \quad 5 \quad \cancel{6} \quad 7 \quad \cancel{8} \quad \cancel{9} \quad \cancel{10}$$
$$11 \quad \cancel{12} \quad 13 \quad \cancel{14} \quad \cancel{15} \quad \cancel{16} \quad 17 \quad \cancel{18} \quad 19 \quad \cancel{20}.$$

The first number after 3 that is not crossed out is 5, so circle 5 and cross out every 5th number after 5. After we do this, the next number after 5 that is not crossed out is 7, so we cross out every 7th number after 7. Listing all numbers up to 50, we now have

$$\begin{array}{cccccccccc}
1 & ② & ③ & \not{4} & ⑤ & \not{6} & ⑦ & \not{8} & \not{9} & \not{10} \\
11 & \not{12} & 13 & \not{14} & \not{15} & \not{16} & 17 & \not{18} & 19 & \not{20} \\
\not{21} & \not{22} & 23 & \not{24} & \not{25} & \not{26} & \not{27} & \not{28} & 29 & \not{30} \\
31 & \not{32} & \not{33} & \not{34} & \not{35} & \not{36} & 37 & \not{38} & \not{39} & \not{40} \\
41 & \not{42} & 43 & \not{44} & \not{45} & \not{46} & 47 & \not{48} & \not{49} & \not{50}.
\end{array}$$

The numbers that remain are 1 and the prime numbers up to 50. We can stop at 7 because of the following.

Proposition 1.8. *If n is composite, then n has a prime factor $p \leq \sqrt{n}$.*

Proof. Since n is composite, we can write $n = ab$ with $1 < a \leq b < n$. Then

$$a^2 \leq ab = n,$$

so $a \leq \sqrt{n}$. Let p be a prime number dividing a. Then $p \leq a \leq \sqrt{n}$. □

The proposition says that the composite numbers up to 50 all have prime factors at most $\sqrt{50} \approx 7.07$, so we could stop after crossing out the multiples of 7. If we want to list the primes up to 1000, we need to cross out the multiples of only the primes through 31 (since $\sqrt{1000} \approx 31.6$).

Why is the process called a sieve? In our example, the multiples of the primes 2, 3, 5, 7 created a net. The numbers that fell through this net are the prime numbers.

CHECK YOUR UNDERSTANDING

4. Use the Sieve of Eratosthenes to compute the prime numbers less than 20.

1.6 The Division Algorithm

If a and b are integers, when we divide a by b we get an integer if and only if $b \mid a$. What can we say when b does not divide a? We can still make a statement using only integers by considering remainders. For example, we can say that 14 divided by 3 is 4 with a remainder of 2. We write this as
$$14 = 3 \cdot 4 + 2$$
to emphasize that our division statement can be written using addition and multiplication of integers. This is just the division with remainder that is taught in elementary school. Our next theorem says that this can always be done.

The Division Algorithm. *Let a and b be integers with $b > 0$. Then there exist unique integers q (the quotient) and r (the remainder) so that*
$$a = bq + r$$
with $0 \leq r < b$.

Proof. Let q be the largest integer less than or equal to a/b, so
$$q \leq a/b < q+1.$$
Multiplying by b yields $bq \leq a < bq + b$, which implies that $0 \leq a - bq < b$. Let $r = a - bq$. Then
$$0 \leq r < b.$$
Since $a = bq + r$, we have proved that the desired q and r exist.
It remains to prove that q and r are unique. If
$$a = bq + r = bq_1 + r_1$$
with $0 \leq r, r_1 < b$ then
$$b(q - q_1) = r_1 - r.$$
Since the left-hand side of this equation is a multiple of b, so is $r_1 - r$. Because $0 \leq r, r_1 < b$, we must have
$$-b < r_1 - r < b. \tag{1.2}$$

The only multiple of b that satisfies equation (1.2) is 0, so $r_1 - r = 0$. Therefore, $r_1 = r$ and the choice of r is unique. Since $b(q - q_1) = r_1 - r$, we now have $b(q - q_1) = 0$. Finally, because $b \neq 0$, we get that $q_1 - q = 0$, so $q_1 = q$ and q is also unique. This completes the proof. □

Examples. (a) Let $a = 27, b = 7$. Then $27 = 7 \cdot 3 + 6$, so $q = 3$ and $r = 6$.
(b) Let $a = -27, b = 7$. Then $-27 = 7 \cdot (-4) + 1$, so $q = -4$ and $r = 1$.
(c) Let $a = 24, b = 8$. Then $24 = 8 \cdot 3$, so $q = 3$ and $r = 0$.
(d) Let $a = 0$ and $b = 5$. Then $0 = 5 \cdot 0 + 0$, so $q = 0$ and $r = 0$.

CHECK YOUR UNDERSTANDING

5. Let $a = 200, b = 7$. Compute q and r such that $a = bq + r$ and $0 \leq r < b$.
6. Let $a = -200, b = 7$. Compute q and r such that $a = bq + r$ and $0 \leq r < b$.

1.6.1 A Cryptographic Application

Here's an amusing cryptographic application of the Division Algorithm. Let's say there is a 6-person committee that has to vote to approve a budget, but they do not want to meet in person. The members prefer to keep their votes anonymous. Here's a mathematical way to have every person vote Yes, vote No, or Abstain, while ensuring that all votes are kept secret.

We'll call the chair A_1 and the other 5 members A_2, A_3, A_4, A_5, A_6. Let's say that A_2, A_5, A_6 are going to vote Yes, A_1, A_4 are going to vote No, and A_3 will Abstain. The chair chooses a large number, say 7923, and e-mails it to A_2. Then A_2 adds 7 for Yes, 1 for No, or 0 for Abstain. Since she's voting Yes, she adds 7 to get $7923 + 7 = 7930$. She e-mails 7930 to A_3. Because A_3 does not know the original number 7923, there is no way for him to know how A_2 voted. Since A_3 is abstaining, he adds 0 to 7930 and e-mails 7930 to A_4. Since A_4 is voting No, she adds 1 to 7930 and e-mails 7931 to A_5. Since A_5 is voting Yes, she adds 7 to 7931 and e-mails 7938 to A_6. Then, A_6 adds 7 to 7938 and e-mails 7945 to the chair, A_1. Now A_1 votes No by adding 1 to 7945 to get

7946. Finally, A_1 subtracts the original number 7923 from 7946 to get 23, which is the amount arising from the sum of the voting. The chair then divides 7 into 23, as in the Division Algorithm, to get

$$23 = 7 \times 3 + 2.$$

The chair announces that 3 people voted Yes, 2 people voted No, and there was 1 abstention. Why? The sum 23 is a sum of three 7s (= Yes votes) and two 1s (= No votes). That accounts for five votes. Therefore, there is also one abstention.

Why do we count a Yes vote as 7 in this example? It's one more than the number of voters. If we used 6 for a Yes vote, then 6 No votes, and one Yes plus 5 abstentions, both give a total of 6.

Let's do another example with 23 people voting. Let's say the chair's random number is 27938. Now, committee members add 24 if they vote Yes and 1 if they vote No. We'll tell you what the votes were so that you can see why the method works. Let's say there are 16 Yes votes, 5 No votes, and 2 abstentions. Then the chair receives the number

$$27938 + 16 \cdot 24 + 5 + 2 \cdot 0 = 27938 + 389 = 28327.$$

Of course, when the chair subtracts 27938 from 28327 the answer is 389, and the Division Algorithm says that

$$389 = 16 \cdot 24 + 5.$$

The voting scheme does have a security flaw. If A_2 and A_4 compare notes, they can figure out how A_3 voted. Therefore, this method should be used only with a friendly committee.

1.7 The Greatest Common Divisor

The divisors of 12 are 1, 2, 3, 4, 6, and 12. The divisors of 18 are 1, 2, 3, 6, 9, and 18. Then $\{1, 2, 3, 6\}$ is the set of common divisors of 12 and 18. Notice that this set has a largest element, 6. If you have any two nonzero integers a and b, you can always form the set of their common divisors. Since 1 is a divisor of every integer, this set is nonempty.

Because this set is finite, it must have a largest element. This idea is so basic, we make special note of it:

Definition 1.9. *Assume that a and b are integers and they are not both zero. Then the set of their common divisors has a largest element d, called the* **greatest common divisor** *of a and b. We write $d = \gcd(a, b)$.*

Examples. $\gcd(24, 52) = 4$, $\gcd(9, 27) = 9$, $\gcd(14, 35) = 7$, $\gcd(15, 28) = 1$.

Definition 1.10. *Two integers a and b are said to be* **relatively prime** *if $\gcd(a, b) = 1$.*

Examples. 14 and 15 are relatively prime. So are 21 and 40.

Remark. If $a \neq 0$, then $\gcd(a, 0) = a$. However, we do not define $\gcd(0, 0)$. Since arbitrarily large integers divide 0, there is no largest divisor. This is the reason we often explicitly write that at least one of a and b is nonzero when we are going to make a statement about $\gcd(a, b)$. In any case, whenever we write $\gcd(a, b)$, it is implicitly assumed that at least one of a and b is nonzero.

We saw that $\gcd(24, 52) = 4$, so 24 and 52 are not relatively prime. If we divide both 24 and 52 by 4, we get 6 and 13, which are relatively prime. This makes sense, since we've divided these numbers by their gcd, which is the largest possible common divisor. We now prove in the following that dividing two integers by their gcd always results in two relatively prime integers.

Proposition 1.11. *If a and b are integers with $d = \gcd(a, b)$, then*

$$\gcd\left(\frac{a}{d}, \frac{b}{d}\right) = 1.$$

Proof. If $c = \gcd(a/d, b/d)$, then $c \mid (a/d)$ and $c \mid (b/d)$. This means that there are integers k_1 and k_2 with

$$\frac{a}{d} = ck_1 \text{ and } \frac{b}{d} = ck_2,$$

which tells us that $a = cdk_1$ and $b = cdk_2$. So, cd is a common divisor of a and b. Since d is the greatest common divisor and $cd \geq d$, we must have $c = 1$. $\square$

Section 1.7   The Greatest Common Divisor

We'll see later that calculating the greatest common divisor has important applications. So, it's natural to ask, how do we go about finding the gcd when the answer is not immediately obvious? One way would be to factor each integer into primes and then take the product of all the primes that they have in common, including repetitions. For example, to find gcd(84, 264), we write

$$84 = 2 \cdot 2 \cdot 3 \cdot 7 \text{ and } 264 = 2 \cdot 2 \cdot 2 \cdot 3 \cdot 11,$$

so their common primes are 2, 2, and 3. We see that $\gcd(84, 264) = 2 \cdot 2 \cdot 3 = 12$. This may seem to be quite efficient, but as we'll see later on, for the numbers of the size (i.e., hundreds of digits) that we'll be interested in, factoring is so slow as to be completely impractical. It's much easier to calculate $\gcd(a, b)$ by the method of the next section.

For reasonably small numbers, Proposition 1.4 is useful for calculating gcd's. For example, suppose we want to calculate $d = \gcd(1005, 500)$. Then $d \mid 1005$ and $d \mid 500$, so $d \mid 1005 - 2 \cdot 500$. Therefore, $d \mid 5$, which means that $d = 1$ or 5. Since $5 \mid 1005$ and $5 \mid 500$, we see that $5 = \gcd(1005, 500)$.

As another example, suppose n is an integer and we want to find all possibilities for $d = \gcd(2n+3, 3n-6)$. By Proposition 1.4,

$$d \mid 2n+3, \quad d \mid 3n-6$$
$$\implies d \mid 3(2n+3) - 2(3n-6) = 21,$$

so $d = 1, 3, 7,$ or 21. In fact, all possibilities occur: when $n = 1$ we have $d = \gcd(5, -3) = 1$, when $n = 3$ we have $d = \gcd(9, 3) = 3$, when $n = 2$ we have $d = \gcd(7, 0) = 7$, and when $n = 9$ we have $d = \gcd(21, 21) = 21$.

CHECK YOUR UNDERSTANDING

7. Evaluate $\gcd(24, 42)$.
8. Find an n with $1 < n < 10$ such that $\gcd(n, 60) = 1$.
9. Let n be an integer. Show that $\gcd(n, n+3) = 1$ or 3, and show that both possibilities occur.

1.8 The Euclidean Algorithm

The Euclidean Algorithm is one of the oldest and most useful algorithms in all of number theory. It is found as Proposition 2 in Book VII of Euclid's *Elements*. One of its features is that it allows us to compute gcd's without factoring. In cryptographic situations, where the numbers often have several hundred digits and are hard to factor, this is very important.

Suppose that we want to compute gcd(123, 456). Consider the following calculation:

$$456 = 3 \cdot 123 + 87$$
$$123 = 1 \cdot 87 + 36$$
$$87 = 2 \cdot 36 + 15$$
$$36 = 2 \cdot 15 + 6$$
$$15 = 2 \cdot 6 + 3$$
$$6 = 2 \cdot 3 + 0.$$

By looking at the prime factorizations of 456 and 123, we see that the last nonzero remainder, namely 3, is the gcd. Let's look at what we did. We divided the smaller of the original two numbers into the larger and got the remainder 87. Then we shifted the 123 and the 87 to the left, did the division, and got a remainder of 36. We continued the "shift left and divide" procedure until we got a remainder of 0.

Let's try another example. Compute gcd(119, 259):

$$259 = 2 \cdot 119 + 21$$
$$119 = 5 \cdot 21 + 14$$
$$21 = 1 \cdot 14 + 7$$
$$14 = 2 \cdot 7 + 0.$$

Again, the last nonzero remainder is the gcd. Why does this work? Let's start by showing why 7 is a common divisor in the second example. The fact that the remainder on the last line is 0 says that $7 \mid 14$. Since $7 \mid 7$ and $7 \mid 14$, the next-to-last line says that $7 \mid 21$, since 21 is a linear combination of 7 and 14. Now move up one line. We have just shown

Section 1.8   The Euclidean Algorithm   31

that $7 \mid 14$ and $7 \mid 21$. Since 119 is a linear combination of 21 and 14, we deduce that $7 \mid 119$. Finally, moving to the top line, we see that $7 \mid 259$ because 259 is a linear combination of 119 and 21, both of which are multiples of 7. Since $7 \mid 119$ and $7 \mid 259$, we have proved that 7 is a common divisor of 119 and 259.

We now want to show that 7 is the largest common divisor. Let d be any divisor of 119 and 259. The top line implies that 21, which is a linear combination of 259 and 119 (namely, $259 - 2 \cdot 119$), is a multiple of d. Next, go to the second line. Both 119 and 21 are multiples of d, so 14 must be a multiple of d. The third line tells us that since $d \mid 21$ and $d \mid 14$, we must have $d \mid 7$. In particular, $d \leq 7$, so 7 is the greatest common divisor, as claimed. We also have proved the additional fact that any common divisor must divide 7.

All of this generalizes to the following:

Euclidean Algorithm. *Let a and b be nonnegative integers and assume that $b \neq 0$. Do the following computation:*

$$a = q_1 b + r_1, \text{ with } 0 \leq r_1 < b$$
$$b = q_2 r_1 + r_2, \text{ with } 0 \leq r_2 < r_1$$
$$r_1 = q_3 r_2 + r_3, \text{ with } 0 \leq r_3 < r_2$$
$$\vdots$$
$$r_{n-3} = q_{n-1} r_{n-2} + r_{n-1}, \text{ with } 0 \leq r_{n-1} < r_{n-2}$$
$$r_{n-2} = q_n r_{n-1} + 0.$$

The last nonzero remainder, namely r_{n-1}, equals $\gcd(a,b)$.

The proof that $r_{n-1} = \gcd(a,b)$ follows exactly the steps used in the example of $7 = \gcd(259, 119)$. Since the last remainder is 0, r_{n-1} divides r_{n-2}. The next-to-last line yields $r_{n-1} \mid r_{n-3}$. Moving up, line by line, we eventually find that r_{n-1} is a common divisor of a and b.

Now suppose that d is a common divisor of a and b. The first line yields $d \mid r_1$. Since $d \mid b$ and $d \mid r_1$, the second line yields $d \mid r_2$. Continuing downwards, line by line, we eventually find that $d \mid r_{n-1}$. Therefore, $d \leq r_{n-1}$, so r_{n-1} is the largest divisor, which means that $r_{n-1} = \gcd(a,b)$. We also obtain the extra fact that each common divisor of a and b divides $\gcd(a,b)$.

There is a geometrical way to view the Euclidean Algorithm. For example, suppose we want to compute $\gcd(48, 21)$. Start at the point $(48, 21)$

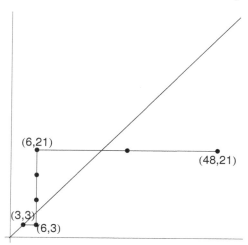

FIGURE 1.1: Computation of gcd(48, 21).

in the plane. Move to the left in steps of size 21 until you land on or cross the line $y = x$. In this case, we take two steps of size 21 and move to $(6, 21)$. Now move downward in steps of size 6 (the smaller of the two coordinates) until you land on or cross the line $y = x$. In this case, we take three steps of size 6 and move to $(6, 3)$. Now move to the left in steps of size 3. In one step we end up at $(3, 3)$ on the line $y = x$. The x-coordinate (also the y-coordinate) is the gcd.

In each set of moves, the number of steps before the next corner is the quotient in the Euclidean Algorithm. Can you see how to determine the remainders in Figure 1.1?

1.8.1 The Extended Euclidean Algorithm

The Euclidean Algorithm yields an amazing and very useful fact: $\gcd(a, b)$ can be expressed as a linear combination of a and b. That is, there exist integers x and y such that $\gcd(a, b) = ax + by$. For example,

$$3 = \gcd(456, 123) = 456 \cdot 17 - 123 \cdot 63$$
$$7 = \gcd(259, 119) = 259 \cdot 6 - 119 \cdot 13.$$

The method for obtaining x and y is called the **Extended Euclidean Algorithm**. Once you've used the Euclidean Algorithm to arrive at

Section 1.8 The Euclidean Algorithm   33

gcd(a, b), there's an easy and very straightforward way to implement the Extended Euclidean Algorithm. We'll show you how it works with the two examples we've just calculated.

When we computed gcd(456, 123), we performed the following calculation:

$$456 = 3 \cdot 123 + 87$$
$$123 = 1 \cdot 87 + 36$$
$$87 = 2 \cdot 36 + 15$$
$$36 = 2 \cdot 15 + 6$$
$$15 = 2 \cdot 6 + 3$$
$$6 = 2 \cdot 3 + 0.$$

We'll form a table with three columns and explain how they arise as we compute them.

We begin by forming two rows and three columns. The first entries in the rows are the numbers we started with. In this case, these numbers are 456 and 123. The columns tell us how to form each of these numbers as a linear combination of 456 and 123. In other words, we will always have

$$\text{entry in first column} = 456x + 123y,$$

where x and y are integers. Initially, this is trivial: $456 = 1 \cdot 456 + 0 \cdot 123$ and $123 = 0 \cdot 456 + 1 \cdot 123$:

	x	y	
456	1	0	($456 = 1 \cdot 456 + 0 \cdot 123$)
123	0	1	($123 = 0 \cdot 456 + 1 \cdot 123$).

Now things get more interesting. If we look at the first line in our gcd(456, 123) calculation, we see $456 = 3 \cdot 123 + 87$. We rewrite this as $87 = 456 - 3 \cdot 123$. Using this as a guide, we compute

$$(\text{1st row}) \; - 3 \cdot (\text{2nd row}),$$

yielding the following:

$$\begin{array}{ccc} & x & y \\ 456 & 1 & 0 \\ 123 & 0 & 1 \\ 87 & 1 & -3 \end{array} \quad \text{(1st row)} - 3\cdot\text{(2nd row)}.$$

The last line tells us that $87 = 456 \cdot 1 + 123 \cdot (-3)$.

We now move to the second row of our gcd calculation. This says that $123 = 1 \cdot 87 + 36$, which we rewrite as $36 = 123 - 1 \cdot 87$. Again, in the column and row language, this tells us to compute (2nd row) − (3rd row). We write this as

$$\begin{array}{ccc} & x & y \\ 456 & 1 & 0 \\ 123 & 0 & 1 \\ 87 & 1 & -3 \\ 36 & -1 & 4 \end{array} \quad \text{(2nd row)} - \text{(3rd row)}.$$

The last line tells us that $36 = 456 \cdot (-1) + 123 \cdot 4$.

Moving to the third row of our gcd calculation, we see that $15 = 87 - 2 \cdot 36 = $ (3rd row)$-2\cdot$(4th row) in our row and column language. This becomes

$$\begin{array}{ccc} & x & y \\ 456 & 1 & 0 \\ 123 & 0 & 1 \\ 87 & 1 & -3 \\ 36 & -1 & 4 \\ 15 & 3 & -11 \end{array} \quad \text{(3rd row)} - 2\cdot\text{(4th row)}.$$

We continue in this way and end when we have $3 = \gcd(456, 123)$ in the first column:

$$\begin{array}{ccc} & x & y \\ 456 & 1 & 0 \\ 123 & 0 & 1 \\ 87 & 1 & -3 \\ 36 & -1 & 4 \\ 15 & 3 & -11 \\ 6 & -7 & 26 \\ 3 & 17 & -63 \end{array} \quad \begin{array}{l} \\ \\ \\ \\ \\ \text{(4th row)} - 2\cdot\text{(5th row)} \\ \text{(5th row)} - 2\cdot\text{(6th row)}. \end{array}$$

Section 1.8 The Euclidean Algorithm

This tells us that $3 = 456 \cdot 17 + 123 \cdot (-63)$.

Notice that as we proceeded, we were doing the Euclidean Algorithm in the first column. The first entry of each row is a remainder from the gcd calculation, and the second and third entries allow us to express the number in the first column as a linear combination of 456 and 123. The quotients in the Euclidean Algorithm told us what to multiply a row by before subtracting it from the previous row.

Here's another example, where we calculate gcd(259, 119). You should go step-by-step to make sure that you understand how we're arriving at the numbers in each row.

	x	y	
259	1	0	
119	0	1	
21	1	−2	(1st row) − 2·(2nd row)
14	−5	11	(2nd row) − 5·(3rd row)
7	6	−13	(3rd row) − (4th row).

The end result is $7 = 259 \cdot 6 - 119 \cdot 13$.

To summarize, we state the following.

Theorem 1.12. *Let a and b be integers with at least one of a, b nonzero. There exist integers x and y, which can be found by the Extended Euclidean Algorithm, such that*

$$\gcd(a, b) = ax + by.$$

Proof. Although it would be fairly straightforward to write a detailed proof that follows the reasoning of the examples, the numerous indices and variables would make the proof rather unenlightening. Therefore, we spare the reader. Instead, we give the following nonconstructive proof that $\gcd(a, b)$ is a linear combination of a and b.

What do we mean by a "nonconstructive proof"? When you read the following, you'll see that we prove the existence of x and y without showing how to find them. In contrast, the Extended Euclidean Algorithm gives a method to find x and y. A proof based on this algorithm would show how to find x and y and therefore would be a "constructive proof."

Let S be the set of integers that can be written in the form $ax+by$ with integers x and y. Since a, b, $-a$, and $-b$ are in S, we see that S contains at least one positive integer. Let d be the smallest positive integer in S (this is an application of the Well Ordering Principle; see Appendix A). Since $d \in S$, we know that $d = ax_0 + by_0$ for some integers x_0 and y_0. We claim that a and b are multiples of d, so d is a *common divisor* of both a and b. To see this, write $a = dq + r$ with integers q and r such that $0 \leq r < d$. Since

$$r = a - dq = a - (ax_0 + by_0)q = a(1 - x_0 q) + b(-y_0 q),$$

we have that $r \in S$. Since d is the smallest positive element of S and $0 \leq r < d$, we must have $r = 0$. This means that $d \mid a$. Similarly, $d \mid b$, so d is a common divisor of a and b.

Now suppose that e is any common divisor of a and b. Proposition 1.4 implies that e divides $ax_0 + by_0 = d$, so $e \leq d$. Therefore, d is the greatest common divisor. By construction, d is a linear combination of a and b. □

Finally, we give a version of Theorem 1.12 that applies to more than two numbers. If $a_1, a_2, \ldots, a_n$ are integers and at least one of them is nonzero, their greatest common divisor is defined to be the largest integer that divides each a_i. For example, $\gcd(30, 42, 70) = 2$.

Theorem 1.13. *Let $n \geq 2$ and let $a_1, a_2, \ldots, a_n$ be integers (at least one of them must be nonzero). Then there exist integers $x_1, x_2, \ldots, x_n$ such that*

$$\gcd(a_1, a_2, \ldots, a_n) = a_1 x_1 + a_2 x_2 + \cdots + a_n x_n.$$

Proof. The proof is a generalization of the proof of Theorem 1.12. Let S be the set of positive integers s that can be written in the form

$$s = a_1 x_1 + a_2 x_2 + \cdots + a_n x_n,$$

where $x_1, x_2, \ldots, x_n$ are integers (positive, negative, or zero). As in the proof of Theorem 1.12, the fact that at least one a_i is nonzero implies that S is nonempty. Let d be the smallest element of S. The fact that d exists is a consequence of the Well Ordering Principle; see Appendix A). Since $d \in S$, we know that

$$d = a_1 y_1 + a_2 y_2 + \cdots + a_n y_n,$$

Section 1.9   Other Bases   37

for some integers $y_1, y_2, \ldots, y_n$.
We claim that $d \mid a_i$ for each i. Fix i and write $a_i = qd + r$ with $0 \le r < d$. Then

$$r = a_i - qd = a_i - q(a_1y_1 + a_2y_2 + \cdots + a_ny_n),$$

which is a linear combination of $a_1, a_2, \ldots, a_n$. If $r > 0$, then $r \in S$, which contradicts the assumption that d is the smallest element of S. Therefore, $r = 0$, so $a_i = qd$, which means that $d \mid a_i$. We have proved that d is a common divisor of all the a_i.

Now suppose that e is a common divisor of all the a_i. Since $e \mid a_i$ for each i, it follows that d divides any linear combination of them, so $e \mid d$. Therefore, $e \le d$, so d is the largest common divisor. This completes the proof. □

Theorem 1.12 and its generalization, Theorem 1.13, are among the most important tools in number theory, and they'll be used to deduce many fundamental properties of the integers. The following is an example.

Proposition 1.14. *Let a, b, c be integers with $a \ne 0$ and $\gcd(a, b) = 1$. If $a \mid bc$, then $a \mid c$.*

Proof. Theorem 1.12 says that we can write $1 = ax + by$ for some integers x and y. Multiply by c to obtain $c = acx + bcy$. Since $a \mid a$ and $a \mid bc$, Proposition 1.4 implies that $a \mid c$. □

CHECK YOUR UNDERSTANDING

10. Compute $\gcd(654, 321)$ without factoring.
11. Find x and y such that $17x + 12y = 1$.

1.9   Other Bases

The numbers that we use in our everyday lives are written using base 10 notation. For example, 783 means $7 \cdot 10^2 + 8 \cdot 10^1 + 3 \cdot 10^0$. The position of each digit tells us what power of 10 that digit will be multiplied by to give us our number, so 58 and 85 represent different numbers because of the positions of the 5 and 8. In the past there have been other ways to represent integers. When Abraham Lincoln wrote the Gettysburg Address, he didn't begin with "Eighty-seven years ago," but with "Four score and seven years ago," using the word *score* (which comes from the Norse *skar*, meaning *mark* or *tall*) for the number 20. In Britain, people still say they weigh 10 stone 7 pounds instead of 147 pounds, using the word stone for 14 from an old unit of measurement.

Our reliance on base 10 is most likely an accident of evolution, and is a reflection of the 10 fingers that we use to count. The Babylonians used a base 60 for their number system, and the Mayans used base 20. (Perhaps they also used their toes.) Computers are based on a binary system and often use base 16 $(= 2^4)$ to represent numbers.

If we have a number in a different base, let's say base 7, then it's easy to rewrite it as a base 10 number. Let's say we have 3524_7, where the subscript 7 means we are working in base 7. Then,

$$3524_7 = 3 \cdot 7^3 + 5 \cdot 7^2 + 2 \cdot 7^1 + 4 \cdot 7^0 = 3 \cdot 343 + 5 \cdot 49 + 2 \cdot 7 + 4 \cdot 1 = 1292_{10}.$$

We can also convert a number from base 10 to any other base with the use of the Division Algorithm.

We give three examples to show how this works.

Example. Convert the base 10 number 21963 to a base 8 number.

We proceed by dividing 21963 by 8, then dividing the quotient by 8, and continuing until the quotient is 0. At the end of the example, we'll show why the process works.

$$\begin{aligned} 21963 &= 2745 \cdot 8 + 3 \\ 2745 &= 343 \cdot 8 + 1 \\ 343 &= 42 \cdot 8 + 7 \\ 42 &= 5 \cdot 8 + 2 \\ 5 &= 0 \cdot 8 + 5. \end{aligned}$$

This tells us that $21963_{10} = 52713_8$. To see why this works, we start from the beginning, making sure to group our factors of 8 together.

$$21963 = 2745 \cdot 8 + 3 = (343 \cdot 8 + 1)8 + 3 =$$
$$343 \cdot 8^2 + 1 \cdot 8 + 3 = (42 \cdot 8 + 7)8^2 + 1 \cdot 8 + 3 =$$

Section 1.9   Other Bases   39

$$42 \cdot 8^3 + 7 \cdot 8^2 + 1 \cdot 8 + 3 = (5 \cdot 8 + 2)8^3 + 7 \cdot 8^2 + 1 \cdot 8 + 3 =$$
$$5 \cdot 8^4 + 2 \cdot 8^3 + 7 \cdot 8^2 + 1 \cdot 8 + 3 =$$
$$52713_8.$$

Example. Convert the base 10 number 1671 to base 2.

$$
\begin{aligned}
1671 &= 835 \cdot 2 + 1 \\
835 &= 417 \cdot 2 + 1 \\
417 &= 208 \cdot 2 + 1 \\
208 &= 104 \cdot 2 + 0 \\
104 &= 52 \cdot 2 + 0 \\
52 &= 26 \cdot 2 + 0 \\
26 &= 13 \cdot 2 + 0 \\
13 &= 6 \cdot 2 + 1 \\
6 &= 3 \cdot 2 + 0 \\
3 &= 1 \cdot 2 + 1 \\
1 &= 0 \cdot 2 + 1.
\end{aligned}
$$

So, $1671_{10} = 11010000111_2$.

Example. It's always a good idea to make sure that any mathematical method works for an example where you already know the answer. This serves as a type of "reality check." So, let's take a base 10 number, say 314159, and use the above algorithm to "convert" it to base 10:

$$
\begin{aligned}
314159 &= 31415 \cdot 10 + 9 \\
31415 &= 3141 \cdot 10 + 5 \\
3141 &= 314 \cdot 10 + 1 \\
314 &= 31 \cdot 10 + 4 \\
31 &= 3 \cdot 10 + 1 \\
3 &= 0 \cdot 10 + 3.
\end{aligned}
$$

It should be reassuring that this gives back the original 314159.

CHECK YOUR UNDERSTANDING

12. Convert 1234_{10} to base 7.
13. Convert 321_5 to base 10.

1.10 Fermat and Mersenne Numbers

When Euclid proved that there are infinitely many primes, he did so by showing that there cannot be a largest prime number. Nevertheless, there is a long history, going back at least 2000 years ago to the Sieve of Eratosthenes, of finding primes, especially large ones. It is remarkable that this area of mathematics, which was long considered a somewhat recreational academic exercise, has now turned out to have important cryptographic uses. We'll discuss these applications in Chapter 7.

Most of the large primes that are found are too large (and not random enough) for cryptographic applications, but they are often a way to test algorithmic advances. Since exponential functions grow very rapidly, it seems reasonable to look for exponential expressions that take on prime values in order to find large prime numbers. This was the motivation behind the work of some sixteenth and seventeenth century mathematicians, whose ideas on generating large primes still spark our interest today.

The most common function to look at was $2^n - 1$. Primes of this form appear in Euclid's *Elements* in his discussion of perfect numbers (see Chapter 13). Many people have thought that this expression is always prime if n is prime, but Hudalricus Regius, in 1536, seems to have been the first person to realize that $2^{11} - 1 = 23 \cdot 89$. (This was before calculators, computers, and the widespread knowledge of algebra. Factoring was very hard and very slow.) Today, numbers of the form

$$M_n = 2^n - 1 \tag{1.3}$$

are called **Mersenne numbers**, after the French monk Marin Mersenne who lived from 1588 to 1648. Mersenne, and others before him, were most likely aware that in order for M_n to be prime, n must be prime. This is the content of the following result.

Proposition 1.15. *If n is composite then $2^n - 1$ is composite.*

Proof. Recall that for every $k \geq 1$,

$$x^k - 1 = (x-1)\left(x^{k-1} + x^{k-2} + x^{k-3} + \cdots + x + 1\right) \tag{1.4}$$

Section 1.10   Fermat and Mersenne Numbers   41

(if we divide both sides of equation (1.4) by $x-1$, we get the well-known formula for a geometric sum; see Appendix A). Since n is composite, $n = ab$ with $1 < a < n$ and $1 < b < n$. Substitute $x = 2^a$ and $k = b$ to get
$$2^{ab} - 1 = (2^a - 1)\left(2^{a(b-1)} + 2^{a(b-2)} + \cdots + 2^a + 1\right).$$
Since $1 < a < n$, we have $1 < 2^a - 1 < 2^n - 1$, so the factor $2^a - 1$ is nontrivial. This means that $2^{ab} - 1$ is composite. □

The proposition tells us that if we want to find n such that M_n is prime, we should look only at prime values of n. This works well at the beginning:
$$2^2 - 1 = 3, \quad 2^3 - 1 = 7, \quad 2^5 - 1 = 31, \quad 2^7 - 1 = 127.$$
All of these are primes. However, Hudalricus Regius's example with $n = 11$ shows that if n is prime, M_n may still be composite. Continuing past the case $n = 11$, we find that
$$2^{13} - 1, \quad 2^{17} - 1, \quad 2^{19} - 1$$
are prime, but $2^{23} - 1$ is composite. As of 2014, there are 48 values of n for which M_n is known to be prime.

In honor of Mersenne, who made some early (and partly incorrect) lists of values of n that make M_n prime, a prime number of the form M_n is called a *Mersenne prime*.

Because of the special form of Mersenne primes, there are very fast tests (see the Lucas–Lehmer Test in Chapter 11) to determine whether M_n is a prime number. Whenever you read that a new "largest" prime has been discovered, you'll likely see that it's a Mersenne prime. In fact, for much of recent history, the largest known prime has been a Mersenne prime. In 1876, Eduoard Lucas proved that M_{127} is prime, and that remained the largest known prime until 1951, when the non-Mersenne $180(M_{127})^2 + 1$ became the champion. In 1952, Mersenne primes took back the lead, with M_{521} being proved prime. Subsequently, larger and larger Mersenne primes were discovered through 1985, when M_{216091} became the leader. Then, in 1989 the non-Mersenne $391581 \times 2^{216193} - 1$ took over and reigned until 1992, when M_{756839} regained the title for the Mersennes. Since then, the largest known prime has always been Mersenne. As of July 2014, the largest known Mersenne prime was

$2^{57885161} - 1$. The base 10 expansion of this number has more than 17 million digits. The Great Internet Mersenne Prime Search (GIMPS) is an Internet-based distributed computing project with numerous volunteers using spare time on their computers to carry out the search for Mersenne primes. The 14 most recent Mersenne primes were found by GIMPS.

Instead of looking at $2^n - 1$, the great French mathematician Pierre de Fermat (1601–1665) examined $2^m + 1$. Fermat realized that the only way for $2^m + 1$ to be prime is for the exponent m to be a power of 2.

Proposition 1.16. *If $m > 1$ is not a power of 2, then $2^m + 1$ is composite.*

Proof. If k is odd then

$$x^k + 1 = (x+1)\left(x^{k-1} - x^{k-2} + x^{k-3} - \cdots - x + 1\right).$$

Since m is not a power of 2, it has a nontrivial odd factor: $m = ab$ with a an odd integer and $a \geq 2$. Let $k = a$ and let $x = 2^b$. Then

$$2^{ab} + 1 = (2^b + 1)\left(2^{b(a-1)} - 2^{b(a-2)} + 2^{b(a-3)} - \cdots - 2^b + 1\right).$$

Since $1 \leq b < m$, we have $1 < 2^b + 1 < 2^m + 1$, so the factor $2^b + 1$ is nontrivial. Therefore, $2^{ab} + 1$ is composite. $\square$

The proposition tells us that if we want to find m such that $2^m + 1$ is prime, we should take m to be a power of 2, leading to the definition of the nth Fermat number:

$$F_n = 2^{2^n} + 1.$$

Fermat incorrectly believed that F_n is prime for every integer $n \geq 0$. Although this is true for small values of n:

$$F_0 = 2^1 + 1 = 3, \quad F_1 = 2^2 + 1 = 5, \quad F_2 = 2^4 + 1 = 17,$$
$$F_3 = 2^8 + 1 = 257, \quad F_4 = 2^{16} + 1 = 65537,$$

Euler showed in 1732 that

$$F_5 = 4294967297 = 641 \cdot 6700417.$$

More recently, F_n has been factored for $6 \leq n \leq 11$, and many more have been proved to be composite, although they are yet to be factored. Many people now believe that F_n is *never* prime if $n \geq 5$.

Fermat primes occur in compass and straightedge constructions in geometry. Using only a compass and a straightedge, it is easy to make an equilateral triangle or a square. It's a little harder to make a regular pentagon, but it's possible. The constructions of equilateral triangles and regular pentagons can be combined to produce a regular 15-gon. Moreover, by bisecting angles, it's easy to double the number of sides of a polygon that is already constructed, so 30-sided, 60-sided, and 120-sided regular polygons can be constructed. What else can be done? In 1796, just before he turned 19, Gauss discovered how to construct a regular 17-gon. Moreover, his methods (completed by Wantzel) yield the following:

Let $n \geq 3$. A regular n-gon can be constructed by compass and straightedge if and only if n is a power of 2 times a product of distinct Fermat primes:
$$n = 2^a F_{n_1} \cdot F_{n_2} \cdots F_{n_r}$$
with $a \geq 0$ and $r \geq 0$.

As mentioned above, the largest known Fermat prime is 65537. Johann Hermes spent 10 years writing down the explicit construction of a 65537-gon, finishing in 1894. The result is stored in a box in Göttingen.

1.11 Chapter Highlights

1. Euclid's theorem: There are infinitely many prime numbers.
2. The Euclidean Algorithm for computing gcd's.
3. If $\gcd(a,b) = 1$, then there exist integers x, y with $ax + by = 1$.

1.12 Problems

1.12.1 Exercises

Section 1.2: Divisibility

1. Show that $5 \mid 120$, $\quad 11 \mid 165$, $\quad$ and $14 \mid 98$.
2. Show that $7 \mid 7$, $\quad 10 \nmid 25$, $\quad$ and $32 \mid -160$.
3. Find all positive divisors of the following integers:
   (a) 20 $\quad$ (b) 52 $\quad$ (c) 195 $\quad$ (d) 203
4. Find all positive divisors of the following integers:
   (a) 12 $\quad$ (b) 13 $\quad$ (c) 15 $\quad$ (c) 16
5. Prove or give a counterexample for the following statements:
   (a) If $10 \mid a$ and $10 \mid b$, then $10 \mid ab$.
   (b) If $10 \mid a$ and $10 \mid b$, then $100 \mid ab$.
   (c) If $10 \nmid a$ and $10 \nmid b$, then $10 \nmid (a+b)$.
6. Prove or give a counterexample for the following statements:
   (a) If $5 \mid a$ and $5 \mid b$, then $5 \mid ab$.
   (b) If $5 \mid a$ and $5 \mid b$, then $25 \mid ab$.
   (c) If $5 \nmid a$ and $5 \nmid b$, then $5 \nmid (a+b)$.
7. Prove or give a counterexample for the following statements:
   (a) If $c \mid a$ and $c \mid b$, then $c \mid ab$.
   (b) If $c \mid a$ and $c \mid b$, then $c^2 \mid ab$.
   (c) If $c \nmid a$ and $c \nmid b$, then $c \nmid (a+b)$.
8. Recall that a number n is even if $n = 2k$ and is odd if $n = 2k+1$. Prove the following: (*Hint:* Look at Section 1.1.)
   (a) The sum of an odd number and an even number is odd.
   (b) The product of an odd number and an even number is even.
   (c) The product of two odd numbers is odd.
9. Find all integers n (positive or negative) such that $n^2 - n$ is prime. (*Hint:* Factor $n^2 - n$).
10. Show that if a is any integer, then 3 divides $a^3 - a$. (*Hint:* Factor $a^3 - a$. What do you notice when you write out the three factors in increasing order?)

11. (a) Find all primes p such that $3p+1$ is a square. (*Hint:* Write $3p+1 = n^2$. Notice that $3p = (n-1)(n+1)$.)
    (b) Find all primes such that $5p+1$ is a square.
    (c) Find all primes such that $29p+1$ is a square.
    (d) If we fix a prime p and ask to find all primes q so that $qp+1$ is a square, what relationship must p and q have if we find a q that works?

12. (a) If $n \geq 2$, show that each of
    $$n!+2,\ n!+3,\ n!+4,\ \ldots,\ n!+n$$
    is composite. (*Hint:* 2 divides $n!+2$ and 3 divides $n!+3$.)
    (b) Find a list of 100 consecutive composite integers.

13. (a) Find all prime numbers that can be written as a difference of two squares. (*Hint:* A difference of two squares can be factored.)
    (b) Find all prime numbers that can be written as a difference of two fourth powers.

Section 1.3: Euclid's Theorem

14. Let p_k denote the kth prime. Show that
    $$p_{n+1} \leq p_1 p_2 p_3 \cdots p_n + 1$$
    for all $n \geq 1$.
    (*Hint:* Let $N = p_1 p_2 p_3 \cdots p_n + 1$. A prime factor of N must be at least as large as p_{n+1}.)

15. Consider the numbers
    $$2+1,\quad 2\cdot 3+1,\quad 2\cdot 3\cdot 5+1,\quad 2\cdot 3\cdot 5\cdot 7+1,\quad \cdots.$$
    Show, by computing several values, that there are composite numbers in this sequence. (This shows that in the proof of Euclid's theorem, these numbers are not necessarily prime, so it is necessary to look at prime factors of these numbers.)

16. The famous Goldbach Conjecture says that every even integer $n \geq 4$ is a sum of two primes. This conjecture is not yet proved. Prove the weaker statement that there are infinitely many even integers that are sums of two primes. (*Hint:* There is an infinite number of primes.)

17. It is conjectured, but not proved, that there is always a prime between every two consecutive squares n^2 and $(n+1)^2$ for $n \geq 1$. Prove the weaker statement that there are infinitely many n such that there is a prime between n^2 and $(n+1)^2$. (*Hint*: There is an infinite number of primes.)

Section 1.6: The Division Algorithm

18. For each of the following, find the quotient and remainder when a is divided by b in the Division Algorithm.
    (a) $a = 43$, $b = 7$ (b) $a = 96$, $b = 11$,
    (c) $a = 140$, $b = 12$ (d) $a = -200$, $b = 21$.

19. For each of the following, find the quotient and remainder when a is divided by b in the Division Algorithm.
    (a) $a = 7$, $b = 7$ (b) $a = 9$, $b = 11$,
    (c) $a = 246$, $b = 10$ (d) $a = 18$, $b = 3$.

20. For each of the following, find the quotient and remainder when a is divided by b in the Division Algorithm.
    (a) $a = 29$, $b = 2$ (b) $a = 53$, $b = 6$,
    (c) $a = 205$, $b = 13$ (d) $a = -87$, $b = 5$.

21. For each of the following, find the quotient and remainder when a is divided by b in the Division Algorithm.
    (a) $a = 12$, $b = 3$ (b) $a = 45$, $b = 97$,
    (c) $a = 138$, $b = 111$ (d) $a = 170$, $b = 20$.

22. When dividing a by b in the Division Algorithm, find the quotient q and remainder r when
    (a) $1 \leq a = b$.
    Hint: What happens if $a = b = 7$? If $a = b = 12$?)
    (b) $a = kb$, k an integer.
    (*Hint*: What happens if $a = 5$ and $b = 15$? If $a = 6$ and $b = 12$?
    (c) $0 \leq a < b$.

23. Find all integers n (positive, negative, or zero) so that $n^2 + 1$ is divisible by $n + 1$. (*Hint*: Subtract a multiple of $n+1$ from $n^2 + 1$. Then, $n + 1$ must divide the remainder. What does that force $n + 1$ to be?)

24. Find all integers n (positive, negative, or zero) so that $n^3 - 1$ is divisible by $n + 1$. (*Hint*: Divide $n^3 - 1$ by $n + 1$. Subtract

Section 1.12   Problems

a multiple of $n+1$ from $n^3 - 1$. Then, $n+1$ must divide the remainder. What does that force $n+1$ to be?)

Section 1.7: The Greatest Common Divisor

25. Evaluate
    (a) $\gcd(6, 14)$
    (b) $\gcd(13, 10)$
    (c) $\gcd(10, 25)$

26. Evaluate
    (a) $\gcd(6, 9)$
    (b) $\gcd(10, 14)$
    (c) $\gcd(8, -9)$

27. Evaluate
    (a) $\gcd(12, 36)$
    (b) $\gcd(3, 0)$
    (c) $\gcd(6, 7)$

28. Show that if n is an integer, then $\gcd(n, n+1) = 1$. (*Hint*: Look at "Check Your Understanding 9.")

29. Show that if n is an integer, then $\gcd(2n-1, 2n+1) = 1$. (*Hint*: Look at "Check Your Understanding 9.")

30. If $\gcd(a, b) = 1$, prove that $\gcd(a+b, a-b) = 1$ or 2. (*Hint*: The first line of your proof should be "Let $d = \gcd(a+b, a-b)$." Then use Corollary 1.12.)

31. Show that if $\gcd(a, b) = 1$ and $c \mid a$, then $\gcd(b, c) = 1$. (*Hint*: Show that $\gcd(b, c)$ divides both a and b.)

32. Let a be an integer. Find all possibilities for $\gcd(5a+1, 12a+9)$ (there are four of them). For each possibility, find an integer a such that the gcd has that value. (*Hint:* Look at the end of Section 1.7.)

33. Show that $n! + 1$ and $(n+1)! + 1$ are relatively prime. (*Hint*: Try multiplying $n! + 1$ by $n+1$ and then using various linear combinations.)

34. Let n be an integer.
    (a) Show that $\gcd(n^2 + n + 6, n^2 + n + 4) = 2$. (*Hint*: Note that $n^2 + n$ is always even and the gcd divides the difference

of the two polynomials.)
(b) Show that $\gcd(n^2 + n + 5, n^2 + n + 3) = 1$.

Section 1.8: The Euclidean Algorithm

35. Evaluate each of the following and use the Extended Euclidean Algorithm to express each $\gcd(a, b)$ as a linear combination of a and b.
    (a) $\gcd(14, 100)$
    (b) $\gcd(6, 84)$
    (c) $\gcd(182, 630)$
    (d) $\gcd(1776, 1848)$

36. Evaluate each of the following and use the Extended Euclidean Algorithm to express each $\gcd(a, b)$ as a linear combination of a and b.
    (a) $\gcd(13, 203)$
    (b) $\gcd(57, 209)$
    (c) $\gcd(465, 2205)$
    (d) $\gcd(1066, 42)$

37. Let n be an integer. Show that $\gcd(n^2, n^2 + n + 1) = 1$. (*Hint*: Use the Euclidean Algorithm to divide $n^2 + n + 1$ by n^2).

38. (a) Compute $\gcd(89, 55)$.
    (b) If F_n denotes the nth Fibonacci number, describe the quotients and remainders in the Euclidean Algorithm for $\gcd(F_{n+1}, F_n)$. (*Hint*: If $F_{n+1} = F_n + F_{n-1}$, the quotient is 1 and the remainder is F_{n-1}.)

39. (a) Use the Euclidean Algorithm to compute $\gcd(13, 5)$.
    (b) Let $n = 1111111111111$ (that's thirteen 1s) and $m = 11111$. Use the Euclidean Algorithm to compute $\gcd(n, m)$.
    (c) Observe the relation between the calculations in (a) and (b). Generalize this to show that if $c = (10^a - 1)/9$ and $d = (10^b - 1)/9$ (so c has a 1s and d has b 1s), then $\gcd(c, d)$ is the number with $\gcd(a, b)$ 1s.

40. Suppose we want to compute the gcd of three numbers, for example, $\gcd(210, 294, 490)$. Do the following computations:
    (a) Compute $d' = \gcd(210, 294)$.
    (b) Compute $d = \gcd(d', 490)$.

(c) It was pointed out in the discussion following the Euclidean Algorithm that every common divisor of a and b divides $\gcd(a,b)$. Use this fact (first applied to d', then to d) to show that every common divisor of 210, 294, and 490 divides d, so d is the largest common divisor of these numbers.

Section 1.9: Other Bases

41. Express the following in base 10:
    (a) 1234_5
    (b) 10101_2
    (c) 111_{11}

42. (a) Convert the base 10 number 54321 to base 6.
    (b) Convert the base 10 number 1000000 to base 2.
    (c) Convert the base 10 number 31416 to base 7.

43. Suppose that a group of 18 billionaires wants to see how many in their group are worth 1 billion, how many are worth 2 billion, etc., but they do not want to reveal their individual wealths. The first person chooses a large integer N (around 100 digits) and computes $N + 20^i$, where i is an integer indicating a net worth of i billions (they round off to the nearest billion). This sum, call it N_1, is given to the second person. This second person computes $N_2 = N_1 + 20^j$, where j is an integer indicating a net worth of j billions. The sum N_2 is given to the third person. This continues until all 18 have added on their numbers. The first person then subtracts N. Show how to determine how many people are worth 0 billion, how many are worth 1 billion, etc. (*Hint*: Write the final number in base 20. What do the coefficients of each power of 20 tell you?)

Section 1.10: Fermat and Mersenne Numbers

44. Show that if $n > 1$ and a is a positive integer with $a^n - 1$ prime, then $a = 2$ and n is prime. (*Hint*: Look at the proof of Proposition 1.15.)

45. Show that if $n > 1$ and $a \geq 2$ is an integer with $a^n + 1$ prime, then $n = 2^k$ for some $k \geq 0$. (*Hint*: Look at the proof of Proposition 1.16.)

1.12.2 Computer Explorations

1. Define a function on positive integers by

$$f(n) = \begin{cases} n/2 \text{ if } n \text{ is even} \\ 3n+1 \text{ if } n \text{ is odd.} \end{cases}$$

   For example, $f(5) = 16$ and $f(6) = 3$. If we start with a positive integer m, we can form the sequence $m_1 = f(m)$, $m_2 = f(m_1)$, $m_3 = f(m_2)$, etc. The *Collatz Conjecture* predicts that we eventually get $m_k = 1$ for some k. For example, if we start with $m = 7$, we get $m_1 = 22$, $m_2 = 11$, $m_3 = 34$, $m_4 = 17$, $m_5 = 52$, $m_6 = 26$, $m_7 = 13$, $m_8 = 40$, $m_9 = 20$, $m_{10} = 10$, $m_{11} = 5, m_{12} = 16$, $m_{13} = 8$, $m_{14} = 4$, $m_{15} = 2$, $m_{16} = 1$.
   (a) Show that the Collatz Conjecture is true for all $m \leq 60$. Which starting value of m required the most steps?
   (b) Suppose you change $3n + 1$ to $n + 1$ in the definition of $f(n)$. What happens? Can you prove this?
   (c) Suppose you change $3n + 1$ to $5n + 1$ in the definition of $f(n)$. Try a few examples and see what happens. Do you see a different behavior for starting values $m = 5$, $m = 6$, and $m = 7$?

2. Let $\pi_{n,a}(x)$ be the number of primes of the form $nk + a$ that are less than or equal to x. For example, $\pi_{4,3}(20) = 4$ since it is counting the primes $3, 7, 11, 19$.
   (a) Let $\pi(x)$ be the number of primes less than or equal to x. Compute the ratios

$$\frac{\pi_{4,1}(x)}{\pi(x)} \text{ and } \frac{\pi_{4,3}(x)}{\pi(x)}$$

   for $x = 1000$, $x = 100000$, and $x = 1000000$. Make a guess as to what each ratio approaches as x gets larger and larger.
   (b) Show that $\pi_{4,1}(x) \leq \pi_{4,3}(x)$ for all integers $x < 30000$ except for two (consecutive) integers x. This comparison is sometimes called a "prime number race." Chebyshev observed in 1853 that usually $\pi_{4,3}(x)$ is ahead of $\pi_{4,1}(x)$, but Littlewood showed in 1914 that the lead changes infinitely often. In 1994, Rubinstein and Sarnak showed (under some yet-to-be-proved assumptions) that $\pi_{4,3}(x)$ is ahead most of the time.

3. Find 20 examples of numbers that are sixth powers. What is true about the remainders when these sixth powers are divided by 7? Can you make a conjecture and then prove it?

4. Write a program that generates 10000 "random" pairs (a, b) of integers, and then use the Euclidean Algorithm (or your software gcd routine) to decide whether $\gcd(a, b) = 1$. If m is the number of pairs that are relatively prime, calculate

$$\frac{m}{10000},$$

which is the fraction of these pairs that are relatively prime. Follow the same procedure for 10^5 pairs and then 10^6 pairs. Can you make a guess as to what number this ratio approaches? (*Hint:* It is an integer divided by π^2.)

5. By Exercise 12 above, there are arbitrarily long sequences of consecutive composite numbers. Find the first string of (at least) 10 consecutive composites, of (at least) 50 consecutive composites, and of (at least) 100 consecutive composites. How do these numbers compare in size to the numbers used in Exercise 12?

6. (a) Show that $n^4 + 4$ is composite for several values of $n \geq 2$. There is a reason for this. Let's find it.
(b) Show that in each example in (a), $n^4 + 4$ always can be factored as $a \times b$, where $a - b = 4n$.
(c) Write $a = x + 2n$ and $b = x - 2n$ for some x. Use $n^4 + 4 = ab$ to find x, and then find the numbers a and b.

1.12.3 Answers to "CHECK YOUR UNDERSTANDING"

1. Since $1001 = 7 \times 143$, we have $7 \mid 1001$.

2. Since $1005 = 7 \times 143 + 4$, 1005 is not a multiple of 7.

3. Suppose $5 \mid 2 \cdot 3 \cdot 5 \cdot 7 + 1$. Since $5 \mid 2 \cdot 3 \cdot 5 \cdot 7$, Corollary 1.5 says that $5 \mid (2 \cdot 3 \cdot 5 \cdot 7 + 1) - (2 \cdot 3 \cdot 5 \cdot 7) = 1$, which is not true. Therefore, $5 \nmid 2 \cdot 3 \cdot 5 \cdot 7 + 1$.

4. We have to use only the primes less than $\sqrt{20} \approx 4.5$, so 2 and 3 suffice. After crossing out 1 and the multiples of 2, the

numbers that remain are 2, 3, 5, 7, 9, 11, 13, 15, 17, 19. Of these, 3 crosses out 9 and 15. The numbers 2, 3, 5, 7, 11, 13, 17, 19 remain and are the primes less than 20.

5. Divide 7 into 200. The quotient is $q = 28$, and the remainder is $r = 4$, so $200 = 7 \cdot 28 + 4$.

6. We get $-200 = 7 \cdot (-28) - 4$, but we want $0 \leq r < 7$. Therefore, we write $-200 = 7 \cdot (-29) + 3$, so $q = -29$ and $r = 3$.

7. The divisors of 24 are 1, 2, 3, 4, 6, 8, 12, 24. The largest of these that divides 42 is 6, so $\gcd(24, 42) = 6$.

8. Since $60 = 2^2 \cdot 3 \cdot 5$, we need n to be odd and not a multiple of 3 or 5. The only n between 1 and 10 that satisfies these requirements is $n = 7$.

9. If $d \mid n$ and $d \mid n+3$, then $d \mid (n+3) - n = 3$. Therefore, $d = 1$ or 3, so any common divisor is either 1 or 3. If $n = 1$, then $\gcd(1, 4) = 1$, so $d = 1$ occurs. If $n = 3$, then $\gcd(3, 6) = 3$, so $d = 3$ occurs.

10. Use the Euclidean Algorithm:

$$654 = 2 \cdot 321 + 12$$
$$321 = 26 \cdot 12 + 9$$
$$12 = 1 \cdot 9 + 3$$
$$9 = 3 \cdot 3 + 0.$$

Therefore, $3 = \gcd(654, 321)$.

11. Use the Extended Euclidean Algorithm:

	x	y	
17	1	0	
12	0	1	
5	1	−1	(1st row) −(2nd row)
2	−2	3	(2nd row) − 2·(3rd row)
1	5	−7	(3rd row) − 2·(4th row).

The end result is $1 = 17 \cdot 5 - 12 \cdot 7$, so $x = 5$ and $y = -7$.

Section 1.12 Problems

12. Use the Division Algorithm:

$$\begin{aligned} 1234 &= 176 \cdot 7 + 2 \\ 176 &= 25 \cdot 7 + 1 \\ 25 &= 3 \cdot 7 + 4 \\ 3 &= 0 \cdot 7 + 3. \end{aligned}$$

Therefore, $1234_{10} = 3412_7$.

13. $321_5 = 3 \cdot 5^2 + 2 \cdot 5 + 1 = 75 + 10 + 1 = 86.$

Chapter 2

Linear Diophantine Equations

2.1 $ax + by = c$

As we mentioned in the introduction, Diophantus lived in Alexandria, Egypt, about 1800 years ago. His book *Arithmetica* gave methods for solving various algebraic equations and had a great influence on the development of algebra and number theory for many years. The part of number theory called Diophantine equations, which studies integer (and sometimes rational) solutions of equations, is named in his honor.

In this section we study the equation

$$ax + by = c,$$

where a, b, and c are integers. Our goal is to find out when **integer** solutions to this equation exist, and when they do exist, to find all of them.

Equations of this form can arise in real life. For example, how many dimes and quarters are needed to pay someone $1.05? This means we have to solve $10x + 25y = 105$. One solution is $x = 3$, $y = 3$. Another solution is $x = 8$, $y = 1$. There are also solutions such as $x = -2$, $y = 5$, which means you pay 5 quarters and get back 2 dimes.

Before we get to the main result of this section, we look at two more examples that will help us understand the general situation. First, consider $6x - 9y = 20$. Notice that 3 must divide the left-hand side, but 3 is not a divisor of the right-hand side. This tells us that this equation can never have an integer solution. To make things notationally simpler, let $d = \gcd(a, b)$. We then see that in order for $ax + by = c$ to have a solution, we must have $d \mid c$. Now let's look at an example where this does occur, say $6x + 9y = 21$. We can divide both sides by 3, giving us $2x + 3y = 7$. After a brief inspection, we see that $x = 2$ and $y = 1$ is a solution. Are there others? It's easy to see that if t is any integer, then $x = 2 + 3t$ and $y = 1 - 2t$ is also a solution. Let's verify this by substituting these expressions for x and y into the original equation, $6x + 9y = 21$:

$$6(2 + 3t) + 9(1 - 2t) = 12 + 18t + 9 - 18t = 21,$$

so our single solution gives rise to an infinite number of them. This can be generalized in the following theorem:

Theorem 2.1. *Assume that a, b, and c are integers, where at least one of a, b is nonzero. Then the equation*

$$ax + by = c \tag{2.1}$$

has a solution if and only if $\gcd(a, b) \mid c$. If it has one solution, then it has an infinite number. If (x_0, y_0) is any particular solution, then all solutions are of the form

$$x = x_0 + \frac{b}{\gcd(a,b)} t, \qquad y = y_0 - \frac{a}{\gcd(a,b)} t \tag{2.2}$$

with t an integer.

Proof. We begin by setting $\gcd(a, b) = d$. We have already seen that if $d \nmid c$, then there are no solutions. Now, assume $d \mid c$. From Theorem 1.12 we know that there are integers r and s so that $ar + bs = d$. Since $d \mid c$, we have that $df = c$ for some integer f. Therefore,

$$a(rf) + b(sf) = df = c.$$

So, $x_0 = rf$ and $y_0 = sf$ is a solution to $ax + by = c$.
Now let

$$x = x_0 + \frac{b}{d} t \text{ and } y = y_0 - \frac{a}{d} t.$$

Then

$$ax + by = a(x_0 + \frac{b}{d}t) + b(y_0 - \frac{a}{d}t) = ax_0 + by_0 + \frac{ab}{d}t - \frac{ba}{d}t = c.$$

This shows that a solution to (2.1) exists (assuming that $\gcd(a,b)$ divides c) and that once we have one solution, we have an infinite number of a specific form.

Next, we need to prove that every solution of Equation (2.1) is of the stated form. Fix one solution x_0, y_0 and let u, v be *any* solution of Equation (2.1). (Any solution continues to mean any integer solution.) Then

$$au + bv = c \tag{2.3}$$

and

$$ax_0 + by_0 = c. \tag{2.4}$$

Subtracting Equation (2.4) from Equation (2.3) gives us

$$a(u - x_0) + b(v - y_0) = 0,$$

so

$$a(u - x_0) = -b(v - y_0) = b(y_0 - v). \tag{2.5}$$

After dividing both sides of Equation (2.5) by d, we get

$$\frac{a}{d}(u - x_0) = \frac{b}{d}(y_0 - v). \tag{2.6}$$

There is a small technicality that needs to be dealt with. If $a = 0$, then we can't say that a/d divides the right-hand side, because we don't allow 0 to divide anything. But if $a = 0$ then our original equation is $by = c$. This means that $v = y_0 = c/b$ and x can be arbitrary, since there is no restriction on x. This is exactly the conclusion of the theorem, which says that all solutions have the form $y = y_0$ and $x = x_0 + t$ (since $\gcd(a,b) = \gcd(0,b) = b$). For the rest of the proof, we now assume that $a \neq 0$.

Equation (2.6) implies that

$$(a/d) \mid (b/d)(y_0 - v).$$

Since $\gcd(a/d, b/d) = 1$, Proposition 1.14 implies that (a/d) divides $(y_0 - v)$. By definition, this means that there is an integer t with

$$y_0 - v = t\frac{a}{d}. \tag{2.7}$$

Substituting the value for $y_0 - v$ from (2.7) into (2.6), we get

$$\frac{a}{d}(u - x_0) = \frac{b}{d}\left(\frac{a}{d}t\right).\tag{2.8}$$

Multiplying both sides by $\dfrac{d}{a}$, we have

$$u - x_0 = \frac{b}{d}t \quad \text{or} \quad u = x_0 + \frac{b}{d}t.\tag{2.9}$$

Combining (2.7) and (2.9), we have

$$u = x_0 + \frac{b}{d}t \quad \text{and} \quad v = y_0 - \frac{a}{d}t.\tag{2.10}$$

Since u and v were arbitrary solutions of (2.1), we have completed the proof. □

In practice, if we want to solve Equation (2.1), we first verify that $d \mid c$. If it doesn't, we're done since there are no solutions. If it does, we divide both sides by d to get a new equation

$$a'x + b'y = c',$$

and in this equation, $\gcd(a', b') = 1$. For example, if we want to solve $6x + 15y = 30$, we divide by 3 and instead solve $2x + 5y = 10$. This means that we will usually be using the following:

Corollary 2.2. *Assume that $a, b,$ and c are integers with at least one of a, b nonzero. If $\gcd(a, b) = 1$, then the equation*

$$ax + by = c$$

always has an infinite number of solutions. If (x_0, y_0) is any particular solution, then all solutions are of the form

$$x = x_0 + bt, \qquad y = y_0 - at$$

with t an integer.

It may seem that we've ignored the problem of actually finding a solution to a linear Diophantine equation; however, the Extended Euclidean Algorithm from the previous chapter provides an efficient method. For example, to solve $13x + 7y = 5$, we write $\gcd(7, 13) = 1$ as a linear combination of 7 and 13 and then multiply our solution by 5. Here's how it works.

We begin by calculating $\gcd(7, 13)$ using the Euclidean Algorithm.

$$13 = 1 \cdot 7 + 6$$
$$7 = 1 \cdot 6 + 1$$
$$6 = 6 \cdot 1 + 0.$$

Now, we use the Extended Euclidean Algorithm to express 1 as a linear combination of 7 and 13:

	x	y	
13	1	0	
7	0	1	
6	1	-1	(1st row) $-$ (2nd row)
1	-1	2	(2nd row) $-$ (3rd row).

We see that $1 = -1 \cdot 13 + 2 \cdot 7$, so that $x = -1$, $y = 2$ is a solution to $13x + 7y = 1$:

$$13(-1) + 7(2) = 1.$$

Multiplying both x and y by 5 gives us $x = -5$, $y = 10$ is the desired solution to the original equation, $13x + 7y = 5$:

$$13(-5) + 7(10) = 5.$$

Theorem 2.1 tells us that all solutions have the form

$$x = -5 + 7t, \quad y = 10 - 13t,$$

where t is an integer.

Here is another example. Let's find all solutions of $10x + 25y = 105$, the equation for paying \$1.05 in dimes and quarters. First, divide by $5 = \gcd(10, 25)$ to get

$$2x + 5y = 21.$$

At this point, you can find a solution by any method. For example, you can try values until something works or use the Extended Euclidean

Algorithm. In any case, one solution is $x_0 = 8$, $y_0 = 1$. The set of all solutions is

$$x = 8 + 5t, \quad y = 1 - 2t.$$

The solution $x = 3$, $y = 3$ given at the beginning of this section is obtained by letting $t = -1$. The solution with $x = -2$, $y = 5$ is obtained by letting $t = -2$.

Now, a warning. It's quite possible that two people working on the same problem may get correct answers that look different. If a problem says find all solutions to $5x - 3y = 1$, you may notice that $(2, 3)$ is a particular solution, so all solutions look like $x = 2 - 3t$, $y = 3 - 5t$. A friend may choose a particular solution to be $(-1, -2)$ and say that all solutions are of the form $x = -1 - 3t, y = -2 - 5t$. These two apparently different sets of solutions are in fact the same, as the following shows:

Solutions of the form $x = 2 - 3t, y = 3 - 5t$:
$$\ldots, (-4, -7), (-1, -2), \mathbf{(2, 3)}, (5, 8), (8, 13), (11, 18), \ldots$$

Solutions of the form $x = -1 - 3t, y = -2 - 5t$:
$$\ldots, (-4, -7), \mathbf{(-1, -2)}, (2, 3), (5, 8), (8, 13), (11, 18), \ldots$$

CHECK YOUR UNDERSTANDING

1. Find all integer solutions to $6x + 8y = 4$.

2.2 The Postage Stamp Problem

If you went to the post office to mail a letter and discovered that they had only three-cent and five-cent stamps, what postage values would you be able to put on your mail? What values are unobtainable from these two stamps? These questions are special cases of what is called the Postage Stamp Problem.

The Postage Stamp Problem: If a and b are positive integers, what

Section 2.2   The Postage Stamp Problem   61

positive integers can be written as $ax + by$ with both x and y nonnegative?

To begin, notice that we want to consider only the case where a and b are relatively prime. If, for example, they were both even, then no odd numbers would ever be expressible as a linear combination of them, and the problem becomes less interesting.

We'll call numbers that can be written as $ax + by$ with both x and y nonnegative **feasible**. For example, $a, b,$ and ab are always feasible since

$$a = 1 \cdot a + 0 \cdot b, \quad b = 0 \cdot a + 1 \cdot b, \text{ and } ab = b \cdot a + 0 \cdot b = 0 \cdot a + a \cdot b.$$

The requirement that x and y both be nonnegative is what makes this an interesting problem. For example, our initial question had three-cent and five-cent stamps, so $a = 3$ and $b = 5$. Since 3 and 5 are relatively prime, if negative coefficients were allowed, then *every* integer could be expressed as a linear combination of them from Theorem 1.12. Let's try to understand which numbers are feasible and which are not by making a chart to see if any patterns occur:

Postage Stamp Problem with $a = 3$ and $b = 5$

Number	1	2	3	4	5	6	7	8	9	10	11	12	13	14
Feasible			✓		✓	✓		✓	✓	✓	✓	✓	✓	✓

An empty space means that the number above it cannot be written as a permissible linear combination of 3 and 5, while a ✓ means that it can be. For example, 13 has a ✓ underneath it because $13 = 1 \cdot 3 + 2 \cdot 5$.

Notice that every number in our list that's greater than 7 is feasible. In fact, once we notice that the three consecutive integers 8, 9, and 10 are feasible, we can show that every number greater than 10 is also feasible. Here's how the argument goes. Every number greater than 10 differs from 8, 9, or 10 by a multiple of 3. (You can use the Division Algorithm to formally prove this.) So, to get any number greater than 10, just add enough three-cent stamps to get the desired amount. For example, suppose you want 92 cents. Since $92 - 8 = 84 = 3 \cdot 28$, take the three-cent and five-cent stamps needed to get eight cents, and add 28 more 3-cent stamps to get 92 cents.

One more thing to point out before we do another example. The last number in our list that is not feasible is 7, which can be written as $3 \cdot 5 - 3 - 5$. Let's see if this pattern holds.

Postage Stamp Problem with $a = 2$ and $b = 7$

Number	1	2	3	4	5	6	7	8	9	10	11	12	13	14
Feasible		✓		✓		✓	✓	✓	✓	✓	✓	✓	✓	✓

Again, $5 = 2 \cdot 7 - 2 - 7$ is the last number that is not feasible. If you try another small example (say, with $a = 4$ and $b = 9$), you will see the same thing occurs, namely every integer greater than $ab - a - b$ can be written as a linear combination of a and b while $ab - a - b$ cannot be. We'll now use two steps to show that this is always the case.

- $ab - a - b$ is not feasible.

- If $m > ab - a - b$, then m is feasible.

We begin by verifying the first step.

Proposition 2.3. *Assume that $a > 1$ and $b > 1$ are positive, relatively prime integers. Then there are no nonnegative integers x and y with $ax + by = ab - a - b$.*

Proof. We start by observing that

$$a(-1) + b(a - 1) = ab - a - b,$$

so $x = -1, y = a - 1$ is a solution to $ax + by = ab - a - b$.
Since $\gcd(a, b) = 1$, we see from Corollary 2.2 that every solution to $ax + by = ab - a - b$ has the form

$$x = -1 + bt, \quad y = a - 1 - at$$

for some integer t. In order to get an x value that's nonnegative, we have to choose a value of t that's positive. But if $t > 0$, then $1 - t \leq 0$ so $y = a - 1 - at = a(1 - t) - 1 \leq -1$. Forcing x to be nonnegative forces y to be negative. This means it is impossible to find a nonnegative solution to $ax + by = ab - a - b$. □

Section 2.2   The Postage Stamp Problem   63

We've now shown that $ab - a - b$ can never be feasible. We next show that every integer larger than $ab - a - b$ is feasible.

Proposition 2.4. *Assume that a and b are positive, with $\gcd(a, b) = 1$. If $n \geq ab - a - b + 1$, there are nonnegative integers x and y with $ax + by = n$.*

Proof. We again appeal to Corollary 2.2. We find a pair of integers (x_0, y_0) with
$$ax_0 + by_0 = n \geq ab - a - b + 1 \qquad (2.11)$$
which enables us to express every solution in the form
$$x = x_0 + bt, \quad y = y_0 - at.$$
It is possible that x_0 or y_0 is negative. What we'll do is find the solution to $ax + by = n$ with the smallest possible nonnegative y, and then show that the corresponding x must be nonnegative. After all, this is our best chance, because making y smaller makes x bigger, so if anything works, it must be the smallest nonnegative value of y. Using the Division Algorithm, we can divide y_0 by a and write $y_0 = at + y_1$, with $0 \leq y_1 \leq a - 1$, for some integer t. This y_1 is our choice of y. Since $y_1 = y_0 - at$, we take $x_1 = x_0 + bt$ as our choice of x. We'll show that $x_1 \geq 0$. If $x_1 \leq -1$, then, because $y_1 \leq a - 1$,

$$\begin{aligned} n &= ax_0 + by_0 \\ &= a(x_1 - bt) + b(y_1 + at) \\ &= ax_1 + by_1 \\ &\leq a(-1) + b(a - 1) \\ &= ab - a - b, \end{aligned}$$

and this contradicts our assumption that $n \geq ab - a - b + 1$. So, (x_1, y_1) is a nonnegative solution. □

Somewhat surprisingly, if there are three or more stamps, there are no known formulas analogous to those we have when there are two stamps.

2.3 Chapter Highlights

1. Finding solutions of linear Diophantine equations $ax + by = c$ using the Extended Euclidean Algorithm.

2.4 Problems

2.4.1 Exercises

Section 2.1: $ax + by = c$

1. For each of the following equations, find all integral solutions or show that it has none.
   (a) $3x + 4y = 10$
   (b) $5x - 7y = 9$
   (c) $9x + 23y = 1$
   (d) $4x + 6y = 11$

2. For each of the following equations, find all integral solutions or show that it has none.
   (a) $8x + 2y = 26$
   (b) $44x - 17y = 9$
   (c) $60x + 9y = 31$
   (d) $60x + 9y = 51$

3. A farmer pays 1770 crowns in purchasing horses and oxen. He pays 31 crowns for each horse and 21 crowns for each ox. How many horses and oxen did he buy? (This problem is from Euler's *Algebra*, which was published in 1770.)

4. You cash a check at the bank, but the teller accidentally switches the dollars and the cents. The amount you are given is 47 cents more than twice the correct amount. What was the amount on the original check? Assume that the number of dollars and the number of cents are less than 100. (*Hint:* If

the original check is for x dollars and y cents, then the check was for $100x + y$. You received $100y + x$.)

5. Let a, b, c be integers, with $\gcd(a,b) = 1$ and with $a > 0$ and $b > 0$. Show that

$$ax - by = c$$

has infinitely many solutions in positive integers x, y.

6. You have nickels, dimes, and quarters, for a total of 20 coins. If the total value of the coins is $2.00, what are the possibilities for the numbers of nickels, dimes, and quarters? (Assume that you have at least one of each type of coin.)

7. An all-you-can-eat restaurant charges $15 for teenagers and $13 for adults. At the end of a day, the restaurant had collected $500. What is the largest number of people who could have eaten there that day? What is the smallest number of people?

Section 2.2: The Postage Stamp Problem

8. (a) Suppose you have only 13-dollar and 7-dollar bills. You need to pay someone 71 dollars. Is this possible without receiving change? If so, show how to do it. If not, explain why it is impossible.
(b) Suppose you need to pay someone 75 dollars. Is this possible? If so, show how to do it. If not, explain why it is impossible.

9. (a) In the Postage Stamp Problem with stamps worth 3 cents and 5 cents, use the fact that 8, 9, and 10 are feasible to show that
   (i) 100 is feasible, (ii) 260 is feasible,
   (iii) 302 is feasible.
(b) If we have stamps of denomination a and b and we can show that the a consecutive numbers $k, k+1, k+2, ..., k+a-1$ are feasible, show that every number greater than $k + a - 1$ must also be feasible.

2.4.2 Answers to "CHECK YOUR UNDERSTANDING"

1. Use Theorem 2.1. Trial and error, for example, yields the solution $(x_0, y_0) = (-2, 2)$. Since $b/\gcd(a,b) = 8/2 = 4$ and $a/\gcd(a,b) = 6/2 = 3$, all solutions are given by

$$x = -2 + 4t, \qquad y = 2 - 3t,$$

where t is an integer (there are similar solutions corresponding to other choices of (x_0, y_0)).

Chapter 3

Unique Factorization

We already know from Lemma 1.7 that every integer greater than 1 has a prime factor, and we'll see that this easily implies that each such integer is prime or a product of primes. The main purpose of this chapter is to prove that this factorization into primes can be done in only one way.

3.1 Preliminary Results

The key to everything we do in this chapter is a consequence of the Division Algorithm. Recall Theorem 1.12: *Let a and b be integers with at least one of a, b nonzero. There exist integers x and y, which can be found by the Extended Euclidean Algorithm, such that*

$$\gcd(a, b) = ax + by.$$

An immediate consequence is the following, which we previously saw was true when we proved that the Euclidean Algorithm gives us the greatest common divisor. However, it is now easy to give another proof.

Corollary 3.1. *If e is a common divisor of a and b, then e divides $\gcd(a, b)$.*

Proof. Both $e \mid a$ and $e \mid b$, and $\gcd(a, b)$ is a linear combination of a and b, so Proposition 1.4 implies that $e \mid \gcd(a, b)$, as desired. $\square$

The next result is the star of this section and is by far the most important property of primes. The theorem makes the transition from a property that can be verified numerically (p doesn't factor) to this important divisibility property for p.

Theorem 3.2. *Let p be prime and let a and b be integers with $p \mid ab$. Then $p \mid a$ or $p \mid b$.*

Proof. Let $d = \gcd(a, p)$. Then $d \mid p$, so $d = 1$ or p. If $d = p$, then the fact that d is a divisor of a implies that $p \mid a$, so we're done in this case. If $d = 1$, we use Theorem 1.12 to write $1 = ax + py$ for some integers x and y. Multiply by b to get

$$b = abx + pby.$$

Since $p \mid ab$ and $p \mid p$, and b is a linear combination of ab and p, Proposition 1.4 implies that $p \mid b$, so we're done in this case, too. □

Remark. If n is composite and $n \mid ab$, we cannot conclude that $n \mid a$ or $n \mid b$. For example, $6 \mid 9 \cdot 8$, but $6 \nmid 9$ and $6 \nmid 8$. In fact, when n is composite, there is always an example: write $n = ab$ with $1 < a, b < n$. Then $n \mid ab$ but $n \nmid a$ and $n \nmid b$.

Corollary 3.3. *Let p be prime and let $a_1, a_2, \ldots, a_r$ be integers such that*

$$p \mid a_1 a_2 \cdots a_r.$$

Then $p \mid a_i$ for some i.

Proof. We'll prove this by induction on r (see Appendix A). The result is clearly true for $r = 1$ since if $p \mid a_1$, then $p \mid a_1$. Suppose it is true for $r = k$ with $k \geq 2$. We'll now prove it must be true for $r = k + 1$. Let $a = a_1 a_2 \cdots a_k$ and let $b = a_{k+1}$. We are assuming that

$$p \mid a_1 a_2 \cdots a_r = a_1 a_2 \cdots a_{k+1} = ab.$$

Theorem 3.2 says that $p \mid a$ or $p \mid b$. If $p \mid a = a_1 a_2 \cdots a_k$, we have $p \mid a_i$ for some i, by the induction assumption. If $p \mid b$, then $p \mid a_{k+1}$ since $a_{k+1} = b$. □

The following consequence of Theorem 1.12 shows, in a sense, that relatively prime integers have little effect on each other when it comes to divisibility. If $\gcd(a, b) = 1$, then (assuming we already know about unique factorization into primes) a and b have no prime factors in common. In Proposition 3.4, all the prime factors of a appear in c and similarly for b. Since $\gcd(a,b) = 1$, there is no overlap in these sets of primes, so all the primes from a and from b occur in c.

Proposition 3.4. *Let a, b, c be integers, with a, b nonzero and $\gcd(a,b) = 1$. If $a \mid c$ and $b \mid c$, then $ab \mid c$.*

Proof. Theorem 1.12 says there exist integers x and y such that $ax + by = 1$. Multiply by c to obtain $acx + bcy = c$. Since $b \mid c$, we have $ab \mid ac$. Since $a \mid c$, we have $ba \mid bc$. Since c is a linear combination of ac and bc, we have $ab \mid c$. □

Proposition 3.4 is false if we don't assume that $\gcd(a,b) = 1$. For example, $6 \mid 30$ and $10 \mid 30$, but $6 \cdot 10 \nmid 30$. The problem is that both 6 and 10 have 2 in their factorizations, and $6 \cdot 10$ contains this 2 twice, which is too much for 30.

CHECK YOUR UNDERSTANDING

1. If p is prime and $p^2 \mid ab$, is it true that $p^2 \mid a$ or $p^2 \mid b$? Give a proof or a counterexample.

3.2 The Fundamental Theorem of Arithmetic

The Fundamental Theorem of Arithmetic says that any positive integer greater than 1 is either prime or can be factored in *exactly one way* as a product of primes. Put more colloquially, this says that every integer greater than 1 is built up out of primes in exactly one way. This type of theorem is called an existence-uniqueness theorem because it makes two assertions. The first, the existence part, is that every integer can be written as a product of prime numbers. This will be relatively easy to show. The second, the uniqueness part, says that this can be done in only one way. This is harder to show and its proof relies on the Euclidean Algorithm.

Although people often say that the Fundamental Theorem of Arithmetic (FTA) was first proved by Euclid, that is not entirely true. There is enough information in Euclid's *Elements* to deduce the FTA, but it's never explicitly stated there. We will have a more detailed discussion of this in the next section.

The Fundamental Theorem of Arithmetic. *Every integer $n > 1$ can be uniquely written as a prime or a product of primes.*

Proof. We begin our proof with the existence part.

Lemma 3.5. *Every integer $n > 1$ is prime or a product of primes.*

Proof. We prove this lemma by assuming there are nonprime integers that cannot be written as products of primes and arriving at a contradiction. Let

$$S = \{\text{integers } n > 1 |\ n \text{ is not prime and}$$
$$n \text{ cannot be written as a product of primes}\}.$$

Assume that S is nonempty. Since all the elements in S are positive integers, S has a smallest element; call it s. (This is an example of the Well-Ordering Principle. See Appendix A.) Now, s cannot be prime, so s must be composite. This means that $s = ab$, where a and b are integers that are greater than 1, so $1 < a < s$ and $1 < b < s$. Since s was the smallest integer that could not be written as a product of primes, we know that a is a product of primes and so is b. This means that $s = ab$ is a product of primes, contradicting that $s \in S$. The only other possibility is that S must be empty, so every integer greater than 1 is either prime or is a product of primes. □

Having shown that every integer $n > 1$ is either prime or a product of primes, we now need to show uniqueness: an integer can be written as a product of primes in only one way. The proof of this relies on Corollary 3.3. We suppose that there is at least one integer that can be written as a product of primes in more than one way. Let n be the smallest of these, so

$$n = p_1 p_2 \cdots p_r = q_1 q_2 \cdots q_s, \qquad (3.1)$$

where the p_i and q_j are primes. Since p_1 divides the left-hand side, it has to divide the right-hand side as well. From Corollary 3.3, we see

Section 3.2   The Fundamental Theorem of Arithmetic

that $p_1 \mid q_i$ for some i. After rearranging terms, we can assume that $q_i = q_1$. Since q_1 is prime, and $p_1 \mid q_1$, we must have that $p_1 = q_1$. So, we write equation (3.1) as

$$n = p_1 p_2 \cdots p_r = p_1 q_2 \cdots q_s$$

and then divide both sides by p_1. This gives us

$$m = p_2 p_3 \cdots p_r = q_2 q_3 \cdots q_s$$

for some integer m. Since the two factorizations of n were different, these must be two different factorizations of m.

Because $m = n/p_1$, we know that $m < n$. This contradicts our assumption that n was the smallest integer that could be factored as a product of primes in more than one way. Therefore, our initial assumption on nonunique factorization was incorrect, and every integer can be factored into a product of primes in exactly one way. □

Throughout the book, we'll see applications of the Fundamental Theorem of Arithmetic. We start by rephrasing divisibility in terms of prime factorizations. Often, we'll write the factorization of an integer as follows:

$$a = 2^{a_2} 3^{a_3} 5^{a_5} \cdots .$$

Each of the exponents is a nonnegative integer, and all but finitely many of the exponents are 0. For example, if $a = 20 = 2^2 5$, then $a_2 = 2, a_3 = 0, a_5 = 1, a_7 = 0, a_{11} = 0, \ldots$.

Proposition 3.6. *Let a and b be positive integers with*

$$a = 2^{a_2} 3^{a_3} 5^{a_5} \cdots \quad \text{and} \quad b = 2^{b_2} 3^{b_3} 5^{b_5} \cdots$$

the prime factorizations of a and b. Then $a \mid b$ if and only if $a_p \leq b_p$ for all p.

Proof. If $a \mid b$, then there exists c with $ac = b$. Write

$$c = 2^{c_2} 3^{c_3} 5^{c_5} \cdots .$$

Then

$$2^{a_2+c_2} 3^{a_3+c_3} 5^{a_5+c_5} \cdots = ac = b = 2^{b_2} 3^{b_3} 5^{b_5} \cdots .$$

By uniqueness of factorization into primes, we must have

$$a_p + c_p = b_p$$

for each p. In particular, $a_p \leq b_p$ for each p.

Conversely, suppose $a_p \leq b_p$ for all p. Let $c_p = b_p - a_p$, so $c_p \geq 0$. Let

$$c = 2^{c_2} 3^{c_3} 5^{c_5} \cdots.$$

Then $ac = b$ so $a \mid b$.

There is one technical point that needs to be mentioned. Only finitely many a_p and b_p are nonzero, so only finitely many c_p are nonzero. Therefore, there are only finitely many factors in the expression for c so c is an integer. (If, for some reason, infinitely many c_p were nonzero, c would not be an integer, since we would have an infinite product.) □

We immediately obtain a method to compute gcd's when the factorizations of a and b are known.

Proposition 3.7. *Let a and b be two positive integers. Let*

$$a = 2^{a_2} 3^{a_3} 5^{a_5} \cdots \text{ and } b = 2^{b_2} 3^{b_3} 5^{b_5} \cdots$$

be the prime factorizations of a and b. Let $d_p = \min(a_p, b_p)$. Then,

$$\gcd(a, b) = 2^{d_2} 3^{d_3} 5^{d_5} \cdots.$$

Example. Let $a = 720 = 2^4 3^2 5$ and $b = 6300 = 2^2 3^2 5^2 7$. Then $d_2 = 2, d_3 = 2, d_5 = 1, d_7 = 0$ (note that the exponent for 7 in a is 0). Therefore, $\gcd(720, 6300) = 2^2 3^2 5 = 180$. This method of finding gcd's is the one many people use before learning the Euclidean Algorithm. It works well for small numbers, but it has a drawback because it requires that you know the factorizations of at least one of a and b. (If you can factor only a, you can test each of the prime factors of a and see if they divide b. The prime factors of a are the only ones that can occur in the gcd, so it doesn't matter if you can't factor b completely and find its other prime factors.)

Proof. Let d be any common divisor and let

$$d = 2^{d_2} 3^{d_3} 5^{d_5} \cdots$$

Section 3.2   The Fundamental Theorem of Arithmetic

be the prime factorization of d.

Since $d \mid a$, we have $d_p \leq a_p$ for each p. Similarly, $d_p \leq b_p$ for each p. The largest common divisor occurs when each exponent d_p is as large as possible. This is when $d_p = \min(a_p, b_p)$ for each p. □

Here is another application of factorization into primes. First, we need a definition.

Definition 3.8. *A positive integer n is* **squarefree** *if all the factors that occur in its prime factorization are distinct. In other words, n is not divisible by a square larger than 1.*

Examples. 165 is squarefree since $165 = 3 \cdot 5 \cdot 11$, but 585 is *not* squarefree since $585 = 3^2 \cdot 5 \cdot 13$. Also, $250 = 2 \cdot 5^3$ is not squarefree since it is divisible by 5^2.

Proposition 3.9. *Let n be a positive integer. Then there is an integer $r \geq 1$ and a squarefree integer $s \geq 1$ such that $n = r^2 s$.*

Proof. Let
$$n = p_1^{a_1} p_2^{a_2} \cdots p_k^{a_k}$$
be the prime factorization of n. If a_i is even, write $a_i = 2b_i$ and if a_i is odd, write $a_i = 2b_i + 1$. (Notice that $b_i = \lfloor a_i/2 \rfloor$, where $\lfloor x \rfloor$ denotes the largest integer less than or equal to the real number x.) Let
$$r = p_1^{b_1} p_2^{b_2} \cdots p_k^{b_k},$$
and let s be the product of the primes p_i for which a_i is odd (if no a_i is odd, let $s = 1$). Then
$$r^2 s = p_1^{2b_1} p_2^{2b_2} \cdots p_k^{2b_k} s = p_1^{a_1} p_2^{a_2} \cdots p_k^{a_k} = n.$$
□

Example. Let $n = 2^5 \cdot 3^2 \cdot 7^3 \cdot 17$. Then $r = 2^2 \cdot 3 \cdot 7$ and $s = 2 \cdot 7 \cdot 17$.

CHECK YOUR UNDERSTANDING

2. Let $m = 2^6 5^3 7^2 17$ and let $n = 2^4 3^4 5^3 11$. Find $\gcd(m, n)$.

3. Write 1440 in the form $r^2 s$, where s is squarefree.

3.3 Euclid and the Fundamental Theorem of Arithmetic

As we said, although Euclid never proved the Fundamental Theorem of Arithmetic, he did come very close to doing so. Here are the relevant propositions from his *Elements*. Note that when he says that a number A measures a number B, he means that A divides B. This is because, if you have a stick of length A, then you can use it to accurately measure any distance B that is a multiple of A.

Book VII, Proposition 30: If two numbers by multiplying one another make some number, and any prime number measures the product, it will also measure one of the original numbers.

This is Theorem 3.2: If p divides the product of two numbers, then it divides at least one of them. This theorem played an essential role in the uniqueness piece of the Fundamental Theorem of Arithmetic.

Book VII, Proposition 31: Any composite number is measured by some prime number.

This says that every composite number is divisible by a prime and tells us that every composite number can be written as a product of primes.

Finally, Euclid does state a unique factorization type of result.

Book IX, Proposition 14: If a number be the least that is measured by prime numbers, it will not be measured by any other prime number except those originally measuring it.

In essence, this says that the least common multiple of prime numbers can be divisible only by those prime numbers. It appears that Euclid was only considering the least common multiple of *distinct* primes, so the least common multiple had to be squarefree.

As you can see, Euclid had all the ingredients necessary to prove the Fundamental Theorem of Arithmetic. Why he never did so explicitly is a mystery.

3.4 Chapter Highlights

1. If p is prime and $p \mid ab$, then $p \mid a$ or $p \mid b$ (Theorem 3.2).
2. The Fundamental Theorem of Arithmetic: Every integer greater than 1 is either prime or a product of primes in exactly one way.

3.5 Problems

3.5.1 Exercises

1. (a) Find the prime factorization of 5625.
   (b) All of the exponents in the factorization in (a) are even. What does this tell you about 5625?

2. (a) Find the prime factorization of 1728.
   (b) All of the exponents in the factorization in (a) are multiples of 3. What does this tell you about 1728?

3. Show that if p is prime and $p \mid a^2$, then $p \mid a$.

4. Suppose that p is prime and $p^2 \mid ab$. Show that if $\gcd(a,b) = 1$, then $p^2 \mid a$ or $p^2 \mid b$.

5. Let $n > 0$. Prove
   (a) If $a^n \mid b^n$, then $a \mid b$.
   (b) If $a^m \mid b^n$ and $m \geq n$, then $a \mid b$.
   (c) Give an example where $a^m \mid b^n$ with $0 < m < n$, and $a \nmid b$.

6. (a) Show that $\gcd(a^2, b^2) = \gcd(a,b)^2$.
   (b) Show that if $n \geq 1$, then $\gcd(a^n, b^n) = \gcd(a,b)^n$.

7. If $\gcd(a,b) = 1$ and $c \mid (a+b)$, show that $\gcd(a,c) = \gcd(b,c) = 1$. (*Hint:* Let $d = \gcd(a,c)$ and show that d divides $\gcd(a,b)$.)

8. Let a and b be integers. Show that if there exist integers x and y such that $ax + by = 1$, then $\gcd(a,b) = 1$.

9. What is the largest number of consecutive squarefree positive integers?

10. What is the largest number of consecutive odd squarefree positive integers?

11. Show that every positive integer n can be uniquely written as $2^r m$ with $r \geq 0$ and m a positive odd integer.

12. If p is a prime and $\gcd(a, b) = p$, what are the possible values of (a) $\gcd(a^2, b)$? (b) $\gcd(a^3, b)$?

13. Assume that a and b are integers and p is a prime. If $\gcd(a, p^2) = p$ and $\gcd(b, p^3) = p^2$, find
    (a) $\gcd(ab, p^4)$? (b) $\gcd(a+b, p^4)$.

14. The **least common multiple** of a and b, written $[a, b]$ or $\mathrm{lcm}(a, b)$, is defined to the smallest positive integer that is divisible by both a and b.
    (a) Find $[15, 21]$, $[30, 40]$, and $[5, 47]$.
    (b) Let
    $$a = 2^{a_2} 3^{a_3} 5^{a_5} \cdots \text{ and } b = 2^{b_2} 3^{b_3} 5^{b_5} \cdots$$
    be the prime factorizations of a and b. Let $c_i = \max(a_i, b_i)$. Show that
    $$[a, b] = 2^{c_2} 3^{c_3} 5^{c_5} \cdots.$$
    (c) By looking at the prime factorizations of a and b, show that
    $$[a, b] \times \gcd(a, b) = ab.$$

3.5.2 Answers to "CHECK YOUR UNDERSTANDING"

1. The statement is false. If $p^2 \mid ab$, we could have one p divide a and one p divide b. For example, $3^2 \mid 6 \cdot 12$, but $3^2 \nmid 6$ and $3^2 \nmid 12$.

2. We take the minimum of the exponents for each prime, so $\gcd(m, n) = 2^4 3^0 5^3 7^0 11^0 17^0 = 2^4 5^3$.

3. Factor 1440 as $2^5 3^2 5$. This equals $(2^2 3)^2 2 \cdot 5$, so $r = 2^2 3 = 12$ and $s = 2 \cdot 5 = 10$.

Chapter 4

Applications of Unique Factorization

4.1 A Puzzle

> There was a young lady named Chris
> Who, when asked her age, answered this.
> Two-thirds of its square
> Is a cube, I declare.
> Now what was the age of the miss?

(This limerick appeared in the *San Francisco Examiner* in the mid-1970s.)

Let A be her age and let

$$A = 2^a 3^b 5^c 7^d \cdots$$

be the prime factorization of A (for convenience, we include all primes and note that only finitely many of the exponents $a, b, c, \ldots$ are nonzero). Then two-thirds of its square is

$$(2/3)A^2 = (2/3)2^{2a} 3^{2b} 5^{2c} 7^{2d} \cdots = 2^{2a+1} 3^{2b-1} 5^{2c} 7^{2d} \cdots.$$

We need the following useful fact.

Proposition 4.1. *Let $k \geq 2$ be an integer and let m be a positive integer. Then m is a kth power if and only if all the exponents in the prime factorization of m are multiples of k.*

77

Examples. We know that $8000 = 2^6 5^3$ is a cube (it equals 20^3) since the exponents 6 and 3 are multiples of 3, but is not a sixth power since the exponent 3 is not a multiple of 6. Similarly, $2000 = 2^4 5^3$ is not a cube since the exponent 4 is not a multiple of 3.

Proof. Suppose
$$m = 2^{y_2} 3^{y_3} 5^{y_5} \cdots.$$
If each exponent y_p is a multiple of k, then we can write $y_p = k z_p$ for each prime p, so
$$m = (2^{z_2} 3^{z_3} 5^{z_5} \cdots)^k,$$
which is a kth power. Conversely, if $m = n^k$, let
$$n = 2^{w_2} 3^{w_3} 5^{w_5} \cdots$$
be the prime factorization of n. Then
$$m = n^k = 2^{kw_2} 3^{kw_3} 5^{kw_5} \cdots.$$
This is the prime factorization of m, and every exponent is a multiple of k. This proves the proposition. $\square$

We now return to the limerick. Since $(2/3)A^2$ is a cube, the proposition implies that
$$2a+1, \quad 2b-1, \quad 2c, \quad 2d, \ldots$$
are multiples of 3. An easy way to accomplish this is to take $a = 1$, $b = 2$, $c = 0$, $d = 0$, ..., which yields $A = 18$. But let's ignore the word "young" and find the general solution.

Since $3 \mid 2c$ and $\gcd(3,2) = 1$, Proposition 1.14 says that $3 \mid c$, so we can write $c = 3c'$. Similarly, d and all later exponents are multiples of 3.

Since $2a+1$ is odd and a multiple of 3, we can write $2a+1 = 3(2j+1)$, which implies $a = 3j+1$ for some j. For b, we have $2b-1 = 3(2k+1)$, hence $b = 3k+2$ for some k. We conclude that
$$A = 2^a 3^b 5^c \cdots = 2 \cdot 3^2 \cdot \left(2^j 3^k 5^{c'} \cdots\right)^3 = 18 B^3$$
for some B. If $B = 1$, we obtain Chris's age as $A = 18$. The next possibility (with $B = 2$) is $A = 18 \cdot 2^3 = 144$.

Section 4.2 Irrationality Proofs   79

CHECK YOUR UNDERSTANDING

1. How does the prime factorization of 5400 tell you that it's not a perfect cube?

4.2 Irrationality Proofs

A real number is **rational** if it can be expressed as the ratio of two integers. For example,

$$3 = 3/1, \quad 3.7 = 37/10, \quad 0 = 0/1, \quad -.001 = -1/1000$$

are rational numbers. A real number is **irrational** if it is not rational. Later, in Chapter 8, we will see that rational numbers are exactly those whose decimal expansions eventually become periodic (that is, repeating the same string of digits forever). For example,

$$\frac{6187}{3300} = 1.874848484848\ldots$$

is rational and the decimal expansion eventually becomes the string 48 repeating forever. Sometimes, irrational numbers are defined as being those whose decimal expansions never become periodic. This is correct, but it is hard to use this definition to prove that a number is irrational. If we compute a billion digits of the decimal expansion and don't find a pattern, how do we know that the repetition doesn't start at the trillionth digit?

Many well-known numbers that one encounters in calculus, for example $e = 2.71828\ldots$ and $\pi = 3.14159\ldots$, are irrational, but we will not consider them here. However, we prove the famous fact that $\sqrt{2}$ is irrational.

Theorem 4.2. $\sqrt{2}$ *is irrational.*

Proof. Suppose instead that $\sqrt{2}$ is rational. Then

$$\sqrt{2} = \frac{a}{b}$$

with some integers a, b. We assume that a/b is a reduced fraction; that is, $\gcd(a, b) = 1$. In particular, we may assume that at least one of a, b is odd.

Squaring the equation $\sqrt{2} = a/b$ yields $2 = a^2/b^2$, and therefore $2b^2 = a^2$. Since a^2 is even (it equals $2b^2$), a must be even. Write $a = 2a_1$ for some integer a_1. Then

$$2b^2 = a^2 = (2a_1)^2 = 4a_1^2.$$

Dividing by 2 yields $b^2 = 2a_1^2$, so b^2 is even. This implies that b is even. We have now shown that both a and b are even. This contradicts the fact that at least one of a, b is odd. Therefore, we conclude that we cannot write $\sqrt{2} = a/b$, so $\sqrt{2}$ is irrational. □

Using unique factorization, we can generalize the theorem.

Theorem 4.3. *Let $k \geq 2$ be an integer. Let n be a positive integer that is not a perfect kth power. Then $\sqrt[k]{n}$ is irrational.*

Examples. $\sqrt{3}$, $\sqrt[7]{16}$, and $\sqrt[3]{5}$ are irrational.

Proof. We show the equivalent statement that if $\sqrt[k]{n}$ is rational, then n is a perfect kth power. Suppose $\sqrt[k]{n} = a/b$, where a, b are positive integers. Then

$$nb^k = a^k.$$

Write

$$n = 2^{x_2} 3^{x_3} 5^{x_5} \cdots$$
$$b = 2^{z_2} 3^{z_3} 5^{z_5} \cdots$$

with exponents $x_p \geq 0$ and $z_p \geq 0$. Then

$$nb^k = 2^{x_2 + kz_2} 3^{x_3 + kz_3} \cdots .$$

Since $nb^k = a^k$ is a perfect kth power, Proposition 4.1 implies that each exponent must be a multiple of k:

$$x_p + kz_p = ky_p,$$

for some y_p. This implies that $x_p = k(y_p - z_p)$, which is a multiple of k. It follows that n is a perfect kth power. □

Section 4.3   The Rational Root Theorem

Note that the irrationality of $\sqrt{2}$ follows immediately from Theorem 4.3 by choosing both k and n to be 2. Since 2 is not a perfect square, $\sqrt{2}\;(=\sqrt[2]{2})$ cannot be a rational number.

CHECK YOUR UNDERSTANDING

2. How do you know that both $\sqrt[3]{28}$ and $\sqrt[9]{50}$ are irrational?

4.3   The Rational Root Theorem

The results of the previous section are all part of a more general result.

Theorem 4.4. (Rational Root Theorem) *Let*

$$P(x) = a_n X^n + a_{n-1} X^{n-1} + \cdots + a_1 X + a_0$$

be a polynomial with integer coefficients (that is, $a_0, a_1, \ldots, a_n$ are integers) such that $a_0 \neq 0$ and $a_n \neq 0$. If $r = u/v$ is a rational number (with $\gcd(u,v) = 1$) and $P(u/v) = 0$, then u divides a_0 and v divides a_n.

Proof. Since $P(u/v) = 0$, we have

$$a_n(u/v)^n + a_{n-1}(u/v)^{n-1} + \cdots + a_0 = 0.$$

Multiply by v^n:

$$a_n u^n + a_{n-1} u^{n-1} v + \cdots + a_1 u v^{n-1} + a_0 v^n = 0.$$

All of the terms except the first are multiples of v, so v must divide $a_n u^n$, too. Since $\gcd(u,v) = 1$, Proposition 1.14 implies that v divides a_n.

Similarly, all the terms except the last are multiples of u, so u divides $a_0 v^n$. Since $\gcd(u,v) = 1$, it follows that u divides a_0.   □

Note that, although the proof of the theorem looks easy, it relies heavily on Proposition 1.14, which was proved as a consequence of Theorem

1.12, which was a key step in proving that factorization into primes is unique.

Examples. Let $P(X) = 2X^3 - 5X^2 + X + 3$. The possible *rational* roots are
$$\pm 3, \quad \pm 1, \quad \pm\frac{3}{2}, \quad \pm\frac{1}{2}.$$
If we try these, we find that $P(3/2) = 0$.

The Rational Root Theorem works well when there are only a few factors of the highest and lowest coefficients. However, it is still possible to use it to advantage when there are many possibilities. Consider the polynomial
$$78X^3 - 547X^2 + 310X - 45.$$

The factors of 45 are
$$\pm 1, \quad \pm 3, \quad \pm 5, \quad \pm 9, \quad \pm 15, \quad \pm 45.$$

The factors of 78 are
$$\pm 1, \quad \pm 2, \quad \pm 3, \quad \pm 6, \quad \pm 13, \quad \pm 26, \quad \pm 39, \quad \pm 78.$$

There are $96 = 12 \cdot 8$ combinations:
$$\pm\frac{1}{1}, \quad \pm\frac{1}{2}, \quad \ldots, \quad \pm\frac{45}{39}, \quad \pm\frac{45}{78};$$
however, some fractions appear more than once. For example, $\frac{45}{39} = \frac{15}{13}$ and $-1/1 = -3/3$. Removing the repeats leaves 64 possible roots. These can be substituted into the polynomial to see if any of them are roots. But it is easier to proceed as follows. Numerical methods (for example, the Newton–Raphson Method or simply using a graphing calculator) yield the roots
$$0.26015\cdots, \quad .34615\cdots, \quad 6.40651\cdots.$$

If a root is rational, the Rational Root Theorem tells us that its denominator divides 78. Therefore, multiply these roots by 78:
$$20.2918\cdots, \quad 27.0000\cdots, \quad 499.7082\cdots.$$

Consequently, the first and third roots are not rational and the second one, if rational, is $27/78 = 9/26$. In fact, substituting $9/26$ into the polynomial gives 0, so we have found a rational root.

The Rational Root Theorem immediately implies that $\sqrt{2}$ is irrational: The only possible rational roots of $X^2 - 2$ are ± 1 and ± 2. Since these are not roots, we see that $X^2 - 2$ has no rational roots. In particular, $\sqrt{2}$, which is a root of this polynomial, is irrational.

Finally, we can use the Rational Root Theorem to give another proof of Theorem 4.3. Let $k \geq 2$ be an integer and let n be a positive integer. Consider the polynomial $X^k - n$. By the Rational Root Theorem, the only possible roots have denominator 1. That is, the only possible rational roots are integers. If m is an integer that is a root, then $m^k - n = 0$, so $n = m^k$ is a perfect kth power. Therefore, if n is not a perfect kth power, then $X^k - n$ has no rational roots, which means that $\sqrt[k]{n}$ is irrational. This is Theorem 4.3.

CHECK YOUR UNDERSTANDING

3. List all the possible rational roots of $2x^3 - 3x^2 + 5x - 10$. Are any of the possibilities actually roots?

4.4 Pythagorean Triples

We all have seen the relation

$$3^2 + 4^2 = 5^2,$$

corresponding to the 3-4-5 right triangle, and maybe also the relations

$$5^2 + 12^2 = 13^2, \qquad 8^2 + 15^2 = 17^2.$$

A triple (a, b, c) of positive integers satisfying

$$a^2 + b^2 = c^2$$

is called a **Pythagorean triple**. If $\gcd(a, b, c) = 1$, then the Pythagorean triple (a, b, c) is called a **primitive Pythagorean triple**. Of course, if $\gcd(a, b, c) = d$, then we may divide a, b, c by d and obtain a primitive triple. For example, $(6, 8, 10)$ is a Pythagorean triple with $\gcd(6, 8, 10) = 2$. Dividing by 2 yields the primitive triple $(3, 4, 5)$.

Questions: Are there infinitely many primitive Pythagorean triples? Is there a way of generating all of them?

There is an easy way of producing many Pythagorean triples. Take an odd number, for example, 9. Square it: $9^2 = 81$. Split the result into two pieces that differ by one: $81 = 40 + 41$. The result produces a Pythagorean triple: $9^2 + 40^2 = 41^2$. This yields many examples:

$$(3, 4, 5), \quad (5, 12, 13), \quad (7, 24, 25), \ldots.$$

Since the last two entries in each triple differ by 1, their gcd is 1, so the triples are primitive.

Algebraically, we have done the following: Let the odd number be $2n+1$. Its square is

$$(2n + 1)^2 = 4n^2 + 4n + 1.$$

Split this into two parts:

$$4n^2 + 4n + 1 = (2n^2 + 2n) + (2n^2 + 2n + 1).$$

Then $(2n + 1, \ 2n^2 + 2n, \ 2n^2 + 2n + 1)$ is a Pythagorean triple. This can be verified using the identity

$$(2n + 1)^2 + (2n^2 + 2n)^2 = (2n^2 + 2n + 1)^2.$$

There is a similar construction that uses a multiple of 4. For example, start with 8. Square it: $8^2 = 64$. Take one-fourth of this square and subtract 1 and add 1:

$$(1/4)\, 8^2 - 1 = 15 \quad \text{and} \quad (1/4)\, 8^2 + 1 = 17.$$

This gives the triple $(8, 15, 17)$. Here are some more examples:

$$(12, 35, 37), \quad (16, 63, 65), \quad \ldots.$$

Since the last two entries in each triple are odd and differ by 2, their gcd is 1, so the triples are primitive. We could do this process starting with any even integer, but the ones that are not multiples of 4 yield triples all of whose entries are even, so they are imprimitive.

We could continue to find series of triples similar to these examples, but instead, we state a general result that gives all primitive triples.

Section 4.4 Pythagorean Triples

Theorem 4.5. *Let (a, b, c) be a primitive Pythagorean triple. Then c is odd and exactly one of a, b is even. Assume that b is even. Then there exist relatively prime integers $m < n$, one even and one odd, such that*

$$a = n^2 - m^2, \quad b = 2mn, \quad c = m^2 + n^2.$$

Examples. If $n = 2$ and $m = 1$, we obtain $(a, b, c) = (3, 4, 5)$. If $n = 7$ and $m = 2$, we obtain the less familiar triple $(a, b, c) = (45, 28, 53)$.

Proof. We start with a very important lemma.

Lemma 4.6. *Let $k \geq 2$, let a, b be relatively prime positive integers such that $ab = n^k$ for some integer n. Then a and b are each kth powers of integers.*

Proof. Factor n into prime powers:

$$n = 2^{x_2} 3^{x_3} 5^{x_5} \cdots$$

with integers $x_p \geq 0$. Then

$$ab = n^k = 2^{kx_2} 3^{kx_3} 5^{kx_5} \cdots.$$

Let p be a prime occurring in the prime factorization of a; let's say p^c is the exact power of p in a. Since $\gcd(a, b) = 1$, the prime p does not occur in the factorization of b. Therefore, p^c is the power of p occurring in ab. But n^k has p^{kx_p} as the power of p in its prime factorization. Since the prime factorization of $ab = n^k$ is unique, we must have $p^c = p^{kx_p}$, so $c = kx_p$. This means that every prime in the factorization of a occurs with an exponent that is a multiple of k, and this implies that a is a kth power. By exactly the same reasoning, b is a kth power. □

We need one more lemma.

Lemma 4.7. *The square of an odd number is 1 more than a multiple of 8. The square of an even number is a multiple of 4.*

Proof. Let n be even. Then $n = 2j$ for some integer j, and $n^2 = 4j^2$, which is a multiple of 4.

Now, let n be odd, so $n = 2k + 1$ for some k. Then

$$n^2 = (2k+1)^2 = 4k(k+1) + 1.$$

Since either k or $k+1$ is even, $k(k+1)$ is a multiple of 2, so $4k(k+1)$ is a multiple of 8. Therefore, n^2 is 1 more than a multiple of 8. □

Remark. Arguments such as the one in the proof of Lemma 4.7 will be much more routine after we develop the machinery of congruences in Chapter 5.

We now can prove the theorem. Suppose

$$a^2 + b^2 = c^2$$

and $\gcd(a, b, c) = 1$.

First, there are some technical details involving parity and gcd's that must be treated — not very exciting, but necessary. Suppose that both a and b are odd. Then a^2 and b^2 are each 1 more than multiples of 8, so $a^2 + b^2$ is 2 more than a multiple of 8. In particular, $a^2 + b^2$ is even but not a multiple of 4. By Lemma 3.7, even squares are multiples of 4, so $a^2 + b^2$ cannot be a square. Therefore, at least one of a, b is even.

If both a and b are even, then $c^2 = a^2 + b^2$ is even, so c is even. But then 2 is a common divisor of a, b, c, so $\gcd(a, b, c) \neq 1$, which is contrary to our assumption that the triple is primitive. It follows that exactly one of a, b is even and the other is odd. Switching the names of a and b if necessary, we may assume that a is odd and b is even. Therefore, $a^2 + b^2$ is an odd plus an even, hence is odd, so c is odd.

Write $b = 2b_1$ so $(c+a)(c-a) = c^2 - a^2 = b^2 = 4b_1^2$. Dividing by 4 yields

$$\left(\frac{c+a}{2}\right)\left(\frac{c-a}{2}\right) = b_1^2.$$

Since c and a are odd, $c+a$ is even, so $(c+a)/2$ is an integer. Similarly, $(c-a)/2$ is an integer.

Let $d = \gcd((c+a)/2, (c-a)/2)$. Suppose $d > 1$. Let p be a prime dividing d. Then

$$c = \frac{c+a}{2} + \frac{c-a}{2} \text{ and } a = \frac{c+a}{2} - \frac{c-a}{2}$$

are multiples of p since these are sums and differences of multiples of p. Therefore, $b^2 = c^2 - a^2$ is a difference of multiples of p, hence b^2 is a multiple of p. Since p is prime and divides b^2, it must divide b, so p is

Section 4.5   Differences of Squares

a common factor of a, b, c. This contradicts the primitivity of (a, b, c). Therefore, there is no prime dividing d, so $d = 1$.

We have now arrived at the heart of the proof. We have two relatively prime positive integers, namely $(c+a)/2$ and $(c-a)/2$, whose product is a square. By Lemma 4.6, each factor is a square:

$$\frac{c-a}{2} = m^2, \quad \frac{c+a}{2} = n^2,$$

for some positive integers m, n. Then

$$c = \frac{c+a}{2} + \frac{c-a}{2} = n^2 + m^2 \text{ and } a = \frac{c+a}{2} - \frac{c-a}{2} = n^2 - m^2.$$

Moreover,

$$b^2 = c^2 - a^2 = (n^2 + m^2)^2 - (n^2 - m^2)^2 = 4m^2n^2,$$

so $b = 2mn$.

Because $(c-a)/2 = m^2$ and $(c+a)/2 = n^2$ are relatively prime, we must have $\gcd(m, n) = 1$. Finally, since $m^2 + n^2 = c$, which is odd, one of m and n is odd and the other is even.   □

CHECK YOUR UNDERSTANDING

4. Using Theorem 4.5, find the Pythagorean triple with $n = 8$ and $m = 3$.

5. Find the values of m and n in Theorem 4.5 that yield the Pythagorean triple $(5, 12, 13)$.

4.5 Differences of Squares

Perhaps you have noticed that odd numbers appear as the differences between successive squares:

$$1^2 - 0^2 = 1, \quad 2^2 - 1^2 = 3, \quad 3^2 - 2^2 = 5, \quad 4^2 - 3^2 = 7, \ldots.$$

This is a consequence of the identity

$$(n+1)^2 - n^2 = 2n + 1.$$

The identity

$$(n+1)^2 - (n-1)^2 = 4n$$

shows that every multiple of 4 is the difference of two squares. Is that all?

Theorem 4.8. *Let m be a positive integer. Then m is the difference of two squares if and only if either m is odd or m is a multiple of 4.*

Proof. We have shown that odd m and multiples of 4 are differences of squares. We need to show the converse. Suppose

$$m = x^2 - y^2 = (x+y)(x-y).$$

Since $x+y$ and $x-y$ differ by the even number $2y$, they are both even or both odd. If they are both even, then $m = (x+y)(x-y)$ is a product of two even numbers, hence a multiple of 4. If they are both odd, then m is a product of two odd numbers, so m is odd. Therefore, either m is odd or m is a multiple of 4. □

A natural question is how many ways can m be written as the difference of two squares? If $m = x^2 - y^2 = (x+y)(x-y)$, then m is the product of two factors $(x+y)$ and $(x-y)$. As we just saw, both factors have the same parity; that is, both are even or both are odd. Conversely, suppose $m = uv$ where u, v have the same parity. Assume $u \geq v$. Let $x = (u+v)/2$ and $y = (u-v)/2$. Then x, y are integers (because u, v have the same parity) and

$$x^2 - y^2 = \frac{(u+v)^2}{4} - \frac{(u-v)^2}{4} = uv = m.$$

Therefore, the ways of writing m as the difference of two squares correspond to the factorizations of m into two factors of the same parity. Both Euclid and Diophantus were aware of this result. Euclid's method appears in his *Elements* (Book II, Propositions 9 and 10) and is geometric. Diophantus's method appears in his *Arithmetica* (Book III, Proposition 19).

Section 4.6   Chapter Highlights

Examples.
Let $m = 15$. Then

$$15 \times 1 \longleftrightarrow 8^2 - 7^2 = 15 \quad \text{(since } 8 + 7 = 15 \text{ and } 8 - 7 = 1\text{)}$$
$$5 \times 3 \longleftrightarrow 4^2 - 1^2 = 15 \quad \text{(since } 4 + 1 = 5 \text{ and } 4 - 1 = 3\text{).}$$

Let $m = 60$. Then

$$30 \times 2 \longleftrightarrow 16^2 - 14^2 = 60$$
$$10 \times 6 \longleftrightarrow 8^2 - 2^2 = 60.$$

Let $m = 144$. Then

$$72 \times 2 \longleftrightarrow 37^2 - 35^2 = 144$$
$$36 \times 4 \longleftrightarrow 20^2 - 16^2 = 144$$
$$24 \times 6 \longleftrightarrow 15^2 - 9^2 = 144$$
$$18 \times 8 \longleftrightarrow 13^2 - 5^2 = 144$$
$$12 \times 12 \longleftrightarrow 12^2 - 0^2 = 144.$$

Note that the second example ($m = 60$) can be obtained from the first ($m = 15$) by multiplying each factor of 15 by 2. As in the above discussion, the factorization $m = uv$ corresponds to $x^2 - y^2 = m$ with $x = (u+v)/2$ and $y = (u-v)/2$.

CHECK YOUR UNDERSTANDING

6. Can 50 be written as the difference of two squares? Explain.
7. How do you know that 12 can be written as the difference of two squares? Find all possible ways to do so.

4.6 Chapter Highlights

1. An integer m is a kth power if and only if all the exponents in its prime factorization are multiples of k.

2. If an integer n is not a kth power, then $\sqrt[k]{n}$ is irrational.
3. All primitive Pythagorean triples (a, b, c) with b even are given by the formula
$$a = n^2 - m^2, \quad b = 2mn, \quad c = m^2 + n^2,$$
for some relatively prime integers m, n, with $n > m$ and $n \not\equiv m \pmod{2}$.

4.7 Problems

4.7.1 Exercises

Section 4.1: A Puzzle

1. Find a positive integer n such that
   (a) $n/2$ is a square and $n/3$ is a cube.
   (b) $n/2$ is a square, $n/3$ is a cube, and $n/5$ is a fifth power.
2. Find a positive integer n such that
   (a) $2n$ is a square and $3n$ is a cube.
   (b) $2n$ is a square, $3n$ is a cube, and $5n$ is a fifth power.
3. Let a and b be positive integers with $\gcd(a, b) = 1$, and let $n \geq ab - a - b + 1$. Show that every nth power of an integer can be written as the product of an ath power and a bth power. (*Hint:* Look at Section 2.2.)
4. (a) Five people, call them A, B, C, D, and E, have a pancake eating contest, and they eat $a, b, c, d,$ and e pancakes, respectively. They each eat a different number of pancakes and all of the ratios
$$b/a, \quad c/b, \quad d/c, \quad \text{and} \quad e/d$$
are integers. If they eat a total of 47 pancakes, how many pancakes did each eat? Justify that the solution you find is the only solution.
   (b) Find a sequence of primes p_1, p_2, p_3, p_4, p_5 such that $p_{i+1} =$

$2p_i + 1$ for $i = 1, 2, 3, 4$.
(c) What is the relation between (a) and (b)?

Section 4.2: Irrationality Proofs

5. For which positive integers k is $\sqrt[k]{64}$ rational?
6. For which positive integers k is $\sqrt[k]{81}$ rational?
7. (a) Show that if x is rational then x^2 is rational.
   (b) Show that $\sqrt{2} + \sqrt{3}$ is irrational.
8. Use the technique of the previous problem to prove that if a and b are positive integers and ab is not a square, then $\sqrt{a} + \sqrt{b}$ is irrational.

Section 4.3: The Rational Root Theorem

9. Find all rational solutions of $2x^3 + 5x^2 - x - 1 = 0$.
10. Show that $x^7 + 3x^3 + 1 = 0$ has no rational solutions.
11. The roots of the polynomial $36x^3 - 81x^2 - 10x + 75$ are -0.8637854, 1.4471187, and 1.6666667 (rounded to 7 decimal places). Use the fact that the rational roots of this polynomial have denominators dividing 36 to find the rational roots of the polynomial.
12. The roots of the polynomial $1092x^3 - 745x^2 - 57x + 70$ are -0.2857143, 0.3846154, and 0.5833333 (rounded to 7 decimal places). Use the fact that the rational roots of this polynomial have denominators dividing 1092 to find the rational roots of the polynomial.
13. (a) Show that $\sqrt{2} + \sqrt{3}$ is a root of $x^4 - 10x^2 + 1$.
    (b) Use the Rational Root Theorem to show that $\sqrt{2} + \sqrt{3}$ is irrational. (Compare with Exercise 7.)

Section 4.4: Pythagorean Triples

14. Find the values of m and n in Theorem 4.5 corresponding to the Pythagorean triples $(7, 24, 25)$ and $(8, 15, 17)$.
15. Find a Pythagorean triple (a, b, c) with $a = 37$.
16. Find all primitive Pythagorean triples (a, b, c) with $b = 44$.

92   Chapter 4   Applications of Unique Factorization

17. Find two primitive Pythagorean triples (a, b, c) with $c = 65$.

18. Find two imprimitive Pythagorean triples (a, b, c) with $c = 65$.

19. Find all (not necessarily primitive) Pythagorean triples (a, b, c) with $b = 18$.

20. Show that $(3, 4, 5)$ is the only primitive Pythagorean triple whose terms are in an arithmetic progression. Then, find *all* Pythagorean triples whose terms are in an arithmetic progression.

21. Show that if $n \geq 3$ is an integer then there is a (possibly imprimitive) Pythagorean triple that has n as one of its numbers. (*Hint:* If n is odd, $n = (2k+1)^2 - k^2$ for some k. If n is even, write $n = 2^r m$ with m odd and separately consider the cases $m = 1$ and $m \geq 3$.)

22. (a) Let $n = 2^j$, where j is an odd positive integer, and let m be an odd positive integer such that $n + m$ is a square. Show that $2n(n + m)$ is a square.
    (b) Find a primitive Pythagorean triangle whose perimeter has length equal to the square of an integer.

23. At the beginning of the section on Pythagorean triples, we saw two easy ways of producing Pythagorean triples, one starting with squaring an odd number and the other starting with squaring a multiple of 4. Find the values of m and n in Theorem 4.5 that correspond to these two families.

Section 4.5: Differences of Squares

24. Show that 7 is the difference of two squares in exactly one way.

25. Let p be an odd prime. Show that p is the difference of two squares in exactly one way.

26. Find all solutions of $x^2 - y^2 = 20$ in positive integers x, y.

27. Find all solutions of $x^2 - y^2 = 15$ in positive integers x, y.

28. Show that if x is an integer then $x^3 + y^2 = z^2$ has a solution in integers y, z. (*Hint:* Rewrite as $x^3 = z^2 - y^2$.)

4.7.2 Computer Explorations

1. The six primes

$$7, \quad 37, \quad 67, \quad 97, \quad 127, \quad 157$$

form an arithmetic progression with common difference 30. Can you find an arithmetic progression of ten primes with common difference 210? There is one whose first prime is less than 300.

Remark: In 2004, Ben Green and Terry Tao proved that there are arbitrarily large arithmetic progressions in the primes. For example, there is a sequence of one million very large primes that form an arithmetic progression. Note, however, that this does not say that there are arithmetic progressions of primes that are infinitely long, just that there exists an arithmetic progression of any desired finite length.

4.7.3 Answers to "CHECK YOUR UNDERSTANDING"

1. The prime factorization of 5400 is $2^3 3^3 5^2$. Since the exponent for 5 is not a multiple of 3, we know that 5400 is not a perfect cube.

2. Since 28 is not a perfect cube, $\sqrt[3]{28}$ is irrational. Since 50 is not a perfect ninth power, $\sqrt[9]{50}$ is irrational.

3. The possible rational roots of $2x^3 - 3x^2 + 5x - 10$ are

$$\pm 10, \quad \pm 5, \quad \pm \frac{5}{2}, \quad \pm 2, \quad \pm \frac{1}{2}, \quad \pm 1.$$

   None of these are actually roots.

4. Since $n = 8$ and $m = 3, a = 8^2 - 3^2 = 55, b = 2 \cdot 3 \cdot 8 = 48$, and $c = 8^2 + 3^2 = 73$.

5. We need $m < n$ and $m^2 + n^2 = 13$. The only solution to $m^2 + n^2 = 13$ with $n > m$ is $n = 3, m = 2$. This produces the triple $(5, 12, 13)$.

6. It's impossible to write 50 as the difference of two squares since 50 is neither odd nor a multiple of 4.

7. $12 = 4^2 - 2^2$. This is the only way to write 12 as the difference of two squares, since $12 = 6 \cdot 2$ is the only way to write 12 as the product of two numbers with the same parity.

Chapter 5

Congruences

5.1 Definitions and Examples

When you classify a number as even or odd, you're using the concept of congruence. In the Division Algorithm, an even number is a number that leaves a remainder of 0 when divided by 2 and an odd number leaves a remainder of 1 when divided by 2. Although nobody would say that 14 and 96 are the same number, they do share the property of having the same remainder when divided by 2. Similarly, when you look at a clock and you notice that it's 10:00 now and think to yourself that in 5 hours it will be 3:00, you're not saying that $10 + 5 = 3$. What you're really doing, perhaps without realizing it, is saying $10 + 5 = 15$, and when I divide 15 by 12, the remainder is 3. Congruences generalize these ideas to all positive integers by saying (as we'll see in a bit) that two numbers are "congruent for the number n" if they leave the same remainder when divided by n. So, to return to our clock example, 3 and 15 are "congruent for the number 12" because they both leave a remainder of 3 when divided by 12. It will take us a few paragraphs to make this precise, and we'll have to develop some notation along the way, so let's begin.

Definition 5.1. *Two numbers a and b are* **congruent mod m**, *written* $a \equiv b \pmod{m}$, *if $a - b$ is a multiple of m. The integer m is called the* **modulus** *of the congruence and is assumed to be positive.*

Examples.

1. $7 \equiv 1 \pmod{2}$, $14 \equiv 0 \pmod{2}$, $19 \equiv 7 \pmod{6}$,
   $-8 \equiv 12 \pmod{5}$, $4 \equiv 4 \pmod{11}$, $6 \equiv 16 \pmod{5}$,
   $34 \equiv 12 \pmod{11}$, $7 \equiv -2 \pmod{3}$, $-11 \equiv 3 \pmod{7}$.

2. We can see how the notion of congruence applies to numbers being even or odd. A number is even if it's congruent to 0 mod 2, and it's odd if it's congruent to 1 mod 2. Using congruence notation, n is even if and only if $n \equiv 0 \pmod{2}$ and n is odd if and only if $n \equiv 1 \pmod{2}$.

3. When we look at a number mod 10, we get its last digit, and if we look at it mod 100, we get its last two digits. So, $2347 \equiv 7 \pmod{10}$, $65931 \equiv 1 \pmod{10}$, $2347 \equiv 47 \pmod{100}$, and $65931 \equiv 31 \pmod{100}$. More generally, two positive integers are congruent mod 10 if their last digits are the same, and they are congruent mod 100 if the last two digits of one are the same as the last two digits of the other.

4. If we consider a modulus of 5, we see that $11 \equiv 1 \pmod{5}$, $27 \equiv 2 \pmod{5}$, and $114 \equiv 4 \pmod{5}$.

As we'll see, congruences behave a lot like equalities, but there are some important differences. For example, $3 = 3$ and 3 is the only number equal to 3. (Yes, we do feel silly writing that.) But, notice that $12 \equiv 4 \pmod{8}$, $20 \equiv 4 \pmod{8}$, and $28 \equiv 4 \pmod{8}$. In fact, there is an infinite number of integers that are congruent to 4 mod 8. That's because of the following.

Proposition 5.2. *$a \equiv b \pmod{m}$ if and only if $a = b + km$ for some integer k.*

Proof. $a \equiv b \pmod{m}$ if and only if $a - b$ is a multiple of m. This means that $a - b = km$ for some integer k, or equivalently that $a = b + km$. $\square$

This proposition gives us an embarrassment of riches. If we're given an integer a and a modulus m, there are infinitely many integers that are congruent to $a \pmod{m}$. Nevertheless, the following lemma tells us that if we're given an a and m as above, we can always single out one.

Proposition 5.3. *If a is an integer and m is a positive integer, then there is a unique integer r with $0 \leq r \leq m - 1$ so that $a \equiv r \pmod{m}$. This integer r is called the **least nonnegative residue** of a mod m.*

Proof. We know from the Division Algorithm that we can write $a = mq + r$ with a unique r satisfying $0 \leq r \leq m - 1$. By the previous lemma, this means that $a \equiv r \pmod{m}$. $\square$

Section 5.1   Definitions and Examples

Examples. The least nonnegative residue of 7 (mod 4) is 3, and the least nonnegative residue of -3 (mod 9) is 6.

Remark. Congruences occurred, at least implicitly, in ancient number theory. Euler, around 1750, was responsible for introducing them into modern number theory. The terminology and notation $a \equiv b \pmod{m}$ was introduced by Gauss in his book *Disquisitiones Arithmeticae* in 1801. The "mod" is short for "modulo," which means "with respect to the modulus" in Latin. The use of parentheses around "mod m" is traditional, but not necessary.

The following is another useful property of congruences. You may recognize this as saying that congruence mod m forms an equivalence relation.

Proposition 5.4. *If a, b, c, and m are integers with $m > 0$, then*
(a) $a \equiv a \pmod{m}$
(b) If $a \equiv b \pmod{m}$, then $b \equiv a \pmod{m}$
(c) If $a \equiv b \pmod{m}$ and $b \equiv c \pmod{m}$, then $a \equiv c \pmod{m}$.

Proof. (a) Since $a = a + 0 \cdot m$, Proposition 5.2 says $a \equiv a \pmod{m}$.
(b) If $a \equiv b \pmod{m}$, then $a = b + km$, for some k. Therefore,

$$b = a + (-k)m,$$

so $b \equiv a \pmod{m}$.
(c) If $a \equiv b \pmod{m}$, then there is an integer k_1 with $a - b = k_1 m$. Similarly, if $b \equiv c \pmod{m}$, then there is an integer k_2 with $b - c = k_2 m$. Then,

$$a - c = (a - b) + (b - c) = k_1 m + k_2 m = (k_1 + k_2)m$$

so $a - c$ is a multiple of m, which means that $a \equiv c \pmod{m}$. □

As we said before, congruences behave a lot like equalities. In Section 5.4, we'll learn how to solve linear congruences using techniques similar to those used to solve linear equations. But before we do that, we have to make sure that the relevant rules still apply. For example, if we want to solve the equation $x - 4 = 10$, we add 4 to both sides and get that $x = 14$. We're adding equal numbers (in this case 4) to both sides of an equality and the result is still equality. For the integers, it's pretty obvious that "equals added to equals are equal," because after all, there's only one number equal to 4. But, as we pointed out, if we fix a modulus, say 15, there are infinitely many integers that are congruent to 4 mod 15, for example (4, 19, 34, 49, ...). Since 49 and 4 are the same mod 15, we need to know that adding 4 to one side and 49 to the other side of a congruence maintains the congruence. In this case, if we want to solve the congruence $x - 4 \equiv 10 \pmod{15}$, if we add 4 to the left side and 49 to the right side, do we get a correct solution? We see that we do: we have $x \equiv 59 \pmod{15}$. Plugging 59 into $x - 4 \equiv 10 \pmod{15}$, we get the valid congruence $55 \equiv 10 \pmod{15}$. The following proposition tells us that congruences obey the same fundamental rules as the integers do when adding, subtracting, and multiplying.

Proposition 5.5. *Assume that a, b, c, d, and m are integers with m positive. If $a \equiv b \pmod{m}$ and $c \equiv d \pmod{m}$, then*
(a) $a + c \equiv b + d \pmod{m}$.
(b) $a - c \equiv b - d \pmod{m}$.
(c) $ac \equiv bd \pmod{m}$.

Proof. Since $a \equiv b \pmod{m}$ and $c \equiv d \pmod{m}$, we know that $a = b + k_1 m$ and $c = d + k_2 m$ for some integers k_1 and k_2. To prove (a), we see that

$$a + c = (b + k_1 m) + (d + k_2 m) = (b + d) + (k_1 + k_2)m \equiv b + d \pmod{m}.$$

The proof of (b) follows in the same way:

$$a - c = (b + k_1 m) - (d + k_2 m) = (b - d) + (k_1 - k_2)m \equiv b - d \pmod{m}.$$

Although the algebra is a little messier, the proof of (c) is very similar to that of (a) and (b):

$$ac = (b + k_1 m)(d + k_2 m) = bd + bk_2 m + dk_1 m + k_1 k_2 m^2$$
$$= bd + (bk_2 + dk_1 + k_1 k_2 m)m \equiv bd \pmod{m}. \qquad \square$$

So, just as we say "equals added to equals are equal" and "equals multiplied by equals are equal," we can say "congruents added to congruents are congruent" and "congruents multiplied by congruents are congruent."

Proposition 5.5 was easy to prove. Nevertheless, it is an extremely powerful and useful result, since it allows us to calculate with congruences

Section 5.1 Definitions and Examples   99

very rapidly. For example, if we want to compute $7^2 \bmod 6$, we just have to notice that
$$7^2 \equiv 7 \cdot 7 \pmod 6,$$
and since $7 \equiv 1 \pmod 6$, Proposition 5.5 allows us to conclude that
$$7^2 \equiv 7 \cdot 7 \equiv 1 \cdot 1 \equiv 1 \pmod 6.$$

Now, it's true that you could have said that $7^2 = 49$ and quickly seen that $49 \equiv 1 \pmod 6$. But the exact same technique works to show that $7^{200} \equiv 1 \pmod 6$, and calculating 7^{200} is much harder. In fact, here is a corollary of Proposition 5.5.

Corollary 5.6. *If $a \equiv b \pmod m$, then $a^n \equiv b^n \pmod m$ for any positive integer n.*

Proof. Since $a \equiv b \pmod m$, we can multiply this congruence by itself to get $a^2 \equiv b^2 \pmod m$. Continuing to multiply, we get $a^n \equiv b^n \pmod m$ for any positive integer n. □

As we'll see later, being able to compute $a^n \pmod m$ quickly is extremely important in cryptographic applications.

Notice that Proposition 5.5 doesn't say anything about dividing in congruences. That's because it's *not* always true that if $ac \equiv bc \pmod m$ then $a \equiv b \pmod m$: You can't always divide a congruence and get a true statement. For example, $4 \cdot 5 \equiv 4 \cdot 2 \pmod{12}$, but if we divide by 4 we see that $5 \not\equiv 2 \pmod{12}$. The problem is that we're dividing by 4, which is a divisor of the modulus of 12. In fact, you can divide only when the number you're dividing by is relatively prime to the modulus. To see this, we recast our congruence in terms of divisibility. If $ac \equiv bc \pmod m$, then by definition, $m \mid (ac - bc)$ so that $m \mid c(a - b)$. We recall from Proposition 1.14 that if c and m are relatively prime, then $m \mid c(a-b)$ implies that $m \mid (a-b)$ so that $a \equiv b \pmod m$. We have proved the following:

Proposition 5.7. *If $ac \equiv bc \pmod m$ and $\gcd(c, m) = 1$, then $a \equiv b \pmod m$.*

Not all is lost if the number you're dividing by and the modulus do have a common factor. We can still get a true statement, but with a smaller modulus.

Proposition 5.8. *If $ac \equiv bc \pmod{m}$ and $\gcd(c, m) = d$, then*
$$a \equiv b \pmod{(m/d)}$$
and
$$a \equiv b + (m/d)k \pmod{m} \text{ with } 0 \le k \le d-1.$$

Proof. If $ac \equiv bc \pmod{m}$, then $m \mid (ac - bc)$ so $m \mid c(a - b)$ and
$$(m/d) \mid (c/d)(a - b).$$
But m/d and c/d are relatively prime from Proposition 1.11. Using Proposition 5.7, we see that
$$a \equiv b \pmod{m/d}.$$
This means that $a = b + (m/d)k$ for some k.

We now need to see what this means as a congruence mod m. Suppose $a_1 = b + (m/d)k_1$ and $a_2 = b + (m/d)k_2$ satisfy $a_1 \equiv a_2 \pmod{m}$. Then
$$a_1 - a_2 = (m/d)(k_1 - k_2) = m\left((k_1 - k_2)/d\right)$$
is a multiple of m. This happens exactly when $(k_1 - k_2)/d$ is an integer, that is, when $k_1 \equiv k_2 \pmod{d}$. Therefore, if we let k run through a complete set of congruence classes mod d, then we get all possibilities for $a = b + (m/d)k \pmod{m}$. □

Here are some examples of the previous two propositions.

1. If $3c \equiv 12 \pmod 5$, we can divide both sides by 3 because 3 and 5 are relatively prime. (We're using Proposition 5.7 again.) After dividing, we get $c \equiv 4 \pmod 5$.

Remark. There's a subtlety that we need to discuss about this example. The congruence $3c \equiv 12 \pmod 5$ is often thought of as an equation in the "mod 5 world." When we solve this, we usually think of having a solution ($c = 4$) in that "mod 5 world." If we changed our perspective to the "integer world," there would be an infinite number of solutions, namely $c = 4, 9, 14, 19, \ldots$. When working with congruences, we will stay in the world we started with unless our situation dictates otherwise.

Section 5.1   Definitions and Examples   101

2. If $10c \equiv 30 \pmod 7$, because 10 and 7 are relatively prime, we can divide both sides by 10 and get $c \equiv 3 \pmod 7$.

3. If $2x \equiv 7 \pmod 9$, we could use guess and check to find that $c = 8$ is a solution. A faster way is to exploit the fact that $9 \equiv 0 \pmod 9$ means that adding multiples of 9 to the right-hand side of this congruence is the same as adding 0. By adding 9 to 7 just once, we get 16 (mod 9), so that $2x \equiv 16 \pmod 9$. Because 2 and 9 are relatively prime, we can divide both sides by 2 to get $x \equiv 8 \pmod 9$, so $x = 8$ is a solution.

4. If $6c \equiv 18 \pmod{21}$, we cannot simply divide both sides by 6, because $\gcd(6, 21) = 3$, so 6 and 21 are *not* relatively prime. In this situation, we use Proposition 5.8, which tells us that after dividing by 6 our modulus changes to $21/3 = 7$. So, we get that $c \equiv 3 \pmod 7$ and $c \equiv 3, 10, 17$ is the list of solutions mod 21.

    We also could have solved this in two steps. First, divide everything by the gcd, namely 3, to get $2c \equiv 6 \pmod 7$. Then solve the new congruence to get $c \equiv 3 \pmod 7$. This gives us the list 3, 10, 17 of solutions mod 21.

Notice that in this example, our equation began in the "mod 21" world. We left this world when we divided by 3, but our final answer was written as a mod 21 result. We see the same situation in the next example as well.

5. If $10c \equiv 80 \pmod{45}$, when we divide by 10 the modulus changes to $45/\gcd(10, 45) = 45/5 = 9$ and we have $c \equiv 8 \pmod 9$. Then $c \equiv 8, 17, 26, 35, 44$ is our list of solutions (mod 45). Notice that if we divided by 10 and forgot to divide the modulus by 5, we would have gotten $c \equiv 8 \pmod{45}$. Then we would have lost 17, 26, 35, 44 as solutions.

6. Suppose we want to solve $15x \equiv 25 \pmod{55}$. Divide everything by 5 to get $3x \equiv 5 \pmod{11}$. Trying a few numbers (or changing 5 (mod 11) to 27 (mod 11)) gives us $x \equiv 9 \pmod{11}$, which is the same as $x \equiv 9, 20, 31, 42, 53 \pmod{55}$.

We conclude this section with some results about congruences when the modulus is prime.

Proposition 5.9. *If p is a prime and $ab \equiv 0 \pmod{p}$, then $a \equiv 0 \pmod{p}$ or $b \equiv 0 \pmod{p}$.*

Proof. If $ab \equiv 0 \pmod{p}$, then $p \mid ab$. By Theorem 3.2, this implies that $p \mid a$ or $p \mid b$, so $a \equiv 0 \pmod{p}$ or $b \equiv 0 \pmod{p}$. □

A useful corollary of this proposition is

Corollary 5.10. *Let p be prime. The congruence*
$$x^2 \equiv 1 \pmod{p}$$
has only the solutions $x \equiv \pm 1 \pmod{p}$.

Proof.
$$x^2 \equiv 1 \pmod{p} \iff x^2 - 1 \equiv 0 \pmod{p}$$
$$\iff (x-1)(x+1) \equiv 0 \pmod{p}. \qquad (5.1)$$

From the previous proposition, congruence (5.1) has a solution if and only if $x - 1 \equiv 0 \pmod{p}$ or $x + 1 \equiv 0 \pmod{p}$, which we can rewrite as $x \equiv \pm 1 \pmod{p}$. □

Remark. Remember Corollary 5.10. You will see it used several times in later chapters.

CHECK YOUR UNDERSTANDING

1. Which of the following are true? (a) $17 \equiv 29 \pmod{4}$, (b) $321 \equiv 123 \pmod{10}$, (c) $321 \equiv 123 \pmod{9}$.
2. Solve $2x \equiv 3 \pmod{11}$.
3. Solve $4x \equiv 6 \pmod{22}$.
4. Solve $(x-1)(x-2) \equiv 0 \pmod{11}$.

5.2 Modular Exponentiation

As we said, being able to calculate $a^n \pmod{m}$ for large values of n and m is extremely important in cryptographic applications. Let's say

Section 5.2   Modular Exponentiation

that we need to calculate

$$3^{385} \pmod{479}. \tag{5.2}$$

We could, of course, compute 3^{385}, divide it by 479, and then obtain the remainder as the answer. Or, we could repeatedly multiply 3 by itself and then reduce mod 479 every time our product exceeds 479. Fortunately, there's a more practical and faster method for modular exponentiation that's based on successive squaring. Here's how it works in equation (5.2). We begin by squaring 3:

$$3^2 \equiv 9 \pmod{479}.$$

We now repeatedly square and reduce mod 479. Doing so gives us the following list of congruences, all of which are taken mod 479.

$$3^4 \equiv 9^2 \equiv 81$$
$$3^8 \equiv 81^2 \equiv 6561 \equiv 334$$
$$3^{16} \equiv 334^2 \equiv 111556 \equiv 428$$
$$3^{32} \equiv 428^2 \equiv 183184 \equiv 206$$
$$3^{64} \equiv 206^2 \equiv 284$$
$$3^{128} \equiv 284^2 \equiv 184$$
$$3^{256} \equiv 184^2 \equiv 326.$$

How does this help? Since the exponent $385 = 256 + 128 + 1$,

$$3^{385} \equiv 3^{256} \cdot 3^{128} \cdot 3^1 \equiv 326 \cdot 184 \cdot 3 \equiv 327.$$

So, $3^{385} \equiv 327 \pmod{479}$.

In order to understand how to use this to calculate $a^m \pmod{n}$ for any $a, m,$ and n, let's look more closely at what we've done. Essentially, there were two steps:

1. Precalculate successive squarings of a, stopping when you exceed the exponent m.
2. Write the exponent m in binary form.

Notice that we did Step 2 when we wrote $385 = 256 + 128 + 1$. This says that the binary form of 385 is 110000001. We saw how to calculate this using successive divisions by 2 in Section 1.9:

$$385 = 2 \cdot 192 + 1$$
$$192 = 2 \cdot 96 + 0$$
$$96 = 2 \cdot 48 + 0$$
$$48 = 2 \cdot 24 + 0$$
$$24 = 2 \cdot 12 + 0$$
$$12 = 2 \cdot 6 + 0$$
$$6 = 2 \cdot 3 + 0$$
$$3 = 2 \cdot 1 + 1$$
$$1 = 2 \cdot 0 + 1,$$

which shows that the base 2, or binary, representation of 385 is 110000001. In practice, the process of modular exponentiation is remarkably fast.

CHECK YOUR UNDERSTANDING

5. Compute $2^{66} \pmod{67}$.

5.3 Divisibility Tests

When you look at a number, you immediately know whether it's divisible by 2 by looking at the last digit. If the last digit of the number is even, it's divisible by 2, and if the last digit is odd, it's not. In this section, we use the language of congruences to develop divisibility tests for 2, 3, 4, 5, 8, 9, 10, and 11. These tests have a long history, first appearing over 1200 years ago in the work of the Persian mathematician al-Khwarizini (whose name when Latinized was *Algoritmi*, giving us the word *algorithm*).

The divisibility tests all exploit the fact that we use a base 10 number system. So, for example, the number 37569 is a shorthand expression of $3 \cdot 10^4 + 7 \cdot 10^3 + 5 \cdot 10^2 + 6 \cdot 10^1 + 9 \cdot 10^0$. In general, every positive integer can be written in an expanded form:

$$n = a_m 10^m + a_{m-1} 10^{m-1} + \cdots + a_1 10 + a_0 \qquad (5.3)$$

Section 5.3 Divisibility Tests   105

Written this way, our observation about divisibility by 2 can be expressed as "A number is divisible by 2 if and only if its last digit, a_0, is divisible by 2."

We will present the other divisibility tests as a series of propositions, referring back to Equation (5.3) when necessary.

Proposition 5.11. *A positive integer n is divisible by 10 if and only if its last digit is 0, and is divisible by 5 if and only if its last digit is 0 or 5.*

Proof. Since $10 \equiv 0 \pmod{10}$, we have $10^k \equiv 0 \pmod{10}$ for every positive integer k, so

$$n = a_m 10^m + a_{m-1} 10^{m-1} + \cdots + a_1 10 + a_0$$
$$\equiv a_m \cdot 0 + a_{m-1} \cdot 0 + \cdots + a_1 \cdot 0 + a_0$$
$$\equiv a_0 \pmod{10}.$$

This means that n is divisible by 10 if and only if $a_0 \equiv 0 \pmod{10}$, which happens exactly when $a_0 = 0$. The proof for divisibility for 5 is similar and rests on the observation that $10 \equiv 0 \pmod 5$. So, we again have $n \equiv a_0 \pmod 5$. This means that 5 divides n if and only if $a_0 \equiv 0 \pmod 5$, which happens if and only if $a_0 = 0$ or 5. □

Examples. Since the last digit of $n = 681094136$ is neither 0 nor 5, n is divisible by neither 5 nor 10. On the other hand, 2107195 is divisible by 5 but not by 10 since the last digit is 5, and 94979420 is divisible by both 5 and 10 since the last digit is 0.

Proposition 5.12. *A positive integer n is divisible by 4 if and only if the number formed by its last two digits is divisible by 4, and n is divisible by 8 if and only if the number formed by its last three digits is divisible by 8.*

Proof. We begin by observing that if $k \geq 2$ then

$$10^k = 10^2 10^{k-2} = 4 \cdot 25 \cdot 10^{k-2} \equiv 0 \pmod 4.$$

Therefore,

$$n = a_m 10^m + a_{m-1} 10^{m-1} + \cdots + a_1 10 + a_0$$
$$\equiv a_m \cdot 0 + a_{m-1} \cdot 0 + \cdots + a_1 \cdot 10 + a_0$$
$$\equiv a_1 10 + a_0 \pmod 4.$$

So, a number is congruent mod 4 to the number formed by its last two digits and it's divisible by 4 if and only if the number formed by its last two digits is divisible by 4. The proof for divisibility by 8 is almost identical: if $k \geq 3$ then

$$10^k = 10^3 10^{k-3} = 8 \cdot 125 \cdot 10^{k-3} \equiv 0 \pmod{8}.$$

This means that a number mod 8 reduces down to its last three digits and is therefore divisible by 8 if and only if the number formed by its last three digits is divisible by 8. □

Examples. 5675972 is divisible by 4 because 72 is, but 5675972 is not divisible by 8 because 972 is not divisible by 8. On the other hand, 32589104 is divisible by both 4 and 8, while 16932594 is divisible by neither 8 nor 4.

Proposition 5.13. *A positive integer is congruent mod 3 (respectively, mod 9) to the sum of its digits mod 3 (respectively, mod 9).*

Proof. We begin with the observation that $10 \equiv 1 \pmod{3}$. Since $1^k = 1$ for every integer k, we have

$$10^k \equiv 1^k \equiv 1 \pmod{3} \text{ for } k \geq 0.$$

This means that when we look at an integer n expanded in its base 10 form mod 3, we get

$$n = a_m 10^m + a_{m-1} 10^{m-1} + \cdots + a_1 10 + a_0$$
$$\equiv a_m + a_{m-1} + \cdots + a_1 + a_0 \pmod{3},$$

where we get the right-hand side of the congruence because each power of 10 is 1 mod 3. We now see that $n \pmod 3$ is the same as the sum of the digits of $n \pmod 3$ and that n is divisible by 3 if and only if the sum of its digits is divisible by 3.

The proof of divisibility by 9 is exactly the same after we note that

$$10 \equiv 1 \pmod{9} \implies 10^k \equiv 1 \pmod{9} \text{ for every integer } k \geq 0.$$

□

Section 5.3   Divisibility Tests

Corollary 5.14. *A positive integer is divisible by 3 if and only if the sum of its digits is divisible by 3. It is divisible by 9 if and only if the sum of its digits is divisible by 9.*

Example. The sum of the digits of 9789042 is $9+7+8+9+0+4+2 = 39$. Since 39 is divisible by 3, so is 9789042. Note that 39 is not divisible by 9, since $39 \equiv 3 \pmod 9$, and when we divide 9789042 by 9, we get a remainder of 3. We can also see this by adding the digits three times:

$$\begin{aligned} 9789042 &\equiv 9+7+8+9+0+4+2 \\ &\equiv 39 \equiv 3+9 = 12 \\ &\equiv 1+2 = 3 \pmod 9. \end{aligned}$$

Remark. We can also prove Proposition 5.13 without congruences in the following elementary way. If we write

$$\begin{aligned} n &= a_0 + a_1(9+1) + a_2(99+1) + a_3(999+1) + \ldots \\ &= a_0 + a_1 + a_2 + \ldots + a_n + 9(a_1 + 11a_2 + 111a_3 + \cdots), \end{aligned}$$

we see that
$$n = a_0 + a_1 + a_2 + \cdots + a_n + 9k,$$

for some integer k. Therefore, n is congruent to the sum of its digits mod 9.

This also shows that n is congruent to the sum of its digits mod 3, since

$$n = a_0 + a_1 + a_2 + \cdots + a_n + 9k = a_0 + a_1 + a_2 + \cdots + a_n + 3 \cdot (3k).$$

The final divisibility test that we'll do is for 11.

Proposition 5.15. *A positive integer is congruent mod 11 to the alternating sum of its digits starting with the ones digit (a_0), subtracting the tens digit (a_1), etc. So,*

$$n = a_m 10^m + a_{m-1} 10^{m-1} + \cdots + a_1 10 + a_0$$

is congruent mod 11 to

$$a_0 - a_1 + a_2 - a_3 + \cdots + (-1)^m a_m.$$

Proof. We begin with the observation that

$$10 \equiv -1 \pmod{11} \implies 10^k \equiv (-1)^k \pmod{11}.$$

This means that when we look at the base 10 expansion of $n \pmod{11}$, we get

$$n \equiv a_0 - a_1 + a_2 - a_3 + \cdots + (-1)^m a_m \pmod{11}.$$

□

Corollary 5.16. *A positive integer is divisible by 11 if and only if the alternating sum of its digits is divisible by 11.*

Examples. If $n = 8563981$ then the alternating sum of the digits of n is $1 - 8 + 9 - 3 + 6 - 5 + 8 = 8$, so $n \equiv 8 \pmod{11}$. But $n = 639162513$ has alternating sum $3 - 1 + 5 - 2 + 6 - 1 + 9 - 3 + 6 = 22 \equiv 0 \pmod{11}$ so 639162513 is divisible by 11. Three-digit numbers such as 264 and 385, where the middle digit is the sum of the first and third digits, are immediately recognized as multiples of 11 because the alternating sum is 0.

CHECK YOUR UNDERSTANDING

6. Show that $987654321 \equiv 0 \pmod{9}$.
7. Show that $343434 \equiv 58 \pmod{11}$.

5.4 Linear Congruences

Let's say that we want to solve $3x \equiv 4 \pmod{7}$. We can try different values of x until we discover a solution. In this case, the smallest positive one is $x = 6$. Now, suppose we want to solve $3x \equiv 6 \pmod{12}$. We find that $x \equiv 2, 6, 10 \pmod{12}$ are the solutions. On the other hand, the equation $2x \equiv 1 \pmod{4}$ has no solutions because $2x$ is always congruent to 0 or 2 mod 4. In general, we have the following.

Section 5.4 Linear Congruences

Theorem 5.17. *Let m be a positive integer and let $a \neq 0$. The congruence*
$$ax \equiv b \pmod{m} \tag{5.4}$$
has a solution if and only if $d = \gcd(a, m)$ is a divisor of b. If $d \mid b$, then there are exactly d solutions that are distinct mod m. In this case, if x_0 is a solution, the solutions mod m are
$$x = x_0 + (m/d)k \text{ for } 0 \le k < d.$$
The solution x_0 can be taken to be any integer satisfying
$$(a/d)x_0 \equiv (b/d) \pmod{m/d}.$$

Proof. We begin by rewriting Equation (5.4) without congruences:
$$ax \equiv b \pmod{m}$$
is equivalent to
$$ax = b + my, \quad \text{or} \quad ax - my = b \tag{5.5}$$
for some integer y. We now need to use Theorem 2.1. Note that this theorem talked about the equation $ax + by = c$, while we're looking at the equation $ax - my = b$, so we replace (a, b, c) in Theorem 2.1 with $(a, -m, b)$.

If $d \nmid b$, Theorem 2.1 tells us that Equation (5.5) has no solutions.

We now assume that $d \mid b$. Theorem 2.1 tells us that the solutions of (5.5) have the form
$$x = x_0 + (m/d)k, \quad y = y_0 + (a/d)k,$$
where k is an integer and (x_0, y_0) is any particular solution of Equation (5.5). Therefore, all solutions of $ax \equiv b \pmod{m}$ have the form $x = x_0 + (m/d)k$. This means that the solutions x are exactly the integers congruent to x_0 mod (m/d). We need to see when these solutions are different mod m.

Let $x_1 = x_0 + (m/d)k_1$ and $x_2 = x_0 + (m/d)k_2$ be two solutions and suppose that $x_1 \equiv x_2 \pmod{m}$. Then $x_1 - x_2 = mk_3$ for some k_3. But
$$\begin{aligned} x_1 - x_2 = mk_3 &\iff (x_0 + (m/d)k_1) - (x_0 + (m/d)k_2) = mk_3 \\ &\iff (m/d)(k_1 - k_2) = mk_3 \\ &\iff k_1 - k_2 = dk_3, \end{aligned}$$

which is the same as saying that $k_1 \equiv k_2 \pmod{d}$. Therefore, we get the distinct solutions mod m by letting $x = x_0 + k(m/d)$ with $0 \le k < d$ (or with k running through any complete set of residue classes mod d). Finally, we need to show that x_0 arises from solving $(a/d)x_0 \equiv (b/d) \pmod{m/d}$. Rewrite this congruence as

$$(a/d)x_0 = (b/d) + (m/d)z,$$

where z is an integer. Multiplying by d yields

$$ax_0 = b + mz,$$

which says that x_0 is a solution of (5.4), and therefore can be used as our particular solution. □

By far, the most common case of Theorem 5.17 is when $d = 1$. Since it is so important, we single it out.

Corollary 5.18. *If* $\gcd(a, m) = 1$, *the congruence* $ax \equiv b \pmod{m}$ *has exactly one solution mod m.*

Proof. Set $d = 1$ in the theorem. Then $d \mid b$, so there is a solution. The distinct solutions mod m are $x_0 + mk$ with $0 \le k < 1$, so only $k = 0$ is allowed, and therefore there is exactly one solution mod m. □

Examples.

- The congruence $6x \equiv 7 \pmod{15}$ has no solutions because $\gcd(6, 15) = 3$ and $3 \nmid 7$.

- The congruence $-10x \equiv 12 \pmod{15}$ has no solutions because we have $\gcd(-10, 15) = 5$ and $5 \nmid 12$.

- The congruence $5x \equiv 6 \pmod{11}$ has $x = 10$ as a solution. Since $\gcd(5, 11) = 1$, this is the only solution mod 11.

- The congruence $3x \equiv 8 \pmod{10}$ has $x = 6$ as a solution, and it is the only solution mod 10.

- In the congruence $9x \equiv 6 \pmod{15}$, we divide by $3 = \gcd(9, 15)$ and get the new congruence $3x \equiv 2 \pmod{5}$. This has $x_0 = 4$ as a solution. The theorem tells us that the solutions to $9x \equiv 6 \pmod{15}$ are obtained by adding multiples of $5 = 15/3$ to 4:

$$x \equiv 4,\ 9,\ 14 \pmod{15}.$$

Section 5.4 Linear Congruences   111

- The congruence $10x \equiv 15 \pmod{45}$ has $5 = \gcd(10, 45)$ solutions. To find them, we divide by 5 and get $2x \equiv 3 \pmod 9$. This has $x_0 = 6$ as a solution. The solutions are obtained by adding multiples of $9 = 45/5$ to 6:

$$x \equiv 6,\ 15,\ 24,\ 33,\ 42 \pmod{45}.$$

There is a subtlety here that you should be aware of. When we had the linear equation $ax - my = b$, we could divide both sides by $\gcd(a, m)$. This made the computations involved simpler and nothing was lost. But if we try the same thing with a linear congruence, we get solutions to a congruence with a new modulus, namely mod m/d, where $d = \gcd(a, m)$. These yield d solutions to the original congruence.

For example, $4x \equiv 8 \pmod{20}$ has 4 solutions mod 20 (namely, $x = 2$, 7, 12, 17). Dividing by $\gcd(4, 20) = 4$ gives us just one solution mod 5, namely $x \equiv 2 \pmod 5$. So we take 2, 2+5, 2+10, 2+15, and stop at $2 + 20 = 22$ since 22 is the same as 2 mod 20. Just remember that if you divide by the gcd, then you need to remember this extra step to get back all solutions for the original modulus.

A linear congruence can be solved using the Extended Euclidean Algorithm by transforming the congruence to an equation. Here's an example.

Example. We'll show how to solve the linear congruence

$$183x \equiv 15 \pmod{31}. \tag{5.6}$$

We begin by reducing 183 mod 31 to get $28x \equiv 15 \pmod{31}$, and then we change this linear congruence to a linear Diophantine equation:

$$28x - 31y = 15.$$

The next step is to use the Euclidean Algorithm to find $\gcd(28, 31)$:

$$31 = 1 \cdot 28 + 3$$
$$28 = 9 \cdot 3 + 1$$
$$3 = 3 \cdot 1 + 0.$$

So $\gcd(28, 31) = 1$. We now use the Extended Euclidean Algorithm:

$$\begin{array}{ccc} & x & y \\ 31 & 1 & 0 \\ 28 & 0 & 1 \\ 3 & 1 & -1 \\ 1 & -9 & 10 \end{array} \quad \begin{array}{l} \\ \\ \text{(1st row)} - \text{(2nd row)} \\ \text{(2nd row)} - 9 \cdot \text{(3rd row)}. \end{array}$$

This tells us that $28 \cdot (10) + 31 \cdot (-9) = 1$ or $28 \cdot 10 - 31 \cdot 9 = 1$. After multiplying both sides by 15, we find

$$28 \cdot 150 - 31 \cdot 135 = 15.$$

Since mod 31 means that we can discard multiples of 31, we obtain $28 \cdot 150 \equiv 15 \pmod{31}$. This tells us that $x = 150$ is a solution to (5.6), and after reducing 150 mod 31, we get that $x = 26$ is a solution. An alternative approach uses inverses, which we now discuss.

Definition 5.19. *We say that integers a and b are* **multiplicative inverses mod m** *if $ab \equiv 1 \pmod{m}$.*

If the context is clear, we drop the word multiplicative and simply say that a and b are inverses $\pmod{m}$.

Examples. Since $2 \cdot 3 \equiv 1 \pmod{5}$, we see that 2 and 3 are inverses mod 5. Similarly, 5 and 7 are inverses mod 34, and 8 and 7 are inverses mod 11 (since $8 \cdot 7 = 56 \equiv 1 \pmod{11}$).

To find the inverse mod m of an integer a, we need to solve the congruence

$$ax \equiv 1 \pmod{m}. \tag{5.7}$$

We immediately see that Theorem 5.17 has the following as a consequence.

Corollary 5.20. *The integer a has an inverse mod m if and only if $\gcd(a, m) = 1$.*

Proof. We know from Theorem 5.17 that $ax \equiv 1 \pmod{m}$ has a solution if and only if $\gcd(a, m) \mid 1$. This is true if and only if $\gcd(a, m) = 1$. □

Example. In order to find the inverse of 13 $\pmod{100}$, we need to solve the congruence

$$13x \equiv 1 \pmod{100}. \tag{5.8}$$

Section 5.4   Linear Congruences

To solve this, we use the same technique used to solve an arbitrary linear congruence. First, we change the congruence into a linear Diophantine equation:
$$13x - 100y = 1.$$
Next, we use the Euclidean Algorithm to find gcd(13, 100):
$$100 = 7 \cdot 13 + 9$$
$$13 = 1 \cdot 9 + 4$$
$$9 = 2 \cdot 4 + 1$$
$$4 = 4 \cdot 1 + 0.$$

So gcd(13, 100) = 1. We now use the Extended Euclidean Algorithm:

	x	y	
100	1	0	
13	0	1	
9	1	−7	(1st row) − 7·(2nd row)
4	−1	8	(2nd row) − (3rd row)
1	3	−23	(3rd row) − 2·(4th row).

Therefore, $1 = 100 \cdot 3 - 13 \cdot 23$, so $13(-23) \equiv 1 \pmod{100}$. The inverse of 13 (mod 100) is $-23 \equiv 77 \pmod{100}$.

Example. To find the inverse of 7 (mod 19), we need to solve the congruence $7x \equiv 1 \pmod{19}$. We can do this by writing this as the linear equation $7x - 19y = 1$ and then solving by using the Extended Euclidean Algorithm as in the previous example. An alternative that is often quicker for small moduli is to manipulate the equation in order to get a coefficient of 1 for x. Here is an example.

$$7x \equiv 1 \pmod{19} \quad \text{Now multiply by 3}$$
$$21x \equiv 3 \pmod{19} \quad \text{Now reduce mod 19}$$
$$2x \equiv 3 \pmod{19} \quad \text{Now add 19 to 3}$$
$$2x \equiv 22 \pmod{19} \quad \text{Now divide by 2}$$
$$x \equiv 11 \pmod{19} \quad \text{We're done!}$$

This says that a solution to $7x \equiv 1 \pmod{19}$ is $x = 11$, so 11 (mod 19) is the inverse to 7 (mod 19). We can write this as $7^{-1} \equiv 11 \pmod{19}$.

Example. Here's how we can solve $7x \equiv 4 \pmod{19}$ by using the (multiplicative) inverse of 7. Multiplying both sides by 11, which we just saw is the inverse of 7 (mod 19), yields the following:

$$7x \equiv 4 \pmod{19} \iff 77x \equiv 44 \pmod{19} \iff x \equiv 6 \pmod{19}$$

since $77 \equiv 1 \pmod{19}$ and $44 \equiv 6 \pmod{19}$. Therefore, $x \equiv 6 \pmod{19}$ is the solution.

In effect, we've solved this equation by multiplying by 11, instead of dividing by 7. This may seem strange, but we do things like this all the time. For example, with real numbers, instead of dividing by 2, we can multiply by 0.50. What we've just done is the modular analogue of this.

Example. Let's try this technique to solve $3x \equiv 4 \pmod{26}$. Since $3 \cdot 9 \equiv 1 \pmod{26}$, we have $3^{-1} \equiv 9 \pmod{26}$. Therefore,

$$3x \equiv 4 \pmod{26} \iff 27x \equiv 36 \pmod{26} \iff x \equiv 10 \pmod{26},$$

so $x \equiv 10 \pmod{26}$ is the solution.

The following are summaries of two important algorithms. Note that the second is a special case of the first and occurs as an intermediate step in the first.

Solve $ax \equiv b \pmod{m}$ when $\gcd(a, m) = 1$

1. Form a linear Diophantine equation: $ax - my = b$.

2. Use the Extended Euclidean Algorithm to find x_0, y_0 with $ax_0 - my_0 = 1$.

3. Compute $x \equiv bx_0 \pmod{m}$.

Find the inverse of $a \pmod{m}$ when $\gcd(a, m) = 1$

1. Form the linear Diophantine equation $ax - my = 1$.

2. Use the Extended Euclidean Algorithm to find x_0, y_0 with $ax_0 - my_0 = 1$.

3. $x_0 \pmod{m}$ is the inverse of $a \pmod{m}$.

Section 5.5   The Chinese Remainder Theorem   115

CHECK YOUR UNDERSTANDING
8. Find the multiplicative inverse of 15 (mod 73).
9. Solve $15x \equiv 11$ (mod 73).

5.5 The Chinese Remainder Theorem

Around 1700 years ago, the Chinese mathematician Sun Tzu asked for a number that leaves a remainder of 2 when divided by 3, leaves a remainder of 3 when divided by 5, and leaves a remainder of 2 when divided by 7. In terms of congruences, this means

$$x \equiv 2 \pmod{3}, \quad x \equiv 3 \pmod{5}, \quad x \equiv 2 \pmod{7}.$$

Similar problems subsequently appeared in various writings in various countries, along with algorithms for solving them. Eventually, the theorem that guarantees (under certain hypotheses) that a solution exists became known as the Chinese Remainder Theorem.

Try to find a solution to Sun Tzu's problem. You'll probably devise methods that are close to some that we discuss in this section.

In the previous section, we learned how to solve a linear congruence. Let's start with asking what happens if we have two (or more) linear congruences with different moduli. For example,

$$x \equiv 2 \pmod{3} \text{ and } x \equiv 4 \pmod{5}. \tag{5.9}$$

We'll begin by solving this by using a somewhat ad hoc method: Start by looking at the congruence that has the larger modulus, in this case $x \equiv 4 \pmod{5}$. Make a list of positive integers that are congruent to 4 mod 5:

$$4, 9, 14, 19, 24$$

and then make a list of these numbers mod 3, stopping when you get a solution to $x \equiv 2 \pmod{3}$:

$$\begin{array}{lccccc} x \equiv 4 \pmod{5}: & 4 & 9 & 14 & 19 & 24 \\ x \pmod{3}: & 1 & 0 & \mathbf{2}. & & \end{array}$$

This tells us that 14 is a solution to both congruences. If we had started our list with the smaller modulus 3, we would have arrived at the answer of 14, but it would have taken us longer to get there since our initial list would be 2, 5, 8, 11, 14.

Here's another example of this method, this time with three congruences.

$$x \equiv 4 \pmod{5}, \quad x \equiv 5 \pmod{7}, \quad \text{and} \quad x \equiv 3 \pmod{11}. \tag{5.10}$$

Since 11 is the largest of the moduli, we begin with a list of integers that are 3 (mod 11):

$$3, 14, 25, 36, 47, 58, 69, 80, \ldots.$$

Next, put each number mod 7 underneath this:

$$\begin{array}{ll} x \equiv 3 \pmod{11}: & 3 \quad 14 \quad 25 \quad 36 \quad 47 \quad 58 \\ x \pmod{7}: & 3 \quad 0 \quad 4 \quad 1 \quad 5 \quad 2. \end{array}$$

This tells us that $x \equiv 47 \pmod{77}$ is the solution to $x \equiv 3 \pmod{11}$ and $x \equiv 5 \pmod{7}$. Now make a list of the numbers that are 47 (mod 77) and look at them mod 5:

$$\begin{array}{ll} x \equiv 47 \pmod{77}: & 47 \quad 124 \quad 201 \quad \cdots \\ x \pmod{5}: & 2 \quad 4 \end{array}$$

So, $x = 124$ is a solution to the three congruences.

Although this method works well for small numbers, if the moduli are large the lists can become very long, which makes this impractical. In the following, we prove a general result that guarantees a solution to simultaneous congruences. Moreover, the proof gives an algorithm for finding the solution.

Theorem 5.21. (Chinese Remainder Theorem) *Let $m_1, m_2, \ldots, m_r$ be positive integers and assume that they are pairwise relatively prime (that is, $\gcd(m_i, m_j) = 1$ if $i \neq j$). Then the system of congruences*

$$\begin{aligned} x &\equiv a_1 \pmod{m_1} \\ x &\equiv a_2 \pmod{m_2} \\ &\vdots \\ x &\equiv a_r \pmod{m_r} \end{aligned}$$

has a unique solution mod $m_1 m_2 \cdots m_r$.

Section 5.5   The Chinese Remainder Theorem

Algorithm for Solving the Chinese Remainder Theorem. Let $m_1, m_2, \ldots, m_r$ and $a_1, a_2, \ldots, a_r$ be as in the statement of the Chinese Remainder Theorem. Let $m = m_1 m_2 \cdots m_r$.

1. Let z_i be the multiplicative inverse of m/m_i mod m_i.
2. Let $x = a_1(m/m_1)z_1 + a_2(m/m_2)z_2 + \cdots + a_r(m/m_r)z_r$.
3. Then $x \equiv a_i \pmod{m_i}$ for each i.

Proof. We prove the theorem and verify the algorithm together. The theorem says that a solution exists and is unique. We first show that the algorithm produces a solution. Then we show the solution is unique.

Note that m/m_i is the product of all m_j except m_i. Because $m_1, m_2, \ldots, m_r$ are pairwise relatively prime, we have $\gcd(m/m_i, m_i) = 1$. Therefore, Corollary 5.20 implies that z_i exists. Second, since m/m_j contains the factor m_i when $i \neq j$,

$$x \equiv a_1(0)z_1 + \cdots + a_i(m/m_i)z_i + \cdots a_r(0)z_r \equiv a_i(m/m_i)z_i \pmod{m_i}.$$

But z_i is the multiplicative inverse of m/m_i mod m_i, so $(m/m_i)z_i \equiv 1 \pmod{m_i}$. Therefore, $x \equiv a_i \pmod{m_i}$. This shows that the solution x exists and verifies that the algorithm works.

The only thing left to prove is the uniqueness of x. Assume that there are an x_1 and an x_2 with

$$\begin{aligned} x_1 &\equiv a_1 \pmod{m_1} & \text{and} && x_2 &\equiv a_1 \pmod{m_1} \\ x_1 &\equiv a_2 \pmod{m_2} & \text{and} && x_2 &\equiv a_2 \pmod{m_2} \\ &\vdots \\ x_1 &\equiv a_r \pmod{m_r} & \text{and} && x_2 &\equiv a_r \pmod{m_r} \end{aligned}$$

with the m_i pairwise relatively prime. Since

$$m_i \mid (x_1 - x_2) \text{ for } 1 \leq i \leq r,$$

the assumption that the m_i are pairwise relatively prime implies that

$$m_1 m_2 \cdots m_r \mid x_1 - x_2.$$

So, $x_1 \equiv x_2 \pmod{m_1 m_2 \cdots m_r}$, which shows the uniqueness of our solution. $\square$

Example. Let's consider the congruences in Equation 5.10:

$$x \equiv 4 \pmod{5}, \quad x \equiv 5 \pmod{7}, \quad \text{and} \quad x \equiv 3 \pmod{11}.$$

Then $m_1 = 5$, $m_2 = 7$, and $m_3 = 11$, so $m = 5 \cdot 7 \cdot 11 = 385$. We have

$$m/m_1 = 77 \quad \text{and} \quad z_1 \equiv 77^{-1} \equiv 3 \pmod{5}$$
$$m/m_2 = 55 \quad \text{and} \quad z_2 \equiv 55^{-1} \equiv 6 \pmod{7}$$
$$m/m_3 = 35 \quad \text{and} \quad z_3 \equiv 35^{-1} \equiv 6 \pmod{11}.$$

Therefore,

$$x \equiv 4(77)3 + 5(55)6 + 3(35)6 \equiv 3204 \equiv 124 \pmod{385}.$$

The Chinese Remainder Theorem gives us a new way of looking at congruences for composite moduli. If we factor the modulus m into distinct prime powers, then either we can look at a congruence mod m, or we can look at a system of congruences mod these prime powers. Sometimes the latter are easier to work with. For example, suppose we want to solve

$$x^2 \equiv 1 \pmod{275}.$$

Since $275 = 5^2 \cdot 11$, this is the same as the system

$$x^2 \equiv 1 \pmod{25}. \quad x^2 \equiv 1 \pmod{11}.$$

The congruence $x^2 \equiv 1 \pmod{25}$ has the solutions $x \equiv 1$ or 24 (mod 25). The congruence $x^2 \equiv 1 \pmod{11}$ has the solutions $x \equiv 1$ or 10 (mod 11). There are four ways to combine these:

$$x \equiv 1 \pmod{25}, \quad x \equiv 1 \pmod{11} \implies x \equiv 1 \pmod{275}$$
$$x \equiv 1 \pmod{25}, \quad x \equiv 10 \pmod{11} \implies x \equiv 76 \pmod{275}$$
$$x \equiv 24 \pmod{25}, \quad x \equiv 1 \pmod{11} \implies x \equiv 199 \pmod{275}$$
$$x \equiv 24 \pmod{25}, \quad x \equiv 10 \pmod{11} \implies x \equiv 274 \pmod{275}.$$

Therefore, the solutions are 1, 76, 199, and 274 mod 275.

A useful general result is the following.

Section 5.6   Fractions Mod m

Proposition 5.22. *Let $n = pq$ be a product of 2 distinct odd primes. Let b be an integer with $\gcd(b,n) = 1$. Then*
$$x^2 \equiv b^2 \pmod{n}$$
has exactly 4 solutions x mod n.

Proof. Suppose x is a solution, so $x^2 \equiv b^2 \pmod{n}$. Then $x^2 \equiv b^2 \pmod{p}$ and $\pmod{q}$. If $x^2 \equiv b^2 \pmod{p}$, then $(x+b)(x-b) \equiv 0 \pmod{p}$, so Proposition 5.9 implies that either $x+b \equiv 0$ or $x-b \equiv 0 \pmod{p}$. Therefore, $x \equiv \pm b \pmod{p}$. Similarly, $x \equiv \pm b \pmod{q}$. Note that $\gcd(b,n)$ implies that $b \not\equiv 0 \pmod{p}$. Since n is odd, $p \neq 2$, which means that $+b$ and $-b$ are distinct mod p, and similarly for q.

There are 4 ways to choose the $\pm$ signs in the simultaneous congruences
$$x \equiv \pm b \pmod{p}, \quad x \equiv \pm b \pmod{q},$$
and each choice yields an x mod n, so we obtain 4 distinct values of x. Moreover, each x satisfies $x^2 \equiv (\pm b)^2 \equiv b^2 \pmod{p}$ and $\pmod{q}$. Therefore, p and q divide $x^2 - b^2$, so $n = pq$ divides $x^2 - b^2$. Consequently, each of the 4 values of x yields a solution of $x^2 \equiv b^2 \pmod{n}$. □

More generally, if n is the product of r distinct odd primes and $\gcd(b,n) = 1$, then $x^2 \equiv b^2 \pmod{n}$ has exactly 2^r solutions mod n. The proof is similar to the proof of Proposition 5.22.

CHECK YOUR UNDERSTANDING

10. Solve $x \equiv 3 \pmod{11}$, $x \equiv 2 \pmod{7}$.

5.6   Fractions Mod m

Let's look at two calculations:

1. *Problem:* Solve $7x = 2$. *Solution:* Multiply by $1/7$ to get $x = 2/7$.

2. *Problem:* Solve $7x \equiv 2 \pmod{19}$. *Solution:* Multiply by 11 to get $77x \equiv 22 \pmod{19}$, which becomes $x \equiv 22 \pmod{19}$ because $77 \equiv 1 \pmod{19}$.

In the second calculation, the role of 1/7 was played by 11, which is the multiplicative inverse 7 in the mod 19 world.

What is 1/7, anyway? It's a symbol with exactly one property: When you multiply it by 7 you get 1 as the answer. In the mod 19 world, if you multiply 11 by 7, you get 77, which is the same as 1, since $77 \equiv 1 \pmod{19}$. Therefore, if you had grown up in the mod 19 world, you would have learned that you could write 11 in place of 1/7. This might seem strange, but you already know a similar situation. You probably grew up in a decimal world, so you don't get upset when someone writes $1/7 = 0.142857\cdots$, even though the second expression does not look much like dividing by 7. In different worlds, 1/7 can be expressed in different ways.

In many situations in number theory, it is useful to be able to work with fractions mod m during intermediate steps in calculations. In this section, we show how this can be done.

Here are two warnings:

1. When working mod m, work only with fractions a/b with $\gcd(b, m) = 1$. This is because such b are the only ones that have multiplicative inverses mod m.

2. At the end of the calculation, change the result back to an integer. This is particularly important when the numbers involved are exponents, since fractional exponents cause problems and you should never have them when you are working mod m.

The fraction a/b is the same as $a(1/b)$. When a fraction a/b occurs mod m, interpret it as $a \cdot b^{-1}$, where $b \cdot b^{-1} \equiv 1 \pmod{m}$; that is, b^{-1} is the multiplicative inverse of b mod m. This is why we need $\gcd(b, m) = 1$.

Example. Since $1/7 \equiv 11 \pmod{19}$, we see that $2/7 \pmod{19}$ is the same as $2 \cdot 11 \pmod{19}$, which is the same as $3 \pmod{19}$.

Here is another example. Suppose we want to compute the sum $x_1 + x_2 + x_3 \pmod{13}$, and we know that $2x_1 \equiv 1 \pmod{13}$, $3x_2 \equiv 1 \pmod{13}$, and $6x_3 \equiv 1 \pmod{13}$. In terms of fractions, we have

$$x_1 \equiv 1/2, \quad x_2 \equiv 1/3, \quad x_3 \equiv 1/6 \pmod{13}.$$

Since
$$\frac{1}{2} + \frac{1}{3} + \frac{1}{6} = 1,$$
$x_1 + x_2 + x_3 \equiv 1 \pmod{13}$. To check this, we compute $x_1 \equiv 7$, $x_2 \equiv 9$, and $x_3 \equiv 11$. Therefore,

$$x_1 + x_2 + x_3 \equiv 7 + 9 + 11 \equiv 1 \pmod{13}.$$

Fractions are convenient when writing certain formulas. For example, the quadratic formula mod an odd prime p (see Section 10.3) says that the solutions to $x^2 + bx + c \equiv 0 \pmod{p}$ are given by

$$x \equiv \frac{-b \pm \sqrt{b^2 - 4c}}{2} \pmod{p}.$$

Of course, we could write this as $(-b \pm \sqrt{b^2 - 4c})(2^{-1})$, where 2^{-1} is the inverse of 2 mod p, but many people find it more convenient to write the formula as a fraction.

As another example, in intermediate steps of some calculations, we start with congruences like $2^x \equiv 2^y \cdot b \pmod{m}$ and then write $2^{x-y} \equiv b \pmod{m}$. We don't want to worry about whether $x - y$ is positive or negative. If it's negative, 2^{x-y} is a fraction, which means that 2^{x-y} is interpreted as the multiplicative inverse of $2^{-(x-y)}$.

5.7 Queens on a Chessboard

The *eight queens problem* is to place eight queens on a chess board so that no queen can attack another. This means that there is exactly one queen in each row and exactly one in each column, and no two can lie on the same diagonal (that is, any line with slope ± 1). One solution is the following:

More generally, the n queens problem is to place n queens on an $n \times n$ checkerboard with the same requirements. It has been proved that there is a solution for every positive integer n except for $n = 2$ and $n = 3$. In the following, we give a simple solution for all n with $\gcd(n, 6) = 1$.

Label the rows and columns from 0 to $n - 1$, so (x, y) denotes the square in column x and row y. Regard the coordinates as numbers mod n. Diagonal lines are given by equations $x + y = c$ or $x - y = c$, where c is a constant.

There are n queens, which we'll number from 0 to $n - 1$. Place the jth queen in square $(j, 2j)$, where $2j$ means $2j \pmod{n}$. Since the x-coordinates take on all n values, there is one queen in each column. If queen i and queen j are in the same row, their y-coordinates are the same, so $2i \equiv 2j \pmod{n}$. Since $\gcd(n, 6) = 1$, we have $\gcd(2, n) = 1$, so we can divide by 2 and obtain $i \equiv j \pmod{n}$. Therefore, there is only one queen in each row.

Now suppose that the ith queen and the jth queen lie on the diagonal $x + y = c$. Since their coordinates are $(i, 2i)$ and $(j, 2j)$ mod n, we have

$$i + 2i \equiv c \pmod{n} \quad \text{and} \quad j + 2j \equiv c \pmod{n}.$$

Therefore, $3i \equiv c \equiv 3j \pmod{n}$. Since $\gcd(3, n) = 1$, we can divide by 3 and obtain $i \equiv j \pmod{n}$. This means that $i = j$, so at most one queen can lie on this diagonal.

Similarly, suppose that the ith queen and the jth queen lie on the diagonal $x - y = c$. Then

$$i - 2i \equiv c \pmod{n} \quad \text{and} \quad j - 2j \equiv c \pmod{n}.$$

Therefore, $-i \equiv c \equiv -j \pmod{n}$. This means that $i = j$, so at most one queen can lie on this diagonal.

Therefore, this placement of the queens solves the problem. Here is an example for $n = 5$. We take the lower left corner as the square $(0,0)$.

		q		
				q
	q			
			q	
q				

It is interesting to observe that our solution works "on a torus." Imagine that the chess board is a torus (that is, the surface of a donut), so the top edge is attached to the bottom edge and the left edge is attached to the right edge. Then the diagonals wrap around. For example, on the 5×5 board, the diagonal that starts at square $(0,1)$ and continues through $(1,2)$, $(2,3)$, and $(3,4)$ is allowed to exit at the top edge and then re-enter at the bottom at $(4,0)$ and continue back to $(0,1)$. In this situation, the diagonal lines are given by congruence $x+y \equiv c \pmod{n}$ and $x - y \equiv c \pmod{n}$ rather than equalities. Our solution shows that the queens cannot attack one another, even if they are allowed to move along these extended diagonals.

In contrast, in the solution for the 8×8 board at the beginning of this section, the queen in position $(2,5)$ (where the lower left corner is $(0,0)$) can attack the queen at $(5,0)$ along an extended diagonal given by $x - y \equiv 5 \pmod{8}$.

5.8 Chapter Highlights

1. If $\gcd(a, m) = 1$, then a has a multiplicative inverse mod m and $ax \equiv b \pmod{m}$ has a unique solution mod m.
2. The Chinese Remainder Theorem.

5.9 Problems

5.9.1 Exercises

Section 5.1: Definitions and Examples

1. Find the least nonnegative residue mod 14 for each of the following:
   (a) 27 (b) 16 (c) −20 (d) 311 (e) −91 (f) 42

2. Find the least nonnegative residue mod 8 for each of the following:
   (a) 11 (b) −5 (c) 121 (d) 2014 (e) −83 (f) 57

3. Verify that each of the following congruences is true:
   (a) $96 \equiv 6 \pmod{10}$ (b) $101 \equiv -9 \pmod{10}$
   (c) $-5 \equiv 13 \pmod{9}$

4. Verify that each of the following congruences is true:
   (a) $77 \equiv 12 \pmod 5$ (b) $136 \equiv 31 \pmod 5$
   (c) $-11 \equiv -60 \pmod 7$

5. Find all positive integers n for which the congruence $123 \equiv 234 \pmod n$ holds.

6. Find all positive integers n for which the congruence $1855 \equiv 1777 \pmod n$ holds.

7. A bug starts at the 12 on a (nondigital) 12-hour clock. It jumps to the 1, then to the 2, then to the 3, etc. After 12345 jumps, what number is it on? Explain your answer in terms of a congruence.

8. Suppose today is Monday. What day of the week will it be in 150 days? Explain your answer in terms of a congruence.

9. Let n be an integer.
   (a) Use congruences to show that $n(n+1)$ is always even.
   (b) Use congruences to show that $n(n+1)(n+2)$ is always a multiple of 6.
   (c) Use the fact that binomial coefficients are integers to prove (a) and (b). (See Appendix A for binomial coefficients.)

Section 5.9   Problems   125

10. Show that if $n \equiv 2 \pmod{4}$, then n cannot be written as a^m with $m > 1$.

11. Show that if you sum the cubes of three consecutive integers, the result is a number that's congruent to 0 $\pmod 9$.

12. Prove that the square of any integer must be a number ending in 0, 1, 4, 5, 6, or 9.

13. If n is an odd integer, what are the possibilities for the last digit of n^4?

14. If n is an integer, what possibilities are there for $n^3 \pmod 9$?

15. Let a, b, d, k, m be positive integers. Decide whether each of the following is true or false. If true, give a proof; if false, give a counterexample.
    (a) If $d \mid m$ and $a \equiv b \pmod m$, then $a \equiv b \pmod d$.
    (b) If $d \mid m$ and $a \equiv b \pmod d$, then $a \equiv b \pmod m$.
    (c) If $a = b$, then $a \equiv b \pmod n$ for every positive integer n.
    (d) If $a \equiv b \pmod n$ for every positive integer n, then $a = b$.
    (e) If $\gcd(k, m) = 1$ and $ka \equiv kb \pmod m$, then $a \equiv b \pmod m$.

16. If (a, b, c) is a Pythagorean triple, prove that
    (a) at least one of $a, b,$ or c is divisible by 3.
    (b) at least one of $a, b,$ or c is divisible by 5.
    (c) we always have abc divisible by 60.
    (*Hint:* Squares are congruent to $0, 1 \pmod 3$ and to $0, 1, 4 \pmod 5$.)

17. Show that if $n \equiv 2 \pmod 3$, then n cannot be written in the form $x^2 + 3y^2$ (originally due to Fermat).

18. Show that the fourth power of any odd integer is of the form $16k + 1$.

19. Suppose that p and q are twin primes, so that p and q are primes whose difference is 2. (Examples are 3 and 5, 5 and 7, 11 and 13, ...)
    (a) Show that if $p, q \geq 5$, then $p + q$ is divisible by 3.
    (b) Show that $p + q$ is a multiple of 4.
    (c) Use (a) and (b) to show that $p + q$ is divisible by 12.

20. (a) Show that a square is congruent to 0, 1, or 4 mod 8.

(b) Show that a sum of two squares is not 3 mod 4.

(c) Show that a sum of three squares is not 7 mod 8.

21. Show that if $9 \mid (a^3 + b^3 + c^3)$, then 3 divides at least one of the integers a, b, or c.

22. (a) Show that if $1 \leq x \leq y \leq z$ and $x! + y! = z!$, then $x = y = 1$ and $z = 2$. (*Hint:* Use congruences mod $y!$.)
    (b) Show that $x!\,y! = z!$ has infinitely many solutions. (*Hint:* For example, $5!\,119! = 120!$.)

23. Alice and Bob have a pile of n markers. They fix an integer m and play the following game. A move consists of removing at least one and at most m markers from the pile. They take turns making moves. Alice plays first.
    (a) Suppose the rules state that the person who removes the last marker wins. Show that Alice has a winning strategy if and only if $n \not\equiv 0 \pmod{m+1}$.
    (b) Suppose the rules state that the person who removes the last marker loses. Show that Alice has a winning strategy if and only if $n \not\equiv 1 \pmod{m+1}$.

24. Let p be an odd prime.
    (a) Let $a \not\equiv 0 \pmod{p}$. Show that $x^2 \equiv a^2 \pmod{p}$ has exactly two solutions.
    (b) Show that there are exactly $(p-1)/2$ nonzero squares mod p.

25. Let F_n be the nth Fibonacci number (see the Appendix). Show that if $n \geq 1$, then $F_n \equiv F_{n+2} \equiv F_{n+3} \pmod{F_{n+1}}$.

26. Let F_n be the nth Fibonacci number (see the Appendix).
    (a) Use induction to show that if $n \geq 1$, then $F_{n+3} \equiv F_n \pmod{2}$.
    (b) Show that if $n \geq 1$, then $F_{n+6} \equiv 5F_n \pmod{8}$.
    (c) Show that if $n \geq 1$, then $F_{n+12} \equiv F_n \pmod{8}$.

Section 5.2: Modular Exponentiation

27. Compute (a) $2^{17} \pmod{13}$, (b) $5^{25} \pmod{11}$, (c) $2^{16} \pmod{10}$.

28. Compute (a) $3^{40} \pmod{100}$, (b) $7^9 \pmod{11}$, (c) $6^{10} \pmod{17}$.

Section 5.3: Divisibility Tests

29. Find the least nonnegative residue mod 9 for each of the following:
    (a) 1453 (b) 1927 (c) 1066 (d) 1855 (e) −4004 (f) −753

30. Find the least nonnegative residue mod 11 for each of the following:
    (a) 1777 (b) −43 (c) 275 (d) 1234567 (e) −83 (f) 235711

31. You calculated
    $$123456789 \times 987654321 = 121932a31112635269,$$
    but your handwriting was so bad that you cannot figure out what digit a represents. Use congruences mod 9 to get the answer.

32. You receive a message from an extraterrestrial alien, who is calculating 43434343^2. The answer is $18865ab151841649$, where the two digits represented as a and b were lost in transmission. Use congruences mod 9 and mod 11 to determine the answer to this fundamental problem.

33. Suppose you know that $12^{10} = 61917ab4224$, where a and b are digits. Use congruences mod 9 and mod 11 to determine a and b.

Section 5.4: Linear Congruences

34. Find all solutions to each of the following:
    (a) $3x \equiv 8 \pmod{11}$ (b) $6x \equiv 7 \pmod 9$
    (c) $4x \equiv 12 \pmod{32}$

35. Find all solutions to each of the following:
    (a) $5x \equiv 19 \pmod{21}$ (b) $8x \equiv 12 \pmod{19}$
    (c) $91x \equiv 3 \pmod{121}$

36. Find all values of n with $0 \le n \le 23$ such that the congruence $10x \equiv n \pmod{24}$ has a solution.

37. Find all values of n with $0 \le n \le 29$ such that the congruence $25x \equiv n \pmod{30}$ has a solution.

38. (a) Solve the congruence $3x \equiv 1 \pmod{17}$.
    (b) Use (a) to solve $3x \equiv 5 \pmod{17}$.

39. (a) Solve $83x \equiv 1 \pmod{100}$.
    (b) Solve $83x \equiv 2 \pmod{100}$.

40. Show that if $7 \mid (a^2 + b^2)$, then 7 divides both a and b. (*Hint:* If $a \not\equiv 0 \pmod 7$, you can divide by a mod 7.)

41. Show that $x^2 - 2y^2 = 10$ has no integer solutions. (*Hint:* Work mod 5.)

Section 5.5: The Chinese Remainder Theorem

42. Solve the system of congruences
$$x \equiv 2 \pmod 3$$
$$x \equiv 1 \pmod 7.$$

43. Solve the system of congruences
$$x \equiv 1 \pmod 3$$
$$x \equiv 2 \pmod 7$$
$$x \equiv 4 \pmod{11}.$$

44. Solve the system of congruences
$$3x \equiv 2 \pmod 5$$
$$4x \equiv 3 \pmod 7$$
$$x \equiv 2 \pmod{11}.$$

45. (Classical problem: Bhāskara, al-Haitham, Fibonacci) When the eggs in a basket are taken out 2 at a time, there is one egg left in the basket. When they are taken out 3, 4, 5, and 6 at a time there are 2, 3, 4, and 5 eggs, respectively, left in the basket. If 7 at a time are taken, then none are left. What is the least number of eggs in the basket?

46. Suppose that you want to calculate
$$1434661 \cdot 3785648743 - 100020304 \cdot 54300201.$$
You are told that the answer is a positive integer less than 90.

Section 5.9   Problems   129

Compute the answer mod 10 and mod 9, then combine to get the answer. *Note:* This technique has often been employed to do calculations with large integers because most of the computations use small integers, with only the Chinese Remainder Theorem step requiring large integers. For example, if the answer is known to be less than 10^{36}, the calculations can be done mod p for each prime less than 100. Since the product of the primes less than 100 is approximately 2.3×10^{36}, the Chinese Remainder Theorem gives the answer exactly.

47. Show that there are 100 consecutive integers that are not squarefree. (*Hint:* Take 100 different primes p and arrange for p^2 to divide an appropriate member of the sequence.)

48. Find an integer n that gives a remainder of 1 when divided by 2, 3, 4, 5, and 6, a remainder of 2 when divided by 7, a remainder of 1 when divided by 8, a remainder of 4 when divided by 9, a remainder of 1 when divided by 10 and a remainder of 0 when divided by 11. (*Hint:* Do not use the Chinese Remainder Theorem. Instead, write $n = 11r$ and determine the smallest possible divisor of r.)

49. Find a positive integer x such that $x \equiv 1 \pmod{2}$, $x \equiv 2 \pmod{3}$, $x \equiv 3 \pmod{4}$, and $x \equiv 4 \pmod{5}$. (*Hint:* First find a negative solution.)

50. If we have the system of congruences

$$x \equiv a \pmod{m} \text{ and } x \equiv b \pmod{n},$$

we know that the Chinese Remainder Theorem guarantees a solution if $\gcd(m, n) = 1$. But what happens otherwise? Show that this system has a solution if and only if $\gcd(m,n) \mid (a-b)$. (*Hint:* If $\gcd(m,n) \mid (a-b)$, express $a-b$ as a linear combination of m and n. Use this to explicitly find a solution to both congruences. If there is a solution to both congruences, show how to express $a - b$ as a linear combination of m and n.)

51. Let m and n be positive integers such that $m \mid n$. Let $\gcd(a, m) = 1$. The Chinese Remainder Theorem says that there exists b with $b \equiv a \pmod{m}$ and $b \equiv 1 \pmod{p}$ for every prime p that divides n but does not divide m. Show

that $\gcd(b,n) = 1$.

This shows that every relatively prime congruence class mod m can be obtained by starting with a relatively prime congruence class mod n and reducing it mod m.

5.9.2 Computer Explorations

1. Find all values of n with $2 \leq n \leq 100$ for which $1^3 + 2^3 + 3^3 + \cdots + (n-1)^3 \equiv 0 \pmod{n}$. If this congruence is true, what congruence conditions does n seem to satisfy? Can you prove this?

2. For which primes p can you solve the congruence $x^2 \equiv -1 \pmod{p}$? Compute several examples and see if you can find the general rule.

3. For which primes p can you solve the congruence $x^2 \equiv 5 \pmod{p}$? Compute several examples and see if you can find the general rule.

4. For which primes p can you solve the congruence $x^2 \equiv 2 \pmod{p}$? Compute several examples and see if you can find the general rule.

5. (a) Solve the simultaneous congruences

$$n \equiv 0 \pmod{4}, \quad n \equiv -1 \pmod{9}, \quad n \equiv -2 \pmod{25},$$
$$n \equiv -3 \pmod{49}, \quad n \equiv -5 \pmod{121}$$

to find six consecutive integers that are not squarefree.
(b) Find the smallest positive integer n such that $n, n+1, n+2, n+3, n+4, n+5$ are not squarefree. The answer should be much smaller than the number from part (a).

5.9.3 Answers to "CHECK YOUR UNDERSTANDING"

1. (a) True: $17 - 29 = -12$, which is a multiple of 4.
   (b) False: $321 - 123 = 198$, which is not a multiple of 10.
   (c) True: $321 - 123 = 198 = 9 \cdot 22$, which is a multiple of 9.

2. Since $3 \equiv 14 \pmod{11}$, we can rewrite the congruence as

$2x \equiv 14 \pmod{11}$. We can divide by 2 because $\gcd(2, 11) = 1$, so $x \equiv 7 \pmod{11}$.

3. We have $\gcd(4, 22) = 2$. Divide the congruence by 2 to get $2x \equiv 3 \pmod{11}$. By the previous problem, $x \equiv 7 \pmod{11}$. Returning to the mod 22 world, we obtain $x \equiv 7, 18 \pmod{22}$.

4. Since 11 is prime, Proposition 5.9 says that either $x - 1 \equiv 0 \pmod{11}$ or $x - 2 \equiv 0 \pmod{11}$. Therefore, $x \equiv 1$ or $2 \pmod{11}$.

5. $2^2 \equiv 4 \pmod{67}$, $2^4 \equiv 16$, $2^8 \equiv 256 \equiv 55$, $2^{16} \equiv 3025 \equiv 10$, $2^{32} \equiv 100 \equiv 33$, $2^{64} \equiv 1089 \equiv 17$, and $2^{66} \equiv 2^{64} 2^2 2^1 \equiv 17 \cdot 4 \equiv 68 \equiv 1$.

6. $987654321 \equiv 9+8+7+6+5+4+3+2+1 \equiv 45 \equiv 4+5 \equiv 0 \pmod 9$.

7. $343434 \equiv 4-3+4-3+4-3 \equiv 3 \pmod{11}$ and $58 \equiv 8-5 \equiv 3 \pmod{11}$, so $343434 \equiv 58 \pmod{11}$.

8. Change the problem to solving $15x - 73y = 1$. Use the Extended Euclidean Algorithm:

$$\begin{array}{ccc} 73 & 1 & 0 \\ 15 & 0 & 1 \\ 13 & 1 & -4 \\ 2 & -1 & 5 \\ 1 & 7 & -34. \end{array}$$

We conclude that $1 = 73(7) - 15(34)$, so $15(-34) \equiv 1 \pmod{73}$. Therefore, the multiplicative inverse of 15 is $-34 \equiv 39 \pmod{73}$.

9. From the previous problem, we know that the multiplicative inverse of 15 (mod 73) is 39. Multiply both sides of $15x \equiv 11 \pmod{73}$ by 39 to get $x \equiv 11 \cdot 39 \equiv 429 \equiv 64 \pmod{73}$.

10. The numbers congruent to 3 mod 11 are 3, 14, 25, 36, 47, 58, ... These are 3, 0, 4, 1, 5, 2 mod 7. Therefore, $x \equiv 58 \pmod{77}$ is the answer.

Chapter 6

Fermat, Euler, Wilson

6.1 Fermat's Theorem

Look at the following statements:

$$2^2 - 2 \equiv 0 \pmod{2}$$

$$2^3 - 2 \equiv 0 \pmod{3}$$

$$2^4 - 2 \not\equiv 0 \pmod{4}$$

$$2^5 - 2 \equiv 0 \pmod{5}$$

$$2^6 - 2 \not\equiv 0 \pmod{6}$$

$$2^7 - 2 \equiv 0 \pmod{7}$$

$$2^8 - 2 \not\equiv 0 \pmod{8}$$

$$2^9 - 2 \not\equiv 0 \pmod{9}$$

$$2^{10} - 2 \not\equiv 0 \pmod{10}$$

$$2^{11} - 2 \equiv 0 \pmod{11}.$$

Do you see a pattern? A reasonable guess is that $2^n - 2 \equiv 0 \pmod{n}$ if and only if n is prime. A computer search soon produces the counterexample $2^{341} - 2 \equiv 0 \pmod{341}$, where $341 = 11 \cdot 31$ is not prime. So part of our conjecture falls apart. But the other part remains, namely that if n is prime then $2^n - 2 \equiv 0 \pmod{n}$. In fact, there is a more general statement.

Theorem 6.1. *(Fermat's Theorem) Let p be prime.*
(a) For every integer b,

$$b^p - b \equiv 0 \pmod{p}.$$

(b) If $b \not\equiv 0 \pmod{p}$, then

$$b^{p-1} \equiv 1 \pmod{p}.$$

Remark. Sometimes Fermat's Theorem is called "Fermat's Little Theorem" in order to distinguish it from Fermat's Last Theorem. However, there is nothing little about it. Fermat's theorem is one of the most important results in number theory and it is essential that you remember it and understand how to use it.

We'll prove the theorem (twice!) at the end of the section. First, we give some examples.

Example. $2^6 = 64 \equiv 1 \pmod{7}$, and $2^7 \equiv 2 \pmod{7}$. Notice that the second statement can be obtained from the first by multiplying by 2.

Example. Fermat's theorem can be used for exponents that are multiples of $p-1$: $3^{28} \equiv (3^4)^7 \equiv 1^7 \equiv 1 \pmod 5$.

Example. Divide 23 into 7^{200}. What is the remainder? By Fermat's theorem, $7^{22} \equiv 1 \pmod{23}$. Therefore,

$$7^{200} \equiv \left(7^{22}\right)^9 \cdot 7^2 \equiv 1^9 \cdot 49 \equiv 3 \pmod{23}.$$

This example demonstrates an important point. The original exponent 200 was reduced to 2 using the relation $200 = 22 \cdot 9 + 2$. In other words, $200 \equiv 2 \pmod{22}$. The general statement is the following.

Corollary 6.2. *Let p be prime and let $b \not\equiv 0 \pmod{p}$. If $x \equiv y \pmod{p-1}$, then*

$$b^x \equiv b^y \pmod{p}.$$

Proof. Since $x \equiv y \pmod{p-1}$, we can write $x = y + (p-1)k$ for some k. Therefore,

$$b^x = b^y (b^{p-1})^k \equiv b^y (1)^k \equiv b^y \pmod{p},$$

where we used Fermat's theorem to change b^{p-1} to 1 mod p. $\square$

Section 6.1 Fermat's Theorem   135

Warning. Be sure to notice that x and y are congruent mod $p-1$, not mod p. In other words, congruences mod $p-1$ in the exponent lead to congruences mod p overall. Do not try to use congruences mod p in the exponent. They lead to wrong answers. For example, $6 \equiv 1 \pmod 5$ but $2^6 = 64$ is not congruent to $2^1 = 2 \bmod 5$. However, $5 \equiv 1 \pmod 4$ and $2^5 \equiv 2^1 \pmod 5$.

Fermat's theorem says that if p is an odd prime, then $2^{p-1} \equiv 1 \pmod p$. The contrapositive of this statement says that if n is odd and $2^{n-1} \not\equiv 1 \pmod n$, then n is not prime. This allows us to prove some integers are composite without factoring. For example, let's show that 77 is composite. We have, using the successive squaring technique from Section 5.2,

$$2^2 \equiv 4 \pmod{77}$$
$$2^4 \equiv (2^2)^2 \equiv 4^2 \equiv 16$$
$$2^8 \equiv (2^4)^2 \equiv 16^2 \equiv 256 \equiv 25$$
$$2^{16} \equiv (2^8)^2 \equiv 25^2 \equiv 625 \equiv 9$$
$$2^{32} \equiv (2^{16})^2 \equiv 9^2 \equiv 4$$
$$2^{64} \equiv (2^{32})^2 \equiv 4^2 \equiv 16$$
$$2^{76} \equiv 2^{64} 2^8 2^4 \equiv 16 \cdot 25 \cdot 16 \equiv 6400 \equiv 9.$$

Since $2^{76} \not\equiv 1 \pmod{77}$, we conclude that 77 is not prime. This means that we have proved that 77 factors without factoring it. This might seem rather silly for a number as small as 77. But the same method can be used to show that much larger numbers are composite, even in cases where their factorization is currently impossible. For example, the 7th Fermat number

$$F_7 = 2^{2^7} + 1$$

was proved to be composite in 1905, but it wasn't until 1970, with the development of modern computer-oriented techniques, that its factorization was found. The Fermat number $F_{24} = 2^{2^{24}} + 1$, a number with more than 5 million digits, is known to be composite, but no prime factors are known.

Now it's time to prove Fermat's theorem. We give two different proofs, each with its own flavor.

Proof 1. We'll prove that $b^p \equiv b \pmod p$ by induction on b. We start with a lemma.

Lemma 6.3. *Let p be a prime. Then the binomial coefficient*

$$\binom{p}{j} = \frac{p!}{j!(p-j)!} \equiv 0 \pmod{p}$$

for $1 \leq j \leq p-1$.

Proof. We have

$$p! = \binom{p}{j}(p-j)!j!.$$

Clearly, p divides the left-hand side, so p must divide one of the factors on the right-hand side. Since all the factors of $(p-j)!$ are between 1 and $p-j$, we see that p does not divide $(p-j)!$. Similarly, p does not divide $j!$ since $j < p$. Therefore, p must divide the binomial coefficient. □

Returning to the proof of Fermat's theorem, we want to prove that $b^p \equiv b \pmod{p}$ for all b. We will do this using induction on b. Note that $0^p \equiv 0 \pmod{p}$ and $1^p \equiv 1 \pmod{p}$, so the statement is true for $b = 0$ and $b = 1$. Now assume that it's true for $b = k$, so $k^p \equiv k$. Then

$$(k+1)^p \equiv k^p + \binom{p}{1}k^{p-1} + \binom{p}{2}k^{p-2} + \cdots + \binom{p}{p-1}k + 1$$
$$\equiv k^p + 1 \equiv k+1 \pmod{p},$$

where we have used the lemma to get rid of the binomial coefficients mod p, and we have used the induction assumption to change k^p to k mod p. Therefore, the statement is true for $b = k+1$. By induction, we conclude that $b^p \equiv b \pmod{p}$ for all positive integers b.

Now let $-c < 0$ be a negative integer. Since c is positive, we already know $c^p \equiv c$, so

$$(-c)^p \equiv -c^p \equiv -c \pmod{p}$$

if p is odd. If $p = 2$, we have $(-c)^2 \equiv +c \equiv -c \pmod{2}$, so the result also holds for $p = 2$.

We have now proved that $b^p \equiv b \pmod{p}$ for all integers b.

If $b \not\equiv 0 \pmod{p}$, we can divide the congruence by b and obtain $b^{p-1} \equiv 1 \pmod{p}$. □

Proof 2. Of course, we start with a lemma.

Section 6.1   Fermat's Theorem

Lemma 6.4. *Let $b \not\equiv 0 \pmod{p}$. Then the set of numbers*
$$b, \ 2b, \ 3b, \ \ldots, \ (p-1)b \pmod{p}$$
contains each nonzero congruence class mod p exactly once.

Example. Let $p = 7$ and $b = 2$. Then the numbers $2, 4, 6, 8, 10, 12$ (mod 7) are the same as $2, 4, 6, 1, 3, 5$ (mod 7), so every nonzero congruence class mod 7 occurs exactly once.

Proof. Let $a \not\equiv 0 \pmod{p}$. Then $bx \equiv a \pmod{p}$ has a solution x, and $x \not\equiv 0$ since $a \not\equiv 0$, so we may assume that $1 \leq x \leq p-1$. Therefore, a occurs in the set $\{b, 2b, \ldots, (p-1)b\}$ mod p. Since a was arbitrary, every nonzero congruence class occurs.

If $bi \equiv bj \pmod{p}$ with $1 \leq i < j \leq p-1$, we can divide by b and get $i \equiv j \pmod{p}$. The bounds on i and j make this impossible, so we conclude that the multiples of b are distinct mod p. Therefore, each nonzero congruence class occurs exactly once among the multiples of b. □

Now we can prove Fermat's theorem. Lemma 6.4 says that
$$\prod_{i=1}^{p-1} i \equiv \prod_{i=1}^{p-1} bi \pmod{p},$$
because the second product is a rearrangement of the first product. The first product is $(p-1)!$. If we factor b^{p-1} from the second product, we see that the second product is $b^{p-1}(p-1)!$. Therefore,
$$(p-1)! \equiv b^{p-1}(p-1)! \pmod{p}.$$
Since p does not divide $(p-1)!$ (because it doesn't divide any of the factors), we can divide by $(p-1)!$ to obtain
$$b^{p-1} \equiv 1 \pmod{p}.$$
Multiplying by b gives the other form: $b^p \equiv b \pmod{p}$ when $b \not\equiv 0$. If $b \equiv 0$, then $b^p \equiv 0^p \equiv 0 \equiv b$, so the congruence holds for all b. □

CHECK YOUR UNDERSTANDING
1. Compute $2^{104} \pmod{101}$.

6.2 Euler's Theorem

In the 1700s, Euler studied Fermat's work on number theory and provided proofs for many of his statements. In addition, he gave a very useful generalization of Fermat's theorem. This is the subject of the present section.

Definition 6.5. *Let n be a positive integer. Define the **Euler ϕ-function** $\phi(n)$ to be the number of integers j with $1 \leq j \leq n$ such that $\gcd(j, n) = 1$.*

Examples.

1. $\phi(3) = 2$ since $j = 1$ and $j = 2$ have $\gcd(j, 3) = 1$.
2. $\phi(4) = 2$.
3. $\phi(12) = 4$ (the numbers are 1, 5, 7, 11).
4. $\phi(1) = 1$ (the only value of j is $j = 1$; this is the reason we have $j \leq n$ rather than $j < n$ in the definition).
5. If p is prime, then $\phi(p) = p - 1$.

The following is a very useful property of ϕ.

Proposition 6.6. *Let m, n be positive integers. If $\gcd(m, n) = 1$ then*

$$\phi(mn) = \phi(m) \cdot \phi(n).$$

We give the proof at the end of the section.

For prime powers, we have the following.

Proposition 6.7. *If p is a prime and $k \geq 1$, then $\phi(p^k) = p^k - p^{k-1}$.*

Proof. There are p^k integers j with $1 \leq j \leq p^k$. Among these, there are p^{k-1} multiples of p, namely, $p, 2p, 3p, \ldots, p^k$. These multiples of p are exactly the values of j with $\gcd(j, p^k) \neq 1$. So we have $p^k - p^{k-1}$ values of j with $\gcd(j, p^k) = 1$. □

Combining the previous two propositions yields the following.

Section 6.2   Euler's Theorem

Theorem 6.8. *Let $n = p_1^{a_1} p_2^{a_2} \cdots p_r^{a_r}$ with distinct primes p_i and with exponents $a_i \geq 1$. Then*

$$\phi(n) = \prod_{i=1}^{r} \left(p^{a_i} - p^{a_i - 1}\right) = n \prod_{p | n} \left(1 - \frac{1}{p}\right),$$

where the second product is over the primes dividing n.

Proof. From Propositions 6.6 and 6.7,

$$\phi(n) = \prod_i \phi(p_i^{a_i}) = \prod_i (p_i^{a_i} - p_i^{a_i - 1}).$$

This is the first formula. For the second, note that $p^a - p^{a-1} = p^a(1 - 1/p)$. Therefore,

$$\phi(n) = \prod_i p_i^{a_i} \left(1 - \frac{1}{p_i}\right) = \left(\prod_i p_i^{a_i}\right) \prod_i \left(1 - \frac{1}{p_i}\right)$$
$$= n \prod_i \left(1 - \frac{1}{p_i}\right).$$

This is the second formula. □

Example. Let's evaluate $\phi(100)$. The first formula says that

$$\phi(100) = \phi(2^2)\phi(5^2) = (2^2 - 2)(5^2 - 5) = 40.$$

We could also use the second formula to obtain

$$\phi(100) = 100 \left(1 - \frac{1}{2}\right)\left(1 - \frac{1}{5}\right) = 100 \left(\frac{1}{2}\right)\left(\frac{4}{5}\right) = 40.$$

Finally, we come to the main result.

Theorem 6.9. *(Euler's Theorem) Let n be a positive integer and let b be an integer with $\gcd(b, n) = 1$. Then*

$$b^{\phi(n)} \equiv 1 \pmod{n}.$$

Remark. Notice that Euler's theorem generalizes Fermat's theorem, since $\phi(p) = p - 1$ and if b is not a multiple of p, then $\gcd(b, p) = 1$.

Proof. We mimic the second proof of Fermat's theorem in order to prove this result. Again, we first need a lemma.

Lemma 6.10. *Let T be the set of j with $1 \leq j \leq n$ and $\gcd(j, n) = 1$. Choose any b with $\gcd(b, n) = 1$, and let $bT \bmod n$ be the set of numbers of the form $bt \bmod n$ with $t \in T$. Then each element of T is congruent to exactly one element of bT.*

Example. Let $n = 12$ and $b = 5$. Then

$$T = \{1, 5, 7, 11\} \text{ and } bT = \{5, 25, 35, 55\} \equiv \{5, 1, 11, 7\} = T.$$

Proof. Let $a \in T$. Since $\gcd(b, n) = 1$, it follows that $bx \equiv a \pmod{n}$ has a solution x. Since $\gcd(a, n) = 1$, we must have $\gcd(x, n) = 1$, so $x \in T$. Therefore, each element of T is congruent to an element of bT. Suppose t_1 and t_2 are two elements in T with $bt_1 \equiv bt_2 \pmod{n}$. Since $\gcd(b, n) = 1$, we can divide by b and get $t_1 \equiv t_2 \pmod{n}$. The elements of T are distinct mod n, which implies that $t_1 = t_2$. This proves that each element of T occurs at most once among the multiples of b. □

Now we can prove Euler's theorem. Let T be as in Lemma 6.10. Then

$$\prod_{i \in T} i \equiv \prod_{i \in T} bi \equiv b^{\phi(n)} \prod_{i \in T} i \pmod{n},$$

because Lemma 6.10 says that the second product is a rearrangement of the first product. There are $\phi(n)$ elements in T, so $b^{\phi(n)}$ factors out to give the third expression. Every factor in the first product is relatively prime to n, so the product is relatively prime to n. Therefore, we can divide by the first product to obtain

$$1 \equiv b^{\phi(n)} \pmod{n}.$$

This is Euler's theorem. □

Examples.

1. $\phi(15) = \phi(3)\phi(5) = 8$, so $2^8 = 256 \equiv 1 \pmod{15}$.

Section 6.2   Euler's Theorem

2. $\phi(10) = 4$ and $3^4 = 81 \equiv 1 \pmod{10}$.

3. What are the last two digits of 3^{84}? The last two digits of a number are given by the number mod 100, so we need to compute $3^{84} \pmod{100}$. Since $\phi(100) = 40$, Euler's theorem tells us that $3^{40} \equiv 1 \pmod{100}$. Therefore,
$$3^{84} \equiv (3^{40})^2 3^4 \equiv (1)^2 3^4 \equiv 81 \pmod{100}.$$
This means that the last two digits are 81.

The last example is a case of a more general phenomenon. Notice that when the modulus n in the following corollary is prime, we get Corollary 6.2.

Corollary 6.11. *Let n be a positive integer, and suppose b is an integer with $\gcd(b, n) = 1$. If $x \equiv y \pmod{\phi(n)}$, then*
$$b^x \equiv b^y \pmod{n}.$$

Proof. Write $x = y + \phi(n)k$ for some integer k. Then
$$b^x \equiv b^y (b^{\phi(n)})^k \equiv b^y (1)^k \equiv b^y \pmod{n},$$
where we used Euler's theorem to change $b^{\phi(n)}$ to 1. □

Examples.

1. Let $n = 15$. Then $\phi(n) = 8$. Since $9 \equiv 1 \pmod{8}$, we have $2^9 \equiv 2^1 \pmod{15}$. (You can check this. It says that $512 \equiv 2 \pmod{15}$.)

2. Let $n = 10$. Then $\phi(n) = 4$. Since $5 \equiv 1 \pmod{4}$, it follows that $b^5 \equiv b \pmod{10}$ when $\gcd(b, 10) = 1$. This says that b^5 and b have the last digit b when b is 1, 3, 7, or 9. A quick check shows that this is also true when $\gcd(b, 10) > 1$ (for example, $2^5 = 32$ has 2 as its last digit).

3. Suppose m is an integer with $\gcd(m, 77) = 1$. Let $c \equiv m^7 \pmod{77}$. What is $c^{43} \pmod{77}$? Since $\phi(77) = \phi(7)\phi(11) = 6 \cdot 10 = 60$, and $301 \equiv 1 \pmod{60}$, we have
$$c^{43} \equiv (m^7)^{43} \equiv m^{301} \equiv m^1 \pmod{77},$$
so $c^{43} \pmod{77}$ gives back the original number m. This is an example of RSA encryption and decryption (see Section 7.5).

4. What is the last digit of 3^{7^5}? (*Note:* a^{b^c} means $a^{(b^c)}$; the other possible interpretation, namely $(a^b)^c$, is better written in the easier form a^{bc}.) We are working mod 10 overall, so we need to work mod $4 = \phi(10)$ in the exponent. That is, we need to know $7^5 \pmod 4$ in order to use Corollary 6.11. In order to do this, we first use Corollary 6.11 to compute $7^5 \pmod 4$. Since $\phi(4) = 2$ and $5 \equiv 1 \pmod 2$, we have

$$7^5 \equiv 7^1 \equiv 3 \pmod 4.$$

We now use Corollary 6.11 again. Since $\phi(10) = 4$ and $7^5 \equiv 3 \pmod 4$, we have

$$3^{7^5} \equiv 3^3 \equiv 27 \equiv 7 \pmod{10}.$$

This means that the last digit is 7.

Finally, let's return to Proposition 6.6 and give its proof. We want to show that if $\gcd(m, n) = 1$ then $\phi(mn) = \phi(m)\phi(n)$.

In order to see how the proof will work, we'll look at an example.

Example. If $m = 4$ and $n = 15$, we'll show that $\phi(4 \cdot 15) = \phi(4)\phi(15) = 2 \cdot 8 = 16$. We form an array of all the integers from 1 to $4 \cdot 15 = 60$ with the numbers from 1 to 15 forming the first column:

$$
\begin{array}{cccc}
1 & 16 & 31 & 46 \\
2 & 17 & 32 & 47 \\
3 & 18 & 33 & 48 \\
4 & 19 & 34 & 49 \\
5 & 20 & 35 & 50 \\
6 & 21 & 36 & 51 \\
7 & 22 & 37 & 52 \\
8 & 23 & 38 & 53 \\
9 & 24 & 39 & 54 \\
10 & 25 & 40 & 55 \\
11 & 26 & 41 & 56 \\
12 & 27 & 42 & 57 \\
13 & 28 & 43 & 58 \\
14 & 29 & 44 & 59 \\
15 & 30 & 45 & 60.
\end{array}
$$

Section 6.2   Euler's Theorem   143

The third, sixth, ninth, twelfth, and fifteenth rows consist of numbers that are multiples of 3, and the fifth, tenth, and fifteenth rows consist of numbers that are multiples of 5. This means that *none* of the numbers in these rows can be relatively prime to 60.

We rewrite the array, eliminating all the rows just mentioned:

$$
\begin{array}{cccc}
1 & 16 & 31 & 46 \\
2 & 17 & 32 & 47 \\
4 & 19 & 34 & 49 \\
7 & 22 & 37 & 52 \\
8 & 23 & 38 & 53 \\
11 & 26 & 41 & 56 \\
13 & 28 & 43 & 58 \\
14 & 29 & 44 & 59.
\end{array}
$$

This leaves only those rows whose numbers are *potentially* relatively prime to 60, namely those relatively prime to both 3 and 5.

This array came about when we considered $3 \cdot 5 = 15$ as a factor of 60. We now focus on the other factor, 4. We need to find the elements remaining in the array that are relatively prime to 4. We again rewrite the array, this time writing each number relatively prime to 4 in bold:

$$
\begin{array}{cccc}
1 & 16 & \mathbf{31} & 46 \\
2 & \mathbf{17} & 32 & \mathbf{47} \\
4 & \mathbf{19} & 34 & \mathbf{49} \\
7 & 22 & \mathbf{37} & 52 \\
8 & \mathbf{23} & 38 & \mathbf{53} \\
11 & 26 & \mathbf{41} & 56 \\
13 & 28 & \mathbf{43} & 58 \\
14 & \mathbf{29} & 44 & \mathbf{59}.
\end{array}
$$

Notice that each of the eight rows has $2 = \phi(4)$ such numbers, leading to a total of $16 = \phi(15)\phi(4)$ numbers relatively prime to 60.

We will use this idea to prove Proposition 6.6 for an arbitrary pair of relatively prime integers. We begin by forming an array of the first mn positive integers:

$$\begin{array}{ccccc} 1 & m+1 & 2m+1 & \cdots & (n\text{-}1)m+1 \\ 2 & m+2 & 2m+2 & \cdots & (n\text{-}1)m+2 \\ \vdots & \vdots & \vdots & & \vdots \\ m\text{-}1 & 2m\text{-}1 & 3m\text{-}2 & \cdots & nm\text{-}1 \\ m & 2m & 3m & \cdots & nm. \end{array}$$

Choose any number in the first column that is *not* relatively prime to m. If we call that number r, then $\gcd(r,m) = d > 1$, and the entries in the rth row are

$$r,\ m+r,\ 2m+r,\ \ldots,\ (n-1)m+r.$$

Since $d \mid m$ and $d \mid r$, by Proposition 1.4, $d \mid (km+r)$ for every integer k. This means that if $\gcd(r,m) > 1$, then none of the entries in the rth row can be relatively prime to m. So, only the $\phi(m)$ rows that begin with a number relatively prime to m need to be considered when calculating $\phi(mn)$.

So, let's focus on a row whose first number s *is* relatively prime to m. Row s looks like

$$s,\ m+s,\ 2m+s,\ \ldots,\ (n-1)m+s$$

and, since $\gcd(m,s) = 1$, every number (not just the first) in this row is relatively prime to m.

We'll now show that the elements of this row mod n are just a rearrangement of $0, 1, 2, \ldots, n-1$. Since there are n numbers in each row, we just have to show that no two of them are congruent mod n. If

$$k_1 m + s \equiv k_2 m + s \pmod{n}, \text{ then } k_1 m \equiv k_2 m \pmod{n}.$$

Since $\gcd(m,n) = 1$, we can divide by m to get $k_1 \equiv k_2 \pmod{n}$. Since $0 \leq k_1, k_2 \leq n-1$, we have $k_1 = k_2$. This shows that the elements of row s are just a rearrangement of $0, 1, 2, \ldots, n-1$. We know that, by definition, $\phi(n)$ of these numbers are relatively prime to n. In addition, since the rows that remain contain only numbers that are relatively prime to m, these $\phi(n)$ numbers must be relatively prime to mn as well.

To summarize, we have $\phi(m)$ rows whose entries are relatively prime to m, and each of these rows contains $\phi(n)$ elements that are relatively

Section 6.3   Wilson's Theorem   145

prime to n. This makes a total of $\phi(m)\phi(n)$ elements of the array relatively prime to mn. Since the array contains all the numbers from 1 to mn, it contains $\phi(mn)$ numbers relatively prime to mn. Therefore, $\phi(mn) = \phi(m)\phi(n)$, which proves the proposition. □

CHECK YOUR UNDERSTANDING

2. Compute 2^{67} (mod 77).

6.3   Wilson's Theorem

In 1770, Edward Waring wrote the book *Meditationes Arithmeticae*, in which he stated a property of prime numbers that his student John Wilson had discovered but neither Waring nor Wilson could prove. Henceforth, it has been known as Wilson's Theorem. However, the result was known to Bhāskara about 1000 years before Wilson, and also to Ibn al-Haytham (around 1000) and to Leibniz (around 1700). The first published proof was by Lagrange in 1771.

Theorem 6.12. *(Wilson's Theorem) Let p be prime. Then*

$$(p-1)! \equiv -1 \pmod{p}.$$

Example. Let $p = 7$. Then

$$(p-1)! = 6! = 720 \equiv -1 \pmod{7}.$$

Proof. Let's see why the example works. Rearrange the factorial as

$$6! \equiv (6)(5 \cdot 3)(4 \cdot 2)(1) \equiv (-1)(1)(1)(1) \pmod{7}.$$

Note that we have paired numbers with their multiplicative inverses. For example, 5 is paired with 3 and $5 \cdot 3 \equiv 1 \pmod{7}$. Since 6 and 1 are their own inverses ($6 \cdot 6 \equiv 1 \pmod{7}$, and $1 \cdot 1 \equiv 1$), they did not get paired with anything else.

The same idea works in general. Start with the integers $1 \leq b \leq p-1$.

Recall that the equation $bx \equiv 1 \pmod{p}$ always has a unique solution x with $1 \leq x \leq p-1$, so we can pair each integer with its multiplicative inverse mod p. If y is the multiplicative inverse of x, then x is the multiplicative inverse of y (this says that $xy \equiv 1 \pmod{p}$ implies $yx \equiv 1 \pmod{p}$). Therefore, if x is paired with y, then y is paired with x. A number b is paired with itself exactly when b is its own inverse, which says that $b \equiv b^{-1}$, hence $b^2 \equiv 1 \pmod{p}$. By Corollary 5.10, the only solutions of this are $b \equiv \pm 1 \pmod{p}$, which means that $b = 1$ and $b = p - 1$ are the only numbers that get paired with themselves. Now rearrange $(p-1)!$ so that each number except 1 and $p-1$ is next to its multiplicative inverse, just as in the example for $p = 7$. All the pairs in the product combine to yield factors of 1, leaving only the unpaired factors 1 and $p - 1$. Therefore,

$$(p-1)! \equiv 1 \cdot (p-1) \equiv p - 1 \equiv -1 \pmod{p}.$$

This completes the proof. □

Example. Here is another example of what happened in the proof of Wilson's theorem. Let $p = 11$. Then

$$\begin{aligned}(p-1)! = 10! &\equiv (10)(9 \cdot 5)(8 \cdot 7)(6 \cdot 2)(4 \cdot 3)(1) \\ &\equiv (10)(1)(1)(1)(1)(1) \equiv 10 \equiv -1 \pmod{11}.\end{aligned}$$

Wilson's theorem yields a primality test. Unfortunately, this test is not practical, since there is no known method for computing factorials efficiently.

Corollary 6.13. *Let $n \geq 2$ be an integer. Then n is prime if and only if $(n-1)! \equiv -1 \pmod{n}$.*

Proof. If n is prime then $(n-1)! \equiv -1 \pmod{n}$ is just Wilson's theorem. Conversely, suppose $(n-1)! \equiv -1 \pmod{n}$. If n is composite, then $n = ab$ with $1 < a < n$. This implies that a is a factor of $(n-1)!$, so $(n-1)! \equiv 0 \pmod{a}$. But a is also a factor of n, so

$$(n-1)! \equiv -1 \pmod{n} \implies (n-1)! \equiv -1 \pmod{a}.$$

The last two sentences tell us that $0 \equiv -1 \pmod{a}$, which is impossible. Therefore, n cannot factor, so n is prime. □

Example. Let $n = 6$. Then

$$(n-1)! = 5! = 120 \equiv 0 \pmod{6},$$

which implies that 6 is not prime.

In general, when n is composite, $(n-1)! \equiv 0 \pmod{n}$ if $n \neq 4$. See Exercise 30.

6.4 Chapter Highlights

1. Fermat's Theorem: If p is prime and $a \not\equiv 0 \pmod{p}$, then $a^{p-1} \equiv 1 \pmod{p}$.

2. Euler's Theorem: If $\gcd(a, n) = 1$, then $a^{\phi(n)} \equiv 1 \pmod{n}$.

3. Wilson's Theorem: If p is prime, then $(p-1)! \equiv -1 \pmod{p}$.

6.5 Problems

6.5.1 Exercises

Section 6.1: Fermat's Theorem

1. Compute $3^{20} \pmod{11}$.
2. Compute $2^{100} \pmod{97}$.
3. Compute $2^{1234} \pmod{11}$.
4. Compute $3^{75} \pmod{73}$.
5. Compute $7^{50} \pmod{47}$.
6. Evaluate $111^{222} + 222^{111} \pmod{13}$.
7. Show that $2222^{5555} + 5555^{2222}$ is a multiple of 7.
8. Let n be an integer. Show that $n^{13} - n$ is a multiple of 2730.

9. If p is a prime number and a is a positive integer, show that $a^{(p-1)!+1} \equiv a \pmod{p}$.

10. Show that $n^{17} - n \equiv 0 \pmod{510}$ for all integers n. (*Hint:* Factor 510 and use Fermat's theorem for each factor)

11. Use Fermat's theorem to show that $1^n + 2^n + 3^n + 4^n$ is divisible by 5 if and only if n is *not* divisible by 4.

12. (a) Show that if $p \equiv 3 \pmod{4}$ is prime then $x^2 \equiv -1 \pmod{p}$ has no solution. (*Hint:* Raise both sides of the congruence to the $(p-1)/2$ power.)
    (b) Suppose p is an odd prime and n is an integer. Show that if $p \mid n^2 + 1$ then $p \equiv 1 \pmod{4}$.
    (c) Show that there are infinitely many primes $p \equiv 1 \pmod{4}$. (*Hint:* Look at $(2p_1 p_2 \cdots p_r)^2 + 1$.)

Section 6.2: Euler's Theorem

13. Compute $5^{60} \pmod{21}$.

14. Compute $3^{240} \pmod{385}$.

15. Compute $2^{35} \pmod{35}$.

16. Find the last two digits of 123^{403}.

17. Find the last three digits of 101^{40404}.

18. (a) Show that $13^{13} \equiv 1 \pmod{4}$.
    (b) Find the last digit of $13^{13^{13}}$.

19. Find the last four digits of $207^{8004002}$.

20. Let n be a positive integer,
    (a) What are the possible values of the last digit of 3^n?
    (b) What are the possible values of the last digit of 4^n?
    (c) What are the possible values of the last digit of 6^n?
    (d) What are the possible values of the last digit of 7^n?

21. Find all solutions to the following. If the equation has no solutions, give an explanation.
    (a) $\phi(x) = 2$
    (b) $\phi(x) = 3$
    (c) $\phi(x) = 4$
    (d) $\phi(x) = 10$
    (e) $\phi(x) = 14$

Section 6.5   Problems   149

22. Decide whether each of the following is true or false. If true, give a proof; if false, give a counterexample.
    (a) If $d \mid n$, then $\phi(d) \mid \phi(n)$.
    (b) If $\phi(d) \mid \phi(n)$, then $d \mid n$.
    (c) If $d \mid n$, then $\phi(dn) = d\phi(n)$.

23. Decide whether each of the following is true or false. If true, give a proof; if false, give a counterexample.
    (a) If n has k distinct prime factors, then $2^k \mid \phi(n)$.
    (b) If n has k distinct odd prime factors, then $2^k \mid \phi(n)$.

24. (a) Let n be an integer with $\gcd(n, 10) = 1$. Show that there is a number $N = 1111\cdots 111$, whose base 10 expression is a string of 1s, such that $n \mid N$.
    (b) Part (a) can be done using the fact that there are only finitely many possibilities mod n for a string of 1s and then subtracting. It can also be done via Euler's theorem. Use a method that you didn't use in (a) to give a second proof of (a).

25. Here is another proof that there are infinitely many primes. Suppose $2, 3, 5, \ldots, p_n$ are all the primes. Let $N = 2 \cdot 3 \cdot 5 \cdots p_n$.
    (a) Show that $\phi(N) \geq 8$.
    (b) Show that if $1 < a \leq N$, then $\gcd(a, N) \neq 1$. (*Hint:* Lemma 1.7.)
    (c) Use (a) and (b) to obtain a contradiction and therefore conclude that there are infinitely many primes.
    Even though this proof looks easy, it uses the multiplicativity of the ϕ-function, which is fairly complicated.

Section 6.3: Wilson's Theorem

26. Show how the numbers match up in the proof of Wilson's theorem for $p = 13$.

27. Let p be prime. Show that $p \mid (p-2)! - 1$.

28. Let p be prime and let $0 < s < p$. Show that
    $$(s-1)!\,(p-s)! \equiv (-1)^s \pmod{p}.$$

29. Let p be prime. Show that if a and b are positive integers such that $a + b \geq 2p - 1$, then $a!\,b! \equiv 0 \pmod{p}$.

30. Let $n \neq 4$ be composite. Show that $(n-1)! \equiv 0 \pmod{n}$ if $n \neq 4$. (*Hint:* Let $1 < b < n$ be a factor of n. Consider the cases $b \neq n/b$ and $b = n/b$ separately. In the latter case, if $b > 2$, then both b and $2b$ occur in $(n-1)!$.)

31. Let p be an odd prime. Let $x = ((p-1)/2)!$.
    (a) Using the fact that $(p-1)! =$
    $$(1(p-1))(2(p-2))\cdots(((p-1)/2)((p+1)/2)),$$
    show that $-1 \equiv (-1)^{(p-1)/2} x^2 \pmod{p}$.
    (b) Suppose that $p \equiv 1 \pmod{4}$. Show that $x^2 \equiv -1 \pmod{p}$.

6.5.2 Computer Explorations

1. (a) A prime p is called a *Wieferich prime* if $2^{p-1} \equiv 1 \pmod{p^2}$. There are only two such primes known. Both are less than 4000. Can you find them?
   (b) A prime p is called a *Mirimanoff prime* if $3^{p-1} \equiv 1 \pmod{p^2}$. There are two such primes known. One is small and the other is slightly larger than 1 million. Can you find them?
   (c) Can you find other examples of primes p where $r^{p-1} \equiv 1 \pmod{p^2}$ for a small prime r?
   Wieferich primes and Mirimanoff primes arose in connection with Fermat's Last Theorem.

2. Let $a > b$ be fixed integers. We say that a prime p is a primitive divisor of $a^n - b^n$ if $p \mid a^n - b^n$, but $p \nmid a^k - b^k$ for $0 \leq k < n$.
   (a) Compute the primitive divisors of $2^n - 1$ for several values of $n \geq 2$. There is one value of n for which there are no primitive divisors. Did you find it?
   (b) Compute the primitive divisors of $3^n - 1$ for several values of $n \geq 1$. Are there any values of n for which there are no primitive divisors?
   (c) Compute the primitive divisors of $3^n - 2^n$ for several values of $n \geq 2$. Are there any values of n for which there are no primitive divisors?

Section 6.5   Problems   151

Remark: Zsigmondy's Theorem says that $a^n - b^n$ has a primitive divisor, except in the case found in (a) and the case where $n = 1$ and $a - b = 1$.

6.5.3 Answers to "CHECK YOUR UNDERSTANDING"

1. Fermat's theorem tells us that $2^{100} \equiv 1 \pmod{101}$. Therefore, $2^{104} \equiv 2^{100} 2^4 \equiv 2^4 \equiv 16 \pmod{101}$.

2. Since $\phi(77) = \phi(7)\phi(11) = 6 \cdot 10 = 60$, Euler's theorem tells us that $2^{60} \equiv 1 \pmod{77}$. Therefore, $2^{67} \equiv 2^{60} 2^7 \equiv 2^7 \equiv 128 \equiv 51 \pmod{77}$.

Chapter 7

Cryptographic Applications

7.1 Introduction

Bob wants to send a message to Alice so that Alice, *and only Alice*, can read it. For example, Alice may own a business and Bob may need to send her his credit card number so that he may make a purchase. This message is usually called the **plaintext**. To keep the message secret, he encrypts it to a secret, unreadable form called the **ciphertext**, which he sends to Alice. Alice then decrypts to obtain the original plaintext.

For a simple example, suppose Bob's original message is *ZEBRA*. He decides to shift each letter forward by four positions to obtain the ciphertext *DIFVE* (note that the *Z* wraps around to the beginning):

$$ZEBRA \longrightarrow DIFVE.$$

Alice receives the ciphertext *DIFVE* and shifts each letter back by four positions to obtain the plaintext, *ZEBRA*:

$$DIFVE \longrightarrow ZEBRA.$$

Of course, Alice needs to know how many positions Bob shifted each letter in order to decrypt. In this example, the number of positions is called the **key**, and it is shared by Alice and Bob. In general, the key is the secret information that is used to encrypt and decrypt the message. In the present case, there are 26 possible keys (this includes no shift, which is the same as sending an unencrypted message). Cryptosystems that are used in real-life situations are much more complicated and have hundreds of millions of possible keys. In fact, most of the systems used in practice today have more than 10^{30} possible keys.

In classical cryptography, Alice and Bob share the same key. This is called a **symmetric cryptosystem**, since the sender and receiver use exactly the same information to encrypt and decrypt. The workhorses of modern cryptography, namely the Data Encryption Standard (DES) and its successor, the Advanced Encryption Standard (AES), are symmetric cryptosystems that are used to transmit massive amounts of data electronically. For these, as with the classical systems, the sender and receiver need to share a key in order to encrypt and decrypt. These systems start with the message written as a sequence of bits or bytes and then use various operations to mix up and modify the bits to obtain the ciphertext. We do not discuss DES and AES in this book, since they have no relation to number theory. If you're interested in seeing how they work, details of these algorithms are described in many books on cryptography.

Symmetric systems work well if Alice and Bob have met previously and have agreed on a key. But what happens if they haven't? For example, suppose Alice is in central Australia (at Alice Springs) and wants to communicate with the Bank of Baltimore (Bob). Alice and Bob cannot agree on a secret key by e-mail or by talking on the telephone, since eavesdroppers could easily intercept their communications. (If they could securely agree on a secret key by mail, e-mail, or telephone, they would have no need of a cryptosystem!) Alice could send a courier to Bob with the secret key locked in a box. But that might take a few days. Fortunately, there is a solution, called **public key cryptography**. Alice generates two keys, one called the **public key** and one called the **private key**. Alice sends the public key to Bob and this allows Bob to encrypt messages. But decryption requires the private key, which only Alice knows. So Bob can send a secret message to Alice that only Alice can read, even though every communication between Alice and Bob is over public channels. Because Alice and Bob have different information about the keys, public key cryptography is an **asymmetric cryptosystem**.

The concept of public key cryptography was invented by James Ellis in 1970, and an implementation was found by Clifford Cocks in 1973. However, they worked for the British cryptographic agency and the discovery was kept secret until 1997. Meanwhile, in the unclassified world,

Section 7.1   Introduction

Whitfield Diffie and Martin Hellman published an influential paper in 1976 describing public key cryptography. Then, in 1977, Ron Rivest, Adi Shamir, and Leonard Adleman discovered the implementation (essentially the same as the one by Cocks) that is known as RSA. This algorithm will be discussed in Section 7.5.

The influence of RSA on number theory has been enormous. It showed that number theory, long assumed to be a purely theoretical branch of mathematics, has important practical applications. For example, since the security of the RSA system is closely tied to the difficulty of factoring, the development of factorization algorithms has received a lot of attention, and computational problems have become much more popular. Of course, you are still permitted to study number theory because it is fun, but now people also study it because it is useful.

One drawback of public key systems is that they are relatively slow, making it impractical to send massive amounts of data. For this reason, the process of sending an encrypted message is usually broken up into two steps. First, a key is transmitted using RSA or another public key method. Once both parties have this key, the much faster symmetric systems such as DES or AES are then used to send the data. For example, public key systems are used to initiate communications whenever you buy things over the Internet. This is natural, since you probably have not met the online company, but you still need to transmit your credit card number or other confidential data.

In this chapter, we discuss some classical symmetric systems (shift and affine ciphers, Vigenère ciphers, Hill ciphers, linear feedback shift registers), that use number theory concepts that we have treated, along with a system (transposition ciphers) that is less number theoretic. We then introduce the RSA cryptosystem, and in Chapter 9, we discuss another method (Diffie-Hellman) for establishing a secret key for Alice and Bob. (*Note:* You may have noticed that many of the cryptosystems are called "ciphers." This is a word that is often used for methods of encrypting messages.)

Not every cryptographic protocol involves sending secret messages. In various sections of this chapter and Chapter 9, we discuss ways to flip coins over the telephone, play mental poker, sign electronic documents, and share secrets. These all use number theory in essential ways.

7.2 Shift and Affine Ciphers

Suppose you have a message that you want to send to someone. That message may be dinner plans, what time your last class is, or your opinion of a song you've just heard. In these cases, you don't really care if anyone else reads your message. But if you want to send the password for your bank account to your spouse or battle plans during wartime to one of your officers in the field, then it becomes very important that the message be read only by the intended recipient. When you have information that you want to distribute selectively, you need to use some method to ensure that only the intended reader can understand it. This is where cryptography comes into play.

Let's say that your online bank account's password is your dog's name, Sparky. You want to e-mail this to your spouse, but are (justifiably) concerned that someone else may read it. (By the way, you should never use your pet's name as your password.) As discussed in the first section, *SPARKY* is the *plaintext*, or the unaltered message, you want to send. We need to encrypt, or disguise, this message by altering it. We'll show one way to do this using a very simple (and very insecure) method, replacing each plaintext letter by the letter immediately to its right. In this case, the key is $+1$. So, S becomes T, P becomes Q, A becomes B, etc., and *SPARKY* becomes *TQBSLZ*.

The cryptosystem discussed above is an example of a **shift cipher**. There's nothing special about shifting by one position to the right. We could have shifted 3 to the right (a key of $+3$) or 5 to the left (a key of -5), or by any predesignated number. In order to make this simpler to discuss, we're going to assign a number to each letter of the alphabet. From a mathematical perspective, it doesn't really matter how we do this. We're going to keep things simple and begin with $A = 0$.

A	B	C	D	E	F	G	H	I	J	K	L	M
0	1	2	3	4	5	6	7	8	9	10	11	12
N	O	P	Q	R	S	T	U	V	W	X	Y	Z
13	14	15	16	17	18	19	20	21	22	23	24	25

Warning: Starting with A as 0 is a standard convention. Since people

Section 7.2   Shift and Affine Ciphers

think of A as being the first letter of the alphabet, they often think of A as being 1. Be careful not to make this mistake.

Using this, $S = 18$, $P = 15$, etc., so we write *SPARKY* as *18 15 00 17 10 24*. Now, we can think of our shift of 1 to the right, or more explicitly as our key being $+1$ in the following way:

PLAINTEXT	S	P	A	R	K	Y
Numerical Translation	18	15	0	17	10	24
Right Shift by One	19	16	1	18	11	25
CIPHERTEXT	T	Q	B	S	L	Z

Mathematically, we've shifted each letter by 1 unit under the rule

$$x \mapsto x + k \pmod{26},$$

where x is the numerical equivalent of the letter.

If we had our key be $k = 7$, we would have

PLAINTEXT	S	P	A	R	K	Y
Numerical Translation	18	15	0	17	10	24
Right Shift by Seven	25	22	7	24	17	5
CIPHERTEXT	Z	W	H	Y	R	F

Once the intended recipient gets the ciphertext, it can be decrypted by subtracting the key.

The Roman historian Suetonius reported that Julius Caesar used a shift by 3 to encrypt, and this shift cipher is often called the **Caesar cipher**. His nephew, Augustus Caesar, used a shift by 1, but apparently didn't know about modular arithmetic, since he used AA instead of A when the shift went past the end of the alphabet.

As we said before, there are 26 possible keys for a shift cipher, making it very easy to break the system. A malicious individual who intercepts the ciphertext could decrypt by trying all 26 keys, one at a time, stopping when a readable message is found. Because of this, we tweak a shift cipher to get a similar cryptosystem, called an **affine cipher**, that has 312 possible keys, making it a little bit more secure. In an affine cipher, instead of simply performing a shift, we first multiply by a nonzero value that is relatively prime to 26 and then shift. Mathematically, an affine cipher works by having

$$x \mapsto ax + b \pmod{26} \tag{7.1}$$

where $\gcd(a, 26) = 1$. (It will soon be clear why this gcd requirement is necessary.) In this situation we will call the ordered pair (a, b) the key for the affine cipher. Since there are twelve numbers (1, 3, 5, 7, 9, 11, 15, 17, 19, 21, 23, 25) relatively prime to 26, we get $\phi(26) = 12$ (see Section 6.2) choices for a and 26 choices for b, and therefore $12 \cdot 26 = 312$ possible keys.

Here's our $SPARKY$ (= 18 15 00 17 10 24) example with key $(a, b) = (3, 5)$:

PLAINTEXT	S	P	A	R	K	Y
Numerical Translation	18	15	0	17	10	24
$3x + 5 \pmod{26}$	7	24	5	4	9	25
CIPHERTEXT	H	Y	F	E	J	Z

The ciphertext is $HYFEJZ$.

For example, we encrypt the letter P by first finding its numerical value of 15. We then take

$$3 \cdot 15 + 5 \equiv 50 \equiv 24 \pmod{26}.$$

Finally, since 24 corresponds to Y, we get that

$$P \mapsto Y$$

under the affine cipher whose key is $(3, 5)$.

Decryption is easy. Let's say the ciphertext of U is received and the key is $(3, 5)$. Since the numerical value of U is 20, we have to solve the congruence

$$3x + 5 \equiv 20 \pmod{26}$$

so $3x \equiv 15 \pmod{26}$. Notice that because $\gcd(3, 26) = 1$, we can divide by 3 and we then get $x = 5$. This means that the plaintext corresponding to U is F, which is the letter corresponding to the number 5.

If we return to our example, let's say that we received the ciphertext

Section 7.2   Shift and Affine Ciphers

HYFEJZ. In order to decrypt this, we first change it into its numerical equivalent: $H = 7$, $Y = 24$, $F = 5$, $E = 4$, $J = 9$, and $Z = 25$. We could then use the method of Section 5.4 to solve the six congruences

$$\begin{aligned} 3x + 5 &\equiv 7 \pmod{26} \\ 3x + 5 &\equiv 24 \pmod{26} \\ 3x + 5 &\equiv 5 \pmod{26} \\ 3x + 5 &\equiv 4 \pmod{26} \\ 3x + 5 &\equiv 9 \pmod{26} \\ 3x + 5 &\equiv 25 \pmod{26}. \end{aligned}$$

However, there is an easier way. Write $y \equiv 3x + 5 \pmod{26}$. Multiply both sides by 9 to get

$$9y \equiv 27x + 45 \equiv x + 19 \pmod{26},$$

since $27 \equiv 1$ and $45 \equiv 19$. This can be rearranged to

$$x \equiv 9y - 19 \equiv 9y + 7 \pmod{26}.$$

Therefore, $9y + 7$ is our decryption function. Note that multiplying by 9 is the same as dividing by 3 in the "mod 26 world," since 9 and 3 are multiplicative inverses mod 26. So our calculation is really the same as changing $y = 3x + 5$ into $x = (1/3)y - (5/3)$ and then converting the fractions to integers mod 26.

The decryption is now easy:

$$\begin{aligned} 7 &\mapsto 9 \cdot 7 + 7 \pmod{26} \equiv 18 = S \\ 24 &\mapsto 9 \cdot 24 + 7 \pmod{26} \equiv 15 = P \\ 5 &\mapsto 9 \cdot 5 + 7 \pmod{26} \equiv 0 = A \\ 4 &\mapsto 9 \cdot 4 + 7 \pmod{26} \equiv 17 = R \\ 9 &\mapsto 9 \cdot 9 + 7 \pmod{26} \equiv 10 = K \\ 25 &\mapsto 9 \cdot 25 + 7 \pmod{26} \equiv 24 = Y \end{aligned}$$

so we recover our plaintext, *SPARKY*.

In general, in order to be able to decrypt a letter that was encrypted using an affine cipher with key (a, b), we have to solve the linear congruence

$$y \equiv ax + b \pmod{26}$$

for x in terms of y. This is equivalent to solving

$$ax \equiv y - b \pmod{26},$$

and in order to be able to solve this uniquely we need to have $\gcd(a, 26) = 1$ so we can divide by a.

Here's what can go wrong if we choose an a with $\gcd(a, 26) > 1$. Let's say our key is $(13, 5)$ so that we encrypt by having

$$x \mapsto 13x + 5 \pmod{26}.$$

Then the ciphertext corresponding to the plaintext *LOVE* is

$$\begin{aligned}
L &= 11 \mapsto 13 \cdot 11 + 5 & \pmod{26} &\equiv 18 = S \\
O &= 14 \mapsto 13 \cdot 14 + 5 & \pmod{26} &\equiv 5 = F \\
V &= 21 \mapsto 13 \cdot 21 + 5 & \pmod{26} &\equiv 18 = S \\
E &= 4 \mapsto 13 \cdot 4 + 5 & \pmod{26} &\equiv 5 = F,
\end{aligned}$$

while the ciphertext corresponding to the plaintext *HATE* is

$$\begin{aligned}
H &= 7 \mapsto 13 \cdot 7 + 5 & \pmod{26} &\equiv 18 = S \\
A &= 0 \mapsto 13 \cdot 0 + 5 & \pmod{26} &\equiv 5 = F \\
T &= 19 \mapsto 13 \cdot 19 + 5 & \pmod{26} &\equiv 18 = S \\
E &= 4 \mapsto 13 \cdot 4 + 5 & \pmod{26} &\equiv 5 = F.
\end{aligned}$$

This could cause some problems on Valentine's Day.

CHECK YOUR UNDERSTANDING

1. The function $9x + 2$ was used to obtain the ciphertext *KMI*. Find the plaintext.

7.3 Vigenère Ciphers

Before introducing the Vigenère cipher, let's take another look at shift ciphers. We know that we can read a ciphertext by simply looking at all shifts of the ciphertext. But we can also decrypt by counting how often

Section 7.3 Vigenère Ciphers

each letter occurs in the ciphertext, a method known as **frequency analysis**. This method is slower, but it allows us to develop an idea that will be useful later.

Here's how frequency analysis works. It is well known that E is the most common letter in standard English texts, and that Z is rather uncommon (unless you have weekly quizzes). Counts of frequencies of letters in English yield the following data:

FIGURE 7.1: Frequencies of letters in English.

For instance, E occurs with frequency around 12.7%, and A occurs around 8.2% of the time.

Let's say we intercept the following ciphertext:

```
YMJRTXYHTRRTSQDZXJIXNCHMFWFHYJWUFXXBTWINXTSJYBTYMW
JJKTZWKNAJXNCNYNXSTYWJHTRRJSIJIYMFYDTZZXJNYFXDTZWX
```

The frequency of letters in this ciphertext is given in Figure 7.2.

For example, of the 100 characters in the ciphertext, 3 are D's, so D has frequency 0.03. Notice that this graph looks similar to the frequency graph for English, but it's shifted 5 units to the right. (Of course, there are variations in height, but that is to be expected in small samples.) So we guess that the ciphertext was obtained by shifting the plaintext by 5 places. If we shift back by 5 places, we obtain (with spaces and punctuation inserted to make reading easier)

> THE MOST COMMONLY USED SIX-CHARACTER PASSWORD IS ONE TWO THREE FOUR FIVE SIX. IT IS NOT RECOMMENDED THAT YOU USE IT AS YOURS.

FIGURE 7.2: Frequencies of letters in ciphertext.

Why did this work? The 100-character plaintext is probably a representative sample of English, so it probably has around 8% A's and 12% E's. When the letters are shifted by 5 places, the A's become F's and the E's become J's. This means that the frequency for the plaintext A is the same as the frequency for the ciphertext F, and the frequency for the plaintext E is the same as the frequency for the ciphertext J. So, the ciphertext frequency distribution is simply a shift 5 places to the right of the plaintext frequency distribution. This is exactly what we saw in Figures 7.1 and 7.2.

Well into the 20th century, sophisticated versions of frequency counts were used to attack many cryptosystems. The **Vigenère cipher** was an attempt to thwart this type of attack. This cryptosystem seems to have been invented around 1553 by Giovan Battista Bellaso, but in the 1800s it was mistakenly named after Blaise de Vigenère (1523–1596), another 16th century cryptographer who devised similar but stronger cryptosystems. It was commonly regarded as unbreakable and was widely used up through World War I, even though the Prussian cryptographer Friedrich Wilhelm Kasiski had published a method for breaking it in 1863. Most likely, Charles Babbage had also developed the method in his work for the British intelligence agency.

In the Vigenère cipher, the key is a sequence of shifts, for example 2, 5, 4, 7. The length of the key could be anything of reasonable length that someone can remember (like today's passwords). Apply these shifts

Section 7.3   Vigenère Ciphers   163

cyclically to the message:

```
T H I S I S A S E C R E T
2 5 4 7 2 5 4 7 2 5 4 7 2
V M M Z K X E M G H V L V
```

The last line is the encrypted message.

If Bob receives this ciphertext, he shifts in the reverse direction to recover the plaintext.

Notice that in the example the letter E is encrypted once to G and once to L. Also, the letter M in the ciphertext comes from both H and I. The overall effect is that the counts for the common letters get spread among several letters and therefore the frequencies of the letters in the ciphertext are more uniform than from a simple shift cipher. This is why the Vigenère cipher was regarded as secure.

The best way to describe how to break the system is to consider an example. Suppose we intercept the following ciphertext:

VHXOEMJOWWSXFFHTTAGPKGPTTAMKOGCNWTETFIGIOYEOWGMXUSTIEL
KSLKMINEBPTAGEQVRXOETPDTVTAGSTOEMKMXKMIQSLKBEGOYVRTPSE
CTBQNNPLXUSMJEDGYBUKGQWGVHXGALGWBVHPJIVJTAGKXAMTABXEHT
PGXFILCNHVHXTPHKNMKNYCVHTOYVHXCDHRTBQNHHTAKSVQDXDYMJOL
GDXUIKKNZVOMTAGUMBVIFROKVAGVMXUSTIELYIMJONVTAGSEKGAVEL
VDTPGXTOYVHXKRFGSLCGXUBXKNZTETFBRROEKTBEAEQRUWSBPELURB
XAEUEME

The first thing we need to know is the key length. One method is due to Kasiski. Observe that the three-letter sequence VHX occurs five times. The first one starts with the V as the first letter, the second has V as the 133rd letter of the ciphertext, the third has V as the 172nd letter, the fourth has V as the 190th letter, and the fifth has V as the 280th letter. Each occurrence of VHX corresponds to three letters of the plaintext, shifted by amounts corresponding to the shifts required by the key. A reasonable guess is that each VHX comes from the same three letters shifted by the same amount. Since the key is applied cyclically, this means that the distances between the occurrences of VHX are multiples of the key length. These distances are

$$133 - 1 = 132, \quad 172 - 133 = 39, \quad 190 - 172 = 18, \quad 280 - 190 = 90.$$

If these are multiples of the key length, then

$$\gcd(132, 39, 18, 90) = 3$$

is a multiple of the key length. Therefore, a good guess for the key length is 3. As we'll see, this guess is correct.

Now that we suspect the key length is three, we know that the letters of the ciphertext in positions 1, 4, 7, 10, 14, ... were all shifted by the same amount. It is reasonable to assume that the corresponding letters of the plaintext represent a random sample of English letters. Therefore, we can guess the shift by the same method as we used at the beginning of this section; namely, we'll use frequency counts.

Extract the letters in positions 1, 4, 7, ... from the ciphertext to get

VOJWF···.

The frequencies for these letters are given in the following graph:

FIGURE 7.3: Frequencies of letters in positions 1, 4, 7, ···.

This is similar to the frequency graph for English given in Figure 7.1, except that it is shifted by 2 places. Therefore, we suspect that the first shift in the key is 2.

Now we do the same for the letters in positions 2, 5, 8, 11, ···. Figure 7.4 gives the graph of frequencies. It is essentially the frequency graph for English, but with no shift. So we suspect that the second shift of the key is 0.

Finally we look at positions 3, 6, 9, 12, ··· and obtain the graph in

Section 7.3   Vigenère Ciphers   165

FIGURE 7.4: Frequencies of letters in positions 2, 5, 8, ⋯.

FIGURE 7.5: Frequencies of letters in positions 3, 6, 9, ⋯.

Figure 7.5. This one is a little harder to analyze, but the best guess is that it is a shift by 19.

We now conjecture that the key is 2, 0, 19. When we realize that this corresponds to the easily-remembered word *CAT*, we're sure that we're right. But the only way to be certain is to decrypt, which we want to do anyway because that was our objective from the beginning.

Shift the ciphertext back by 2, 0, 19, applied cyclically. The result is (with spaces and punctuation added)

THE METHOD USED FOR THE PREPARATION AND READ-
ING OF CODE MESSAGES IS SIMPLE IN THE EXTREME

AND AT THE SAME TIME IMPOSSIBLE OF TRANSLATION UNLESS THE KEY IS KNOWN. THE EASE WITH WHICH THE KEY MAY BE CHANGED IS ANOTHER POINT IN FAVOR OF THE ADOPTION OF THIS CODE BY THOSE DESIRING TO TRANSMIT IMPORTANT MESSAGES WITHOUT THE SLIGHTEST DANGER OF THEIR MESSAGES BEING READ BY POLITICAL OR BUSINESS RIVALS ETC.

This text is from a 1917 *Scientific American* article describing the advantages of using the Vigenère cipher. As we have demonstrated, the author's analysis was not quite correct.

After World War I, the weaknesses inherent in the Vigenère cipher became more well known and its use decreased markedly.

7.4 Transposition Ciphers

Can you read the following ciphertext?

```
         IARATICNEDHS.
```

This was encrypted by a method that has been around for thousands of years and is often called zig-zag writing or a rail fence cipher. Here's how to read it. Split it in half to get `IARATI   CNEDHS`, and put the first half in one column and the second half in another column:

```
         IC
         AN
         RE
         AD
         TH
         IS.
```

You get the message by reading the letters row-by-row (the path you follow is definitely a zig-zag):

```
         I CAN READ THIS.
```

Section 7.4 Transposition Ciphers

It was encrypted by writing the plaintext in a series of rows and then reading the columns.

More generally, suppose Alice starts with a plaintext such as

THIS IS THE PLAINTEXT THAT WE ARE ENCRYPTING.

Alice decides to use five columns. The message has 37 characters, so she writes it in rows of 5 letters until the last row, which gets only 2 letters:

THISI
STHEP
LAINT
EXTTH
ATWEA
REENC
RYPTI
NG

To get the ciphertext, she reads this column-by-column:

TSLEARRNHTAXTEYGIHITWEPSENTENTIPTHACI.

Bob receives this ciphertext and counts that there are 37 characters. The *key* that Alice and Bob share beforehand is the number of columns. Since he knows that there were five columns, he computes $37/5 = 7.4$, which means he wants to group the letters into groups of 8 and 7 (because when 37 letters are put into 5 columns, some columns will have 8 letters and some will have 7). In fact, two 8s and three 7s yields 37. Therefore, he writes the message in 5 columns of lengths 8, 8, 7, 7, 7:

THISI
STHEP
LAINT
EXTTH
ATWEA
REENC
RYPTI
NG

Reading row by row yields the message.

Suppose Eve intercepts the ciphertext but does not know the number of columns. She knows that there will be an r-by-c array with r rows and c columns, and that $r \times c \geq 37$. Some possibilities are 3×13, 4×10, 8×5, and 7×6. For example, if Eve guesses that there are 3 rows and 13 columns, then she puts the ciphertext into the array

```
TERTTGIEEEIHI
SANAEITPNNPA
LRHXYHWSTTTC.
```

This does not yield a meaningful plaintext, so she tries other possibilities for the number of columns until she gets to 5 columns, which gives her the correct answer.

To increase the difficulty of Eve's attack, Alice can permute the columns according to some prearranged key word. For example, suppose the *key word* for the day is *water*. First notice that w is the first letter in water, a is the second letter, t the third, e the fourth, and r is the fifth. Next, put the letters of the key in alphabetical order: *aertw*. Finally, Alice puts the columns in the order 2, 4, 5, 3, 1 corresponding to the alphabetization of *water* together with where each letter occurs in the key word. For example, since a is the second letter of water and the first letter in the alphabetical order, she moves the second column to the first position. Similarly, since e is the 4th letter in *water* and is the 2nd letter in alphabetical order, she moves the 4th column to the 2nd position. This yields the following array:

```
HSIIT
TEPHS
ANTIL
XTHTE
TEAWA
ENCER
YTIPR
G   N
```

The columns yield the ciphertext:

HTAXTEYGSENTENTIPTHACIIHITWEPTSLEARRN.

This is much harder for Eve to break. Not only does she need to guess how many columns there are, but she also needs to guess the order of the columns. Otherwise, she doesn't know which are missing letters, and this affects where the breaks are between columns. In our example, if she guesses that the order was 2, 4, 5, 3, 1, then, using the fact that there are two columns of 8 and three of seven, she deduces that the plaintext should be arranged into columns of lengths 8, 7, 7, 7, 8. Once this is done, the columns can be rearranged to yield the plaintext by reversing the moves that put them in order 2, 4, 5, 3, 1.

The above procedure can be applied again to the ciphertext to yield a double encryption of the original plaintext. This was used briefly by the German military in World War I (until the French broke the system) and by the Dutch Resistance in World War II. Breaking this double encryption system is much more difficult than breaking a single encryption.

The above methods are called **columnar transpositions**. More generally, a **transposition cipher** keeps the letters of the plaintext unchanged, but puts them in a different order. Since the frequency counts for the text do not change, a cryptanalyst should notice that the frequency counts resemble those for the language used by the sender, which will lead to a suspicion that such a transposition cipher is being used.

CHECK YOUR UNDERSTANDING

2. The ciphertext `MMSTIEEALSEIIOTNNU` was encrypted by the method of this section. Decrypt it.

7.5 RSA

Let's say that Alice owns a bookstore and, in order to generate revenue, needs to accept online orders. Bob sees a book he wants, but in order to purchase it, he needs to send Alice his credit card number. He's concerned that if he does so, someone may intercept his communication with Alice's store and steal his number. Being well read, he knows that in symmetric key cryptosystems, the key used to encrypt a message is essentially the same as the key used to decrypt the message. This means that the person who wants to send a secret message has to protect the encryption key as well as the decryption key. Since Alice wants lots of people to have her encryption key, this can cause logistical problems no matter how strong a symmetric key cryptosystem she possesses. This dilemma was addressed in the groundbreaking paper by Diffie and Hellman, which set the stage for the rise of practical and commercially feasible public key cryptosystems.

All public key cryptosystems have a common theme: A public encryption key is produced using secret or private numbers. This key is distributed without any concerns about concealment. It could be printed in every local and national newspaper or distributed to every e-mail address in the world. With this key, anyone can encrypt a message and send it to you. The security of this system rests on the difficulty of finding the secret numbers that were used to create it, because they are the numbers used to decrypt.

The most widely used public key cryptographic system is RSA, named after Ron Rivest, Adi Shamir, and Len Adleman. It was developed in 1977, a year after the Diffie-Hellman paper appeared. The public encryption key uses the product of two large primes, and its security rests on the difficulty of being able to factor that product and retrieve the primes. The following is a step-by-step description of how Alice uses RSA in order to allow Bob and others to communicate securely with her.

RSA Setup

1. Alice chooses two primes, p and q, and calculates $n = pq$ and $\phi(n) = (p-1)(q-1)$.

2. Alice chooses an e (the encryption key) so that $\gcd(e, \phi(n)) = 1$.

3. Alice calculates d (the decryption key) with the property that $ed \equiv 1 \pmod{\phi(n)}$.

4. Alice makes n and e public and keeps d, p, q secret.

Section 7.5   RSA   171

Public Information: n and e
Private Information: $p, q,$ and d

RSA Encryption

1. Bob looks up Alice's n and e.
2. Bob writes his message as $m \pmod{n}$.
3. Bob computes $c \equiv m^e \pmod{n}$.
4. Bob sends c to Alice.

RSA Decryption

1. Alice receives c from Bob.
2. Alice computes $m \equiv c^d \pmod{n}$.

Example. Alice, realizing that she can't make money selling books, decides to become a spy. She needs to be able to communicate securely with the other spies in her network. To do so, she chooses two primes p and q and then multiplies them together to get her modulus n:

$$p = 3598279, \quad q = 7815629, \text{ and } n = p \cdot q = 28122813702491.$$

She next chooses an encryption exponent e, which has to be relatively prime to $\phi(n) = (p-1)(q-1)$, and calculates her decryption exponent d. The number d is a solution to the congruence $ed \equiv 1 \pmod{(p-1)(q-1)}$ and is kept secret. Alice's choice of

$$e = 233$$

forces

$$d = 27519308677241,$$

which she calculates using the Extended Euclidean Algorithm. Alice makes n and e public while keeping p, q, and d private.

Bob needs to set up a meeting with Alice. Out of their prearranged spots, he decides that they'll meet at his car. (They're not very good spies.) So, he needs to send the secret message CAR. In order to do so, he converts each letter to a number. Previously, we started with $a = 0$. We don't want to do that here because any message starting with the letter A would then start with the number 0 and the 0 would get dropped during the RSA process. So, he starts with $A = 01$, $B = 02$,..., $Z = 26$. This makes his message (CAR) have the numerical value

$$m = 030118 = 30118$$

since $C = 03$, $A = 01$, and $R = 18$. Bob calculates his ciphertext with the following calculation:

$$c \equiv m^e \equiv 30118^{233} \equiv 21666077416496 \pmod{28122813702491}.$$

After Alice receives the ciphertext $c = 21666077416496$, she decrypts it with her secret exponent d:

$$m \equiv c^d \equiv 21666077416496^{27519308677241}$$
$$\equiv 30118 \pmod{28122813702491}$$

(a calculation such as this is done with a modular exponentiation algorithm; see Section 5.2). Alice now has Bob's original message and knows to meet him at the car.

Remarks:

1. In Step 1, both of the two primes chosen should be large. For secure cryptographic applications, they are more than 100 digits long. In Exercise 35, you'll see that computing $\phi(n)$ is equivalent to factoring n.

2. In Step 2, somewhat surprisingly, e does not have to be large. Sometimes $e = 3$ has been used, although it is usually recommended that larger values be used. In practice, the most common value is $e = 65537$. The advantages are that 65537 is a reasonably large prime, so it's easier to arrange that $\gcd(e, \phi(n)) = 1$, and that it is one more than a power of

Section 7.5   RSA   173

2, so raising a number to the 65537th power consists mostly of squaring, plus only one multiplication, when the algorithm of Section 5.2 is used.

3. In Step 3, d is calculated using the Extended Euclidean Algorithm and is kept secret. To do this, Alice uses her knowledge of p and q to compute $\phi(n) = (p-1)(q-1)$, which allows her to solve $de \equiv 1 \pmod{\phi(n)}$ by the Extended Euclidean Algorithm.

4. We will see in Section 11.3 that if someone knows e, d, and n, then it is easy to factor n. Therefore, finding d is equivalent to factoring n. If we believe that factoring n is hard, then we conclude that there is no easy way for an eavesdropper to find d. Therefore, attacks on an RSA system often try to decrypt a plaintext without finding d. One such attack is given later in this section.

5. If Bob's message is larger than n, then it has to be broken up into pieces (or blocks), with each piece smaller than n. Since we're primarily concerned with the mathematics and not with cryptographic implementation issues, we'll ignore this and assume that $0 \leq m < n$.

Your first question might be, "Why does $c^d \pmod n$ give back m?" This "magic" is a consequence of Euler's theorem, as the following proposition shows.

Proposition 7.1. *Let $n = pq$ be the product of two distinct primes, and let d, e be positive integers satisfying $ed \equiv 1 \pmod{(p-1)(q-1)}$. Then, for all integers m,*

$$m^{ed} \equiv m \pmod{n}.$$

Therefore, if $c \equiv m^e \pmod n$, we have $m \equiv c^d \pmod n$.

Proof. First, we'll treat the case where $\gcd(m, n) = 1$. Since $ed \equiv 1 \pmod{\phi(n)}$, we can write

$$ed = 1 + k\phi(n) = 1 + k(p-1)(q-1)$$

for some k. By Euler's theorem,

$$m^{\phi(n)} \equiv 1 \pmod{n},$$

which implies that

$$m^{ed} \equiv m^{1+k\phi(n)} \equiv m\left(m^{\phi(n)}\right)^k \equiv m(1)^k \equiv m \pmod{n}.$$

This is the heart of the proof, but we still need to treat the cases where $\gcd(m,n) > 1$. In practice, n is the product of two extremely large primes, so the chances of this actually occurring are infinitesimally small. Since $\gcd(m,n)$ is a divisor of $n = pq$, we then must have $\gcd(m,n) = p$, q, or pq.

The case $\gcd(m,n) = pq$ is easy. In this case $n \mid m$, so $m \equiv 0 \pmod{n}$. Therefore, $m^{ed} \equiv 0^{ed} \equiv 0 \equiv m \pmod{n}$, as desired.

Now assume that $\gcd(m,n) = p$. Then $p \mid m$, so $m \equiv 0 \pmod{p}$. Therefore, $m^{ed} \equiv 0^{ed} \equiv 0 \equiv m \pmod{p}$. But $q \nmid m$, so $m^{q-1} \equiv 1 \pmod{q}$ by Fermat's theorem. Therefore,

$$m^{ed} \equiv m^{1+k(p-1)(q-1)} \equiv m\left(m^{q-1}\right)^{k(p-1)}$$
$$\equiv m(1)^{k(p-1)} \equiv m \pmod{q}.$$

We have now proved that

$$p \mid m^{ed} - m, \text{ and } q \mid m^{ed} - m.$$

Therefore, $pq \mid m^{ed} - m$, which is the same as $m^{ed} \equiv m \pmod{pq}$.

The case where $\gcd(m,n) = q$ uses the same argument, with p and q switched. Since we have treated all cases, the proof is complete. □

Why are people so confident in the security of this system? If we can factor n, then we can use the knowledge of p and q to compute d. In fact, we'll show (Section 11.3) that knowing d is computationally equivalent to factoring n. Factoring integers has been a topic of research for hundreds of years. In our situation, where the number that needs to be factored is the product of two very large primes, this is generally believed to be computationally infeasible.[1] Therefore, the public information does not allow an attacker to find d.

One of the uses of RSA is to transmit keys for use in symmetric cryptosystems. In this case, n could be a few hundred digits, while the

---
[1] This term is rather time dependent. It means that the computation would take thousands of years on the fastest computers currently available.

Section 7.5   RSA   175

message could have only a few digits. For example, suppose the message m has 16 or fewer digits. In this situation, with a relatively short message, the plaintext can be retrieved without factoring n or finding d. Here's how to accomplish this. An eavesdropper intercepts the ciphertext c, which probably has approximately as many digits as n. Next, the eavesdropper makes two lists:

1. $x^e \pmod{n}$ for all x with $1 \leq x \leq 10^9$.
2. $cy^{-e} \pmod{n}$ for all y with $1 \leq y \leq 10^9$.

Suppose there is a match between the two lists: $x^e \equiv cy^{-e} \pmod{n}$. Then
$$(xy)^e \equiv c \pmod{n}.$$
Since $1 \leq xy \leq 10^{18} < n$, the original message must have been $m = xy$.

Why can we hope to get a match? Since m is at most a 16-digit number, there is a reasonable chance that it factors as the product of two numbers of at most 9 digits (this is certainly not guaranteed, but it happens some of the time). If $m = xy$ with $1 \leq x, y < 10^9$, then $c \equiv m^e \equiv (xy)^e$, so $x^e \equiv cy^{-e}$. This yields a match. The way to avoid this "small message attack" is to append some random digits at the beginning and end of the message to make it into a much larger number.

Example. Let's take another look at the previous example in this section. We know that the modulus is $n = 28122813702491$, the encryption exponent is $e = 233$, and the ciphertext is $c = 21666077416496$. We suspect that the message satisfies $m < 10^5$, and we hope that it factors as the product of two numbers less than 1000. We make two lists,

1. $x^{233} \pmod{28122813702491}$ for $1 \leq x < 1000$
2. $21666077416496 \cdot y^{-233} \pmod{28122813702491}$
   for $1 \leq y < 1000$,

and discover that there is a match:

$$74^{233} \equiv 21666077416496 \cdot 407^{-233} \pmod{28122813702491}.$$

Therefore,

$$21666077416496 \equiv (74 \cdot 407)^{233} \equiv 30118^{233},$$

which means (because we can decrypt uniquely) that

$$m = 30118 \quad (= \ car).$$

However, suppose Bob changes his message to $m = 111030118111$. Then m cannot factor as $m = xy$ unless we allow x or y to have at least 6 digits, so this attack cannot succeed without longer lists. Bob can even tell everyone that he has added the extra digits.

In real-life situations, if someone knows that the message satisfies $m < 10^5$, then an eavesdropper simply encrypts all $m < 10^5$ and sees which one gives the ciphertext. This takes a very short amount of time to do. That is why we talked about a 16-digit message when describing the method. However, a 16-digit message could also be discovered by encrypting all $m < 10^{16}$. But it would have to be a very important message before this would be worthwhile because an m of this size would take a relatively long time to discover with this technique. Making and comparing two lists of length 10^9 is much more feasible.

CHECK YOUR UNDERSTANDING

3. Suppose an RSA public key is $n = 55$ and $e = 27$. If the ciphertext is $c = 4$, find the plaintext (you may use the factorization $55 = 5 \cdot 11$).

7.6 Stream Ciphers

A stream cipher changes a plaintext message one bit (or one letter) at a time. Let's say that Alice writes her message as a sequence of 0s and 1s. There are many ways of doing this. One of the most basic methods is called *ASCII*, which, for example, represents the letter A as 1000001 and B as 1000010. We won't worry about what method is used. The key is also a sequence of 0s and 1s and has the same length as the plaintext. To encrypt, Alice adds the key to the plaintext, bit by bit mod 2. This process is called **XOR** and is often denoted $\oplus$. For example, if the plaintext is 101100111 and the key is 001110101, the ciphertext 100010010 is obtained as follows:

$$101100111$$
$$\underline{001110101}$$
$$100010010.$$

Note that this uses the mod 2 addition laws $0+0 = 0$, $0+1 = 1$, $1+1 = 0$ without any carrying. All the operations are done one bit at a time.

Because adding mod 2 and subtracting mod 2 are the same, decryption is the exact same procedure with the exact same key. In our example,

$$100010010 \oplus 001110101 = 101100111,$$

so XORing the key with the ciphertext yields the plaintext.

A stream cipher encryption machine XORs the key bits one at a time with the plaintext bits, thus producing the ciphertext bits. Here's a picture of what's happening:

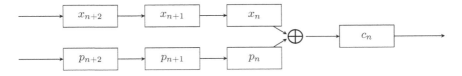

FIGURE 7.6
Stream cipher encryption: p_n = plaintext bit, x_n = key bit, c_n = ciphertext bit.

This encryption method is called a **stream cipher** since it uses a stream of 0s and 1s to encrypt. Stream ciphers are used in many situations. Some are very sophisticated and have high security. In the following, we describe two methods, one which is very secure and the other a simple one that is not very secure.

One-time pad

For a **one-time pad**, Alice uses a random process to generate a random sequence of 0s and 1s as a key. This is then used to encrypt by XORing it with the plaintext, as described above. If the key bits are truly random, this encryption method is completely secure. For example, suppose Eve intercepts the ciphertext 100010010 and is trying to decide whether the plaintext is 101100111 or 111110000. The first possibility corresponds to the key 001110101, and the second corresponds to the key 011100010. Since the key was chosen randomly, both possibilities are equally likely. In this way, we see that all possible plaintexts are equally likely. The ciphertext gives no information about the plaintext.

During the Cold War, the hotline between Washington, D.C., and Moscow used a one-time pad. The random sequences of 0s and 1s were written on paper tape and carried securely by couriers (probably as in the movies, with an attache case handcuffed to the courier's wrist). The random sequence was then available for use whenever needed.

Although the one-time pad is very secure, it is very expensive to use. Massive numbers of random 0s and 1s need to be transmitted in advance. Therefore, in practice it is desirable to use methods that quickly generate random-looking sequences but which can be transmitted more efficiently between Alice and Bob. One such method is described next.

Linear Feedback Shift Registers (LFSR)

Our goal is to create a key for a stream cipher. Here's an example of one way to do this.

We start off with two bits — we'll call them x_1 and x_2 — and arbitrarily choose $x_1 = x_2 = 1$. The values of x_1 and x_2 are called the seed because we'll grow the rest of the bits used in the key from these seed values. From these two bits we need to create a third, x_3. We do so by adding x_1 and x_2 (mod 2):

$$x_3 \equiv x_1 + x_2 \equiv 1 + 1 \equiv 0.$$

We now need to get the fourth key bit, x_4. We do this by adding x_2 and x_3 (mod 2):

$$x_4 \equiv x_2 + x_3 \equiv 1 + 0 \equiv 1 \pmod{2}.$$

In general, we have

$$x_{n+2} \equiv x_n + x_{n+1} \pmod{2}.$$

So, we see that

$$x_5 \equiv x_3 + x_4 \equiv 0 + 1 \equiv 1 \quad \text{and} \quad x_6 \equiv x_4 + x_5 \equiv 1 + 1 \equiv 0 \pmod{2}.$$

Section 7.6   Stream Ciphers

In this way, we generate a sequence

$$1, 1, 0, 1, 1, 0, \ldots.$$

(It's the Fibonacci sequence mod 2.) These 0s and 1s can be used as the key for a stream cipher.

The equation

$$x_{n+2} \equiv x_n + x_{n+1} \pmod{2}$$

is called a **recurrence relation**. In general, we choose a length ℓ, and then coefficients $c_0, c_1, \ldots, c_{\ell-1} \in \{0, 1\}$, and form the recurrence relation

$$x_{n+\ell} \equiv c_0 x_n + c_1 x_{n+1} + \cdots + c_{\ell-1} x_{n+\ell-1} \pmod{2}.$$

In the example above, we had $\ell = 2$, $c_0 = 1$, $c_1 = 1$.

As another example, suppose we have the recurrence relation

$$x_{n+5} \equiv x_n + x_{n+3} \pmod{2},$$

so $\ell = 5$, $c_0 = 1$, $c_1 = 0$, $c_2 = 0$, $c_3 = 1$, $c_4 = 0$. If we choose initial seed values $x_1 = 0, x_2 = 1, x_3 = 0, x_4 = 1, x_5 = 1$, then we can calculate $x_6, x_7, x_8, \ldots$. For example,

$$x_6 \equiv x_1 + x_4 \equiv 0 + 1 \equiv 1.$$

Similarly, we find that

$$x_7 = 0, \quad x_8 = 1, \ldots.$$

We again get a sequence of 0s and 1s.

If we have initial values $x_1, x_2, \ldots, x_\ell$, then we can use a recurrence relation to produce a sequence of bits to use in a stream cipher. Such a sequence is called a **Linear Feedback Shift Register sequence**, or **LFSR sequence**.

The reason for the name is the way a machine can produce the LFSR bits. Look at Figure 7.7. The boxes, which are called registers, are labeled x_n, x_{n+1}, etc. They contain bits. For example, suppose $n = 7$, so the bits in the boxes are x_{10}, x_9, x_8, and x_7. At the click of the machine's clock, the entries of a register are shifted to the right. The bit x_7 moves to the output position, and x_8 moves so that it occupies the space previously occupied by x_7, and similarly for the other bits. Moreover, x_7 and x_9 are also fed back into the system; namely, they follow the vertical arrows and are added together mod 2, and the result, which we'll call x_{11}, is put into the place previously occupied by x_{10}. The rectangles then contain x_{11}, x_{10}, x_9, and x_8. At the next click of the clock, x_8 is outputted, and an x_{12} is put into the leftmost box. Since the recurrence relation tells us that $x_{11} \equiv x_7 + x_9$ and $x_{12} \equiv x_8 + x_{10}$, we see that the feedback mechanism is designed so that the new bit that enters at the left is exactly the one corresponding to the recurrence relation.

FIGURE 7.7: LFSR for $x_{n+4} \equiv x_n + x_{n+2} \pmod{2}$.

An advantage of an LFSR sequence is that only a few bits, namely the coefficients and the initial values, need to be sent to produce a long sequence of bits. The disadvantage is that the method is not very secure. For example, suppose Eve knows the initial several plaintext bits and the corresponding ciphertext bits. She XORs these to get several key bits. Suppose, for example, that these key bits are

$$0, 1, 1, 1, 0, 0, 1, 0, 1, 1, 1, 0, 0, 1, 0.$$

She guesses that the length ℓ is 3, which means she guesses that the recurrence relation is

$$x_{n+3} \equiv c_0 x_n + c_1 x_{n+1} + c_2 x_{n+2} \pmod{2}.$$

Letting $n = 1, 2, 3$, and substituting in the values of the x_i, she obtains

$$\begin{aligned} 1 &\equiv c_0 \cdot 0 + c_1 \cdot 1 + c_2 \cdot 1 \\ 0 &\equiv c_0 \cdot 1 + c_1 \cdot 1 + c_2 \cdot 1 \\ 0 &\equiv c_0 \cdot 1 + c_1 \cdot 1 + c_2 \cdot 0. \end{aligned}$$

For example, the first line is simply the relation $x_4 \equiv c_0 x_1 + c_1 x_2 + c_2 x_3$. Adding the second and third lines yields $c_2 = 0$. Then the first line yields $c_1 = 1$ and the second line yields $c_0 = 1$. So the recurrence relation is

$$x_{n+3} \equiv x_n + x_{n+1} \pmod{2}.$$

Once Eve knows this, she can use the x_i's she knows as initial values and then generate as many terms of the sequence as she desires. In our example, she used only the first six members of the sequence to obtain the relation. She can check that the relation generates the remaining key bits that she knows. If it does not, then she guesses a different length and again solves for the coefficients. In this way, Eve eventually can find the key stream and thereby decrypt the remainder of the message. Therefore, this method of LFSR sequences should be used only for very low-level security. In practice, when high security is desired, much more complicated methods are used to produce the key stream.

CHECK YOUR UNDERSTANDING

4. An LFSR machine produces the key bits 110110110110. You receive the ciphertext 011101011101, which was encrypted with a stream cipher using these key bits. What is the plaintext?

7.7 Block Ciphers

As we discussed, stream ciphers encrypt bit by bit. **Block ciphers**, which form the second type of symmetric key cryptosystems, encrypt several bits at a time. More precisely, the plaintext is divided into blocks of a fixed length. Each block of plaintext is then encrypted, yielding a block of ciphertext. In the simplest version of a block cipher, Alice breaks her plaintext into blocks, all of the same length, and encrypts each one by the same encryption algorithm. The ciphertext is therefore a series of blocks of characters. More complicated implementations use techniques such as feedback, so that the encryption of one block depends on the encryption of preceding blocks. See Exercise 49.

For example, if the plaintext is

ATTACK AT DAWN,

and our block length is 2, we can break up the message into

AT TA CK AT DA WN,

and encrypt each pair of letters separately. If, instead, our block length is five, our plaintext blocks would be

ATTAC KATDA WNXXX,

where we have filled in the empty spaces with X's.

One of the earliest block ciphers, the **Hill cipher**, was invented by Lester Hill in 1929. The key is an $n \times n$ integer matrix M that has an inverse mod 26. (For a brief review of matrices, see the Appendix.) Although the Hill cipher works for any block size n, to simplify matters we look only at $n = 2$, so all our matrices are 2×2. For example, suppose

$$M = \begin{pmatrix} 5 & 7 \\ 2 & 11 \end{pmatrix}.$$

Let

$$N = \begin{pmatrix} -1 & 3 \\ 12 & 9 \end{pmatrix}.$$

Then

$$MN = \begin{pmatrix} 5 & 7 \\ 2 & 11 \end{pmatrix} \begin{pmatrix} -1 & 3 \\ 12 & 9 \end{pmatrix} = \begin{pmatrix} 79 & 78 \\ 130 & 105 \end{pmatrix} \equiv \begin{pmatrix} 1 & 0 \\ 0 & 1 \end{pmatrix} \pmod{26},$$

so N is the inverse of M mod 26. Therefore, M is a suitable key for the Hill cipher.

How did we find N? The formula for the inverse of a 2×2 matrix is

$$\begin{pmatrix} a & b \\ c & d \end{pmatrix}^{-1} = \frac{1}{ad - bc} \begin{pmatrix} d & -b \\ -c & a \end{pmatrix}.$$

For our example, we obtain

$$\begin{pmatrix} 5 & 7 \\ 2 & 11 \end{pmatrix}^{-1} = \frac{1}{41} \begin{pmatrix} 11 & -5 \\ -2 & 5 \end{pmatrix}.$$

Since we are working mod 26, we replace 1/41 with 7, since 7 is the multiplicative inverse of 41 mod 26. (We discussed multiplicative inverses mod n and how to calculate them in Section 5.4.)

$$7 \begin{pmatrix} 11 & -5 \\ -2 & 5 \end{pmatrix} \equiv \begin{pmatrix} -1 & 3 \\ 12 & 9 \end{pmatrix},$$

Section 7.7 Block Ciphers

which is N. In general, the formula yields an inverse matrix whenever $\gcd(ad - bc, 26) = 1$. In fact, it can be shown that $\gcd(ad - bc, 26) = 1$ if and only if the matrix has an inverse mod 26.

Let's continue with the encryption. Alice breaks her ciphertext into blocks of n letters (where n is the size of the matrix) and converts each letter to a number using $A = 0, B = 1, \ldots, Z = 25$. Each block of n letters is then regarded as an n-dimensional vector. The encryption multiplies the matrix M times the vectors mod 26. These are then converted back to letters.

For example, suppose the plaintext is *HIDETHIS*. This becomes the vectors

$$\begin{pmatrix} 7 \\ 8 \end{pmatrix}, \begin{pmatrix} 3 \\ 4 \end{pmatrix}, \begin{pmatrix} 19 \\ 7 \end{pmatrix}, \begin{pmatrix} 8 \\ 18 \end{pmatrix}$$

(the first vector uses $H = 7, I = 8$, for example). The encryption is

$$\begin{pmatrix} 5 & 7 \\ 2 & 11 \end{pmatrix} \begin{pmatrix} 7 \\ 8 \end{pmatrix} \equiv \begin{pmatrix} 13 \\ 24 \end{pmatrix}$$

$$\begin{pmatrix} 5 & 7 \\ 2 & 11 \end{pmatrix} \begin{pmatrix} 3 \\ 4 \end{pmatrix} \equiv \begin{pmatrix} 17 \\ 24 \end{pmatrix}$$

$$\begin{pmatrix} 5 & 7 \\ 2 & 11 \end{pmatrix} \begin{pmatrix} 19 \\ 7 \end{pmatrix} \equiv \begin{pmatrix} 17 \\ 24 \end{pmatrix}$$

$$\begin{pmatrix} 5 & 7 \\ 2 & 11 \end{pmatrix} \begin{pmatrix} 8 \\ 18 \end{pmatrix} \equiv \begin{pmatrix} 10 \\ 6 \end{pmatrix}.$$

Using $13 = N$, $24 = Y$, etc., converts this back to the letters *NYRYOLKG*, which is the ciphertext.

When Bob receives the ciphertext, he decrypts by multiplying by the inverse of M mod 26. In our example, this is the matrix N. So Bob computes

$$\begin{pmatrix} -1 & 3 \\ 12 & 9 \end{pmatrix} \begin{pmatrix} 13 \\ 24 \end{pmatrix} \equiv \begin{pmatrix} 7 \\ 8 \end{pmatrix}$$

$$\begin{pmatrix} -1 & 3 \\ 12 & 9 \end{pmatrix} \begin{pmatrix} 17 \\ 24 \end{pmatrix} \equiv \begin{pmatrix} 3 \\ 4 \end{pmatrix}$$

$$\begin{pmatrix} -1 & 3 \\ 12 & 9 \end{pmatrix} \begin{pmatrix} 17 \\ 24 \end{pmatrix} \equiv \begin{pmatrix} 19 \\ 7 \end{pmatrix}$$

$$\begin{pmatrix} -1 & 3 \\ 12 & 9 \end{pmatrix} \begin{pmatrix} 10 \\ 6 \end{pmatrix} \equiv \begin{pmatrix} 8 \\ 18 \end{pmatrix}.$$

Bob changes this to letters and obtains *HIDETHIS*.

If Eve discovers the plaintext that corresponds to some ciphertext, she can set up linear congruences and solve for the entries in the matrix. Because of this, the Hill cipher should not be used in situations that require high levels of security. But the real reason it was never used widely is that encryption and decryption are rather slow, especially if larger matrices (for example, 6×6) are used.

The block ciphers that are used for situations where security is important use bigger blocks and are faster than Hill ciphers. One of the most widely used block ciphers was DES (Data Encryption Standard), which was invented in the 1970s and was the standard cryptographic block cipher for more than 25 years. It used plaintext blocks of 64 bits and produced ciphertext blocks also of 64 bits. In 2001, AES (Advanced Encryption Standard) was introduced as a replacement for DES. It uses blocks of 128 bits. Both DES and AES have been widely used in Internet commerce.

CHECK YOUR UNDERSTANDING

5. Use the Hill cipher with matrix $\begin{pmatrix} 3 & 2 \\ 13 & 1 \end{pmatrix}$ to encrypt the message *GOLD*.

7.8 Secret Sharing

You are the manager of a bank and you want to skip work tomorrow. Even though you won't be there, the bank's safe will still need to be opened. But, you don't trust any individual employee enough to allow one person to have the lock's combination. So, you need a scheme where any two employees together can deduce the combination, but any single employee obtains no information. Your first idea is to give the first number of the combination to half the employees and the last two numbers to the other half. This works well if a first-number employee gets together with a second/third-number employee, but it doesn't work with two first-number employees. So you need something else. While

Section 7.8 Secret Sharing

musing over the requirement

> two people determine a secret,

you think back to your high school geometry days and recall that

> two points determine a line.

Aha! If you represent the combination as the slope of a line and give each employee a point on the line, then one employee alone cannot determine the slope, but any two employees can find the slope and then determine the combination.

Let's try an example. The lock's combination is 17-34-02, which you write as 173402. We could work with real numbers, but, to avoid slight problems with round-off errors, it is better to work mod a prime. Let's assume that the largest possibility for each number of the combination is 40, so we need a prime larger than 404040. We choose the prime $p = 404051$. Now we choose a line. The slope will be 173402. Choose a random intercept, say 141421, and form the line

$$L: \quad y \equiv 173402x + 141421 \pmod{404051}.$$

Let's say you have five employees: A, B, C, D, E. Give them the following points on L:

$$
\begin{aligned}
A &: \ (1, 314823) \\
B &: \ (2, 84174) \\
C &: \ (3, 257576) \\
D &: \ (4, 26927) \\
E &: \ (5, 200329).
\end{aligned}
$$

For example, B's point is calculated by putting $x = 2$ into the equation for L and getting $y \equiv 173402 \times 2 + 141421 \equiv 84174 \pmod{404051}$.

Suppose that C and D want to open the safe. The slope of the line through their points is

$$\frac{26927 - 257576}{4 - 3} \equiv -230469 \equiv 173402 \pmod{404051}.$$

They have recovered the combination!

Let's try another pair. Suppose that B and E want to open the safe. The slope of the line through their points is

$$\frac{200329 - 84174}{5 - 2} \equiv \frac{116155}{3} \pmod{404051}.$$

What do we do now? Since we need to get rid of the 3 in the denominator, we use the Extended Euclidean Algorithm to find the multiplicative inverse of 3 mod 404051:

$$\frac{1}{3} \equiv 134684 \pmod{404051}.$$

Then the slope is

$$\frac{116155}{3} \equiv 116155 \times 134684 \equiv 173402 \pmod{404051}.$$

Again, they have recovered the combination.

In general, suppose you have a secret s expressed as an integer. Choose a prime p larger than the largest possibility for the s. (Note that p does not need to be a cryptographic-sized prime of 300 digits.) Pick a random integer r and form the line $L: y \equiv sx + r \pmod{p}$. Give each person one of the points $(1, s+r), (2, 2s+r), (3, 3s+r), \ldots$. Then any two people together can determine the secret s. However, one point gives no information about the slope, so one person does not obtain any information.

This method can be generalized. If we replace the line L with a parabola $y \equiv ax^2 + bx + s \pmod{p}$ with randomly chosen coefficients a and b and give each person a point on this parabola, then any three people can determine s, but two people obtain no information about s (as long as no one is given a point with x-coordinate 0). More generally, we can use polynomials of degree k with one coefficient equal to the secret and the other coefficients randomly chosen. This yields a scheme where any $k+1$ people can determine a secret but k people do not obtain any information.

CHECK YOUR UNDERSTANDING

6. In a secret sharing scheme mod 13 where the slope is the secret, person A has the point (2,4), person B has (5,0), and person C has (8,9). Find the secret.

7.9 Chapter Highlights

1. Shift and affine ciphers
2. The Vigenère cipher
3. The RSA algorithm
4. Stream and block ciphers
5. Secret sharing

7.10 Problems

7.10.1 Exercises

Section 7.2: Shift and Affine Ciphers

1. The plaintext $ATDAWN$ was encrypted using the affine function $9x + 13$. What is the ciphertext? What is the decryption function? Verify that it works.

2. The plaintext $THEBEACH$ was encrypted using the affine function $7x + 23$. What is the ciphertext? What is the decryption function? Verify that it works.

3. The ciphertext $QBULF$ was encrypted using the affine function $3x + 11$. What is the plaintext?

4. The ciphertext $QPMZ$ was encrypted using the affine function $17x + 9$. What is the plaintext?

5. The ciphertext $YRFE$ was encrypted using the affine function $5x + 17$. What is the plaintext?

6. The ciphertext $LFICJBOPS$ was encrypted using the affine function $7x + 5$. What is the plaintext?

7. You intercept the ciphertext $KRON$, which was encrypted using an affine function. You know that the plaintext starts DO. Find the encryption function and find the plaintext.

8. You intercept the ciphertext *YFWD*, which was encrypted using an affine function. You know that the plaintext starts *ST*. Find the encryption function and find the plaintext.

9. Suppose you try to use the affine function $2x + 1$ to encrypt. Find two letters that encrypt to the same ciphertext letter.

10. Suppose you try to use the affine function $13x + 7$ to encrypt. Find two different three-letter words that encrypt to *HHH*.

11. If you receive a message encrypted by a shift cipher, usually you can decrypt it by trying all possibilities.
    (a) Decrypt *ITSJ*, which is a ciphertext for a shift cipher.
    (b) Show that *LCPYL* has two different decryptions (it is the longest known example).
    (c) Find some two-letter ciphertexts that have more than one decryption to English words.

Section 7.4: Transposition Ciphers

12. The following were encrypted by a columnar transposition cipher with 3 columns, with no permutation of the columns. Read the messages.
    (a) MHFAIUTSN.
    (b) ERNNSAVYEOTTEOKWH.

13. The following were encrypted by a columnar transposition cipher, with no permutation of the columns. Read the messages.
    (a) IMNOENEEDY.
    (b) TEECHIDUISILSMFTOOFNRI.

14. Suppose Alice decides to use a columnar transposition cipher with only one column. What can you say about the ciphertext?

15. The following were encrypted by columnar transportation ciphers, with no permutation of the columns. Read the messages.
    (a) TROAHSONBTETBOEIQIETTSUOOTHTEN.
    (b) TNESEOOTTSBTHHTETAEIOOTQORBIUN.

16. The following was encrypted by a columnar transportation cipher, with no permutation of the columns. Read the message.
    ICASKPAFADTATPYNTHCOHORHARGEUEINYRR.

Section 7.10   Problems   189

17. The following was encrypted by a columnar transportation cipher, with no permutation of the columns. Read the message.
TIEEWHSNREADAGTOIOTFNLHTBDEHO.

Section 7.5: RSA

18. In each of the following, you are given two primes p and q to use for RSA. Decide whether the given encryption exponent can be used. If it can be used, find the decryption exponent d. If it cannot be used, explain why.
    (a) $p = 7, q = 11, e = 6$.
    (b) $p = 11, q = 13, e = 7$.
    (c) $p = 13, q = 23, e = 22$.
    (d) $p = 13, q = 29, e = 11$.

19. Alice chooses primes $p = 149$ and $q = 317$ and encryption exponent $e = 71$. What public modulus does she publish? What is her decryption exponent?

20. Alice chooses primes $p = 3491$ and $q = 8219$ and encryption exponent $e = 97$. What public modulus does she publish? What is her decryption exponent?

21. For the following, encrypt the message m using RSA with modulus $n = pq$ and exponent e to obtain the ciphertext c. Then find the decryption exponent d, and verify that $c^d \equiv m \pmod{n}$:
    (a) $p = 7, q = 13, e = 5, m = 5$.
    (b) $p = 7, q = 13, e = 7, m = 7$.
    (c) $p = 3, q = 19, e = 7, m = 5$.
    (d) $p = 11, q = 17, e = 7, m = 9$.
    (e) $p = 11, q = 17, e = 13, m = 3$.

22. A general at the Pentagon tells the lieutenant colonels (labeled (a), (b), (c), (d)) in the field to move m tanks into position for an attack by using RSA with modulus pq. The lieutenant colonels know their values of p and q and e the encryption exponent. The ciphertext is c. Use this information to find the decryption exponent d and the number of tanks for each lieutenant colonel.
    (a) $p = 3, q = 11, e = 3, c = 7$.
    (b) $p = 5, q = 11, e = 3, c = 9$.

(c) $p = 3, q = 17, e = 3, c = 5$.
(d) $p = 13, q = 17, e = 11, c = 7$.

23. In the following, a message m was encrypted using RSA with modulus pq and encryption exponent e, yielding the ciphertext c. Find m.
    (a) $p = 3, q = 23, e = 3, c = 9$.
    (b) $p = 3, q = 29, e = 3, c = 5$.
    (c) $p = 5, q = 17, e = 5, c = 3$.

24. Romeo sends Juliet the hour at which they will meet. He uses RSA with $p = 3, q = 17, e = 5$. The ciphertext is $c = 3$. When will they meet?

The following 7 problems are easiest to do with computer software that can handle up to 25-digit numbers and calculate mods.

25. Alice chooses primes $p = 8923$ and $q = 6581$. She publishes $n = p \cdot q = 58722263$ and her encryption exponent $e = 307$. Using this, Bob sends her the message 21511484. What did Bob tell her?

26. Let's say that Bob wants to send Alice the message *HELP*. Alice chooses her secret primes $p = 43753$ and $q = 87149$ and encryption exponent $e = 6043$. She calculates $n = p \cdot q = 3813030197$ and sends Bob n and e. We'll help Bob get started by writing out the number *HELP* corresponds to. Since $H = 08$, $E = 05$, $L = 12$, and $P = 16$ (remember that $A = 01$ when we use RSA), his message $m = 8051216$.
    (a) What ciphertext c does Alice receive?
    (b) What is Alice's decryption exponent d?
    (c) Verify that Alice can decrypt Bob's message.

27. Let's say that Bob wants to send Alice the message *HELP*. Alice chooses her secret primes $p = 658943$ and $q = 357389$ and encryption exponent $e = 2347$. She calculates $n = p \cdot q = 235498979827$ and sends Bob n and e.
    (a) What ciphertext c does Alice receive?
    (b) What is Alice's decryption exponent d?
    (c) Verify that Alice can decrypt Bob's message.

28. Bob wants to send Alice the message *GOOD LUCK* using her encryption key $(e, n) = (5, 21631)$. He writes his message as

a single number: $m = 715150412210311$ and encrypts it using Alice's public key.
(a) What message does Bob send to Alice?
(b) What are Alice's "secret" p and q? What is her decryption exponent?
(c) What message does Alice receive after she decrypts? What went wrong?
(d) Can you come up with two ways to fix the problem?

29. When RSA is used with encryption key $(e, n) = (17, 31439)$, the ciphertext received is

$$3807 \quad 17063 \quad 25522 \quad 12107 \quad 26663$$

(each block was encrypted separately). What is the plaintext?

30. When RSA is used with encryption key $(e, n) = (5, 5352499)$, the ciphertext received is

$$4784648 \quad 1933497 \quad 4437506$$

(each block was encrypted separately). If $n = 1237 \cdot 4327$, find the decryption key and then the plaintext.

31. Sometimes, when decrypting a message, a cryptographer can speed up the process by decrypting a piece of the ciphertext and then making educated guesses about what follows. For example, if a letter is decrypted as Q, we can be pretty certain that the next letter is U. Suppose RSA is being used with encryption key $(e, n) = (23, 58951)$. The plaintext is encrypted in blocks of two letters at a time and the received ciphertext is the following sequence of blocks:

$$40658||35980||31131||02677||12202||37240||52650.$$

You are told that the plaintext begins with

$$2015||0205||1518.$$

(a) Can you guess the rest of the message? (If you can't, you must decide whether 'tis nobler to factor n or keep trying.)
(b) Show that your guess is correct by encrypting it and comparing with the ciphertext.

32. Explain why it would be silly to use an RSA encryption exponent of 1. Why would an encryption exponent of 2 never be used?

33. Alice and Alicia each set up an RSA public key cryptosystem with the same modulus n but different, relatively prime, encryption exponents, so Alice's key is (e_1, n) and Alicia's is (e_2, n). Bob encrypts the same message, sending m^{e_1} (mod n) to Alice and m^{e_2} (mod n) to Alicia. Show that if Eve intercepts both ciphertexts she can find the plaintext m.

34. Choose two primes p and q for your RSA cryptosystem and an encryption exponent e and let $\lambda(n) = \phi(n)/\gcd(p-1, q-1)$. Find d' by solving

$$d' \cdot e \equiv 1 \pmod{\lambda(n)}.$$

Show that d' can also be used as the private decryption exponent.

35. Suppose that you know n can be factored as pq, where p and q are prime numbers, but you are not given p and q. In this exercise, we show that knowledge of both n and $\phi(n)$ allows you to factor n.
(a) Let $r = n - \phi(n) + 1$. Show that $r = p + q$.
(b) Show that p and q are the two solutions to the quadratic equation $x^2 - rx + n = 0$.
(c) If $n = pq = 218957$ and $\phi(n) = 217980$, find p and q using the quadratic formula.

36. Suppose Alice chooses three distinct primes p, q, r and computes $n = pqr$. She chooses encryption and decryption exponents e and d. What requirements should d and e satisfy in order for Alice to encrypt and decrypt using $c \equiv m^e$ (mod n) and $m \equiv c^d$ (mod n) ?
(There seems to be no advantage to using three primes in RSA. The difficulty of factoring n often depends on the smallest prime factor of n, so using three primes might make factoring n easier than factoring a standard RSA modulus of the same size.)

37. Suppose that Eve wants Alice to decrypt c but Alice refuses. Eve computes $c_1 \equiv 123^e c$ (mod n) and gives c_1 to Alice. Alice

decrypts c_1 to get m_1 and, since the message m_1 does not appear to be anything meaningful, she gives m_1 to Eve. How does Eve recover m from m_1?

38. Suppose that Bob knows about the short message attack. He has a message m with $m < 10^{12}$, so he appends 75 zeros at the end of m to get $m_1 = 10^{75}m$. He then encrypts m_1 to get c_1. Show how Eve can still use the short message attack to obtain m.

Section 7.6: Stream Ciphers

39. An LFSR machine produces the key bits 10101010. You encrypt the plaintext 11111111 using these key bits. What is the ciphertext?

40. An LFSR machine produces the key bits 10101010. You receive the ciphertext 11111111, which was encrypted with a stream cipher using these key bits. What is the plaintext?

41. An LFSR machine uses the recurrence relation $x_{n+4} \equiv x_n + x_{n+1}$ (mod 2) and initial values $x_1 = x_2 = x_3 = x_4 = 1$. Find $x_5, x_6, \ldots, x_{10}$.

42. An LFSR machine uses the recurrence relation $x_{n+3} \equiv x_n + x_{n+1} + x_{n+2}$ (mod 2) and initial values $x_1 = x_2 = 1$, $x_3 = 0$. Find $x_4, x_5, \ldots, x_{10}$.

43. The sequence $0, 1, 1, 1, 0, 1, 0, 0, 1, 1, 1, 0$ was produced by a recurrence relation $x_{n+3} \equiv c_0 x_n + c_1 x_{n+1} + c_2 x_{n+2}$ (mod 2). Find c_0, c_1, c_2.

44. The sequence $0, 1, 0, 0, 1, 0, 0, 1$ was produced by a recurrence relation $x_{n+3} \equiv c_0 x_n + c_1 x_{n+1} + c_2 x_{n+2}$ (mod 2). Find c_0, c_1, c_2.

Section 7.7: Block Ciphers

45. Use the Hill cipher with matrix $\begin{pmatrix} 3 & 2 \\ 12 & 9 \end{pmatrix}$ to encrypt the message *HILL*.

46. Use the Hill cipher with matrix $\begin{pmatrix} 1 & 7 \\ 4 & 11 \end{pmatrix}$ to encrypt the message *BUSY*.

47. A Hill cipher with matrix $\begin{pmatrix} 11 & 2 \\ 12 & 3 \end{pmatrix}$ was used to encrypt a message. The ciphertext is $WYED$. What is the plaintext?

48. A Hill cipher with matrix $\begin{pmatrix} 2 & 3 \\ 5 & 12 \end{pmatrix}$ was used to encrypt a message. The ciphertext is $ALNS$. What is the plaintext?

49. A common method of using block ciphers to encrypt is called *cipher block chaining*, or *CBC*. Suppose Bob has an encryption machine that takes an input B of 128 bits and encrypts it to a block C of 128 bits. We'll write this as $C = E(B)$. Alice knows how to decrypt E. That is, Alice knows the function D such that $D(C) = B$. Bob wants to send a message m to Alice. He breaks m into 128-bit blocks $B_1, B_2, B_3, \ldots$. Then he chooses a random 128-bit seed C_0. He computes $C_1 = E(B_1 \oplus C_0)$ to get the first block of ciphertext. Then he computes $C_2 = E(B_2 \oplus C_1)$ to get the second block. In general, he computes $C_n = E(C_{n-1} \oplus B_n)$. The initial seed and the ciphertext blocks $C_0, C_1, C_2, \ldots$ are sent to Alice. When Alice receives this ciphertext, how does she decrypt it in order to recover $B_1, B_2, \ldots$?

Section 7.8: Secret Sharing

50. Suppose we have a secret sharing scheme that works mod 53 and where A, B, C, D get the points $(1, 6)$, $(2, 48)$, $(3, 37)$, $(4, 26)$, respectively. The slope of the line through these points is the secret. Find the secret.

51. Suppose we have a secret sharing scheme that works mod 13 and where A, B, C, D get the points $(1, 12)$, $(2, 4)$, $(3, 9)$, $(4, 1)$, respectively. The slope of the line through these points is the secret. Find the secret.

52. Suppose we have a secret sharing scheme that works mod 37 and where A, B, C, D get the points $(1, 19)$, $(2, 30)$, $(3, 22)$, $(4, 15)$, respectively. The slope of the line through three of these points is the secret, but the remaining point is a forgery and is not correct. Find which point is the forgery and find the secret.

53. Suppose we have a secret sharing scheme that works mod 37

and where A, B, C, D get the points $(1, 29)$, $(2, 15)$, $(3, 28)$, $(4, 9)$, respectively. The slope of the line through three of these points is the secret, but the remaining point is a forgery and is not correct. Find which point is the forgery and find the secret.

54. You have a secret sharing scheme mod 29, as in Section 7.8. Your point is $(2, 18)$ and the secret is $s = 13$. Person Q's point is $(17, *)$. Find the y-coordinate of B's point.

7.10.2 Computer Explorations

1. (a) For RSA, let $p = 167$ and $q = 251$, so $n = 41917$. Let $e = 3$. Encrypt $ban = 20114$, $bat = 20120$, and $bay = 20125$. Can you tell from the ciphertexts that only one letter has been changed in the plaintexts? (In a good cryptosystem, a small change in the plaintext makes a large change in the ciphertext.)
    (b) Find the decryption exponent d.
    (c) Bob tries to send you the ciphertext $c = 27120$, but there is a transmission error and you receive 27121. Do you obtain anything close to the intended message?

2. Suppose Bob sends the same message m to three different people. The ciphertexts are

    $$c_1 = 257261 \pmod{303799}, \quad c_2 = 117466 \pmod{289279},$$
    $$c_3 = 260584 \pmod{410503},$$

    and their RSA moduli are

    $$n_1 = 303799, \quad n_2 = 289279, \quad n_3 = 410503.$$

    Each person has encryption exponent $e = 3$ (so $m^3 \equiv c_1 \pmod{n_1}$, and similarly for the other two).
    (a) Find an x with $0 \leq x < n_1 n_2 n_3$ and

    $$x \equiv c_1 \pmod{n_1}, \quad x \equiv c_2 \pmod{n_2}, \quad x \equiv c_1 \pmod{n_2}.$$

    (b) Show that $0 \leq m^3 < n_1 n_2 n_3$. (Hint: $m < n_i$.)
    (c) Show that $x = m^3$ (not just congruent, but actually equal).
    (d) Find the message m.
    This shows a disadvantage of using a small encryption exponent. (An advantage is that encryption is fast.)

7.10.3 Answers to "CHECK YOUR UNDERSTANDING"

1. To obtain the decryption function, solve $y \equiv 9x + 2$ for x in terms of y. Since 3 is the inverse of 9 mod 26, we have

   $$y \equiv 9x+2 \implies x \equiv (1/9)(y-2) \equiv 3(y-2) \equiv 3y-6 \pmod{26}.$$

   (We could also use $3y+20$, but using $3y-6$ lets us use smaller numbers.) *KMI* is 10, 12, 8 in numbers. These decrypt as follows:

   $$K = 10 \mapsto 3(10) - 6 = 24 \mapsto Y$$
   $$M = 12 \mapsto 3(12) - 6 \equiv 4 \mapsto E$$
   $$I = 8 \mapsto 3(8) - 6 \equiv 18 \mapsto S.$$

   The plaintext is *YES*.

2. There are 18 letters in the ciphertext, so there could be 2, 3, 4, 5, ... columns. Try them. Eventually you get to 4 columns, which yields

   ```
   MEET
   MEIN
   SAIN
   TLOU
   IS
   ```

   So the plaintext is MEET ME IN SAINT LOUIS.

3. First, we need the decryption key. To do this, solve $de \equiv 1 \pmod{(p-1)(q-1)}$, which means solve $27d \equiv 1 \pmod{40}$. The solution (find it by trial and error or by the Extended Euclidean Algorithm) is $d = 3$. Now we can decrypt:

   $$m \equiv c^d \equiv 4^3 \equiv 64 \equiv 9 \pmod{55}.$$

   The plaintext is 9 (if you want to change it to letters, the plaintext could be *I*).

4. To get the plaintext, compute ciphertext $\oplus$ key = 011101011101 $\oplus$ 110110110110 = 101010101010. Therefore, 101010101010 is the plaintext.

5. Break the plaintext into 2-letter blocks: *GO LD*. Change this to numbers: (6, 14), (11, 3). Now compute

$$\begin{pmatrix} 3 & 2 \\ 13 & 1 \end{pmatrix} \begin{pmatrix} 6 \\ 14 \end{pmatrix} = \begin{pmatrix} 46 \\ 92 \end{pmatrix} \equiv \begin{pmatrix} 20 \\ 14 \end{pmatrix} \pmod{26}$$

$$\begin{pmatrix} 3 & 2 \\ 13 & 1 \end{pmatrix} \begin{pmatrix} 11 \\ 3 \end{pmatrix} = \begin{pmatrix} 39 \\ 146 \end{pmatrix} \equiv \begin{pmatrix} 13 \\ 16 \end{pmatrix} \pmod{26}.$$

Change this back to letters to obtain the ciphertext: *UONQ*.

6. Choose any two points, for example, (5,0) and (8,9). The line through these points has slope $(9-0)/8-5) = 3$. So the secret is 3. If we had chosen (2,4) and (8,9), we would obtain slope $(9-4)/(8-2) = 5/6 \equiv 3 \pmod{13}$.

Chapter 8

Order and Primitive Roots

8.1 Orders of Elements

Let n be a positive integer and let $\gcd(a, n) = 1$. The **order** of a mod n, denoted $\text{ord}_n(a)$, is the smallest positive integer m such that

$$a^m \equiv 1 \pmod{n}.$$

Example.

$$2^6 \equiv 1 \pmod 9, \quad 2^i \not\equiv 1 \pmod 9 \text{ for } 1 \leq i \leq 5,$$

so $\text{ord}_9(2) = 6$.

Here's another way to look at the order of an element. The powers of 2 mod 9 are

$$2^0 \equiv 1, \ 2^1 \equiv 2, \ 2^2 \equiv 4, \ 2^3 \equiv 8, \ 2^4 \equiv 7, \ 2^5 \equiv 5,$$
$$2^6 \equiv 1, \ 2^7 \equiv 2, \ 2^8 \equiv 4, \ldots.$$

The powers form a periodic sequence of period 6 (that is, they repeat after 6 steps). In general, the powers of a mod n are periodic with period equal to $\text{ord}_n(a)$.

Before we go any farther, there's a fundamental question. How do we know the order of an element exists? In other words, is there *always* some exponent $m \geq 1$ such that $a^m \equiv 1 \pmod n$? The answer is "Yes," because Euler's theorem says that $a^{\phi(n)} \equiv 1 \pmod n$. Therefore, the order exists and is less than or equal to $\phi(n)$.

One of the most useful results about the order of an element is the following.

Theorem 8.1. *Let n be a positive integer, let a be an integer with $\gcd(a, n) = 1$, and let m be an integer. Then*

$$a^m \equiv 1 \pmod{n} \iff \text{ord}_n(a) \mid m.$$

Proof. For ease of notation, let $m_0 = \text{ord}_n(a)$. If $m_0 \mid m$, then $m = m_0 k$ for some k, so

$$a^m \equiv (a^{m_0})^k \equiv (1)^k \equiv 1 \pmod{n},$$

where we have used the fact that $a^{m_0} \equiv a^{\text{ord}_n(a)} \equiv 1$, by the definition of $\text{ord}_n(a)$.

Conversely, suppose $a^m \equiv 1 \pmod{n}$. Write $m = m_0 q + r$ with $0 \leq r < m_0$. Then

$$a^r \equiv a^r (1)^q \equiv a^r (a^{m_0})^q \equiv a^{m_0 q + r} \equiv a^m \equiv 1 \pmod{n}.$$

Since m_0 is the smallest positive exponent that yields 1 and r is smaller than m_0, we must have $r = 0$. Therefore, $m = m_0 q$, so $m_0 \mid m$. □

Corollary 8.2. *(a) Let p be prime and let a be an integer with $a \not\equiv 0 \pmod{p}$. Then $\text{ord}_p(a) \mid p - 1$.*
(b) Let n be a positive integer and let a be an integer with $\gcd(a, n) = 1$. Then $\text{ord}_n(a) \mid \phi(n)$.

Proof. Part (a) is a special case of (b), but we single it out since it is so useful. However, it suffices to prove (b). By Euler's theorem, $a^{\phi(n)} \equiv 1 \pmod{n}$. The theorem with $m = \phi(n)$ implies that $\text{ord}_n(a) \mid \phi(n)$. □

Example. Let's compute $\text{ord}_{23}(3)$. We could compute $3^i \pmod{23}$ for $i = 1, 2, 3, \ldots$ until we get 1, but it's faster to use the fact that $\text{ord}_{23}(3)$ divides 22 by Corollary 8.2. Therefore, $\text{ord}_{23}(3) = 1, 2, 11, 22$. Since $3^1 \not\equiv 1$ and $3^2 \not\equiv 1 \pmod{23}$, the next possibility is 3^{11}, which is 1 $\pmod{23}$. So $\text{ord}_{23}(3) = 11$.

CHECK YOUR UNDERSTANDING:
1. Compute $\text{ord}_{13}(5)$.

Section 8.1   Orders of Elements

8.1.1 Fermat Numbers

We claimed that Theorem 8.1 is useful. Let's back up our claim. Back in the 1600s, Fermat looked at the *Fermat numbers*

$$F_n = 2^{2^n} + 1.$$

The first few are $F_0 = 3$, $F_1 = 5$, $F_2 = 17$, $F_3 = 257$, $F_4 = 65537$. Fermat observed that all of these are prime and conjectured that F_n is prime for all $n \geq 0$. But in 1732, Euler showed that

$$F_5 = 4294967297 = 641 \times 6700417.$$

Here's how he did this.

Proposition 8.3. *Let $n \geq 2$ and let p be a prime dividing F_n. Then $p \equiv 1 \pmod{2^{n+2}}$.*

Proof. If $p \mid 2^{2^n} + 1$, then $2^{2^n} \equiv -1 \pmod{p}$. Squaring yields

$$\left(2^{2^n}\right)^2 = 2^{2^{n+1}} \equiv (-1)^2 \equiv 1 \pmod{p}.$$

By Theorem 8.1, $\operatorname{ord}_p(2) \mid 2^{n+1}$, which means that $\operatorname{ord}_p(2) = 2^j$ for some $j \leq n+1$. If $j \leq n$, then

$$2^{2^n} \equiv \left(2^{2^j}\right)^{2^{n-j}} \equiv 1^{2^{n-j}} \equiv 1 \pmod{p}.$$

Since $2^{2^n} \equiv -1$, this is a contradiction. Therefore, $j > n$, so $j = n+1$ and $\operatorname{ord}_p(2) = 2^{n+1}$. By Corollary 8.2, $2^{n+1} \mid p-1$.

We now have $p \equiv 1 \pmod{2^{n+1}}$. But we want $p \equiv 1 \pmod{2^{n+2}}$. First, $p \equiv 1 \pmod 8$, since $n \geq 2$. We'll show in Exercise 29 and in Chapter 10 that there exists b such that $b^2 \equiv 2 \pmod{p}$ when $p \equiv 1 \pmod 8$. We now repeat the above steps with b in place of 2:

$$b^{2^{n+1}} \equiv \left(b^2\right)^{2^n} \equiv 2^{2^n} \equiv -1 \pmod{p},$$
$$b^{2^{n+2}} \equiv \left(b^2\right)^{2^{n+1}} \equiv 2^{2^{n+1}} \equiv 1 \pmod{p}.$$

This means that $\operatorname{ord}_p(b)$ divides 2^{n+2} and does not divide 2^{n+1}, so $\operatorname{ord}_p(b) = 2^{n+2}$. Corollary 8.2 tells us that $2^{n+2} \mid p-1$. □

Here's the method Euler used to factor F_5. Let's say we're looking for prime factors of F_5. By Proposition 8.3, any such prime must be congruent to 1 mod 128. The first few primes satisfying $p \equiv 1 \pmod{128}$ are

$$257, \quad 641, \quad 769, \quad 1153, \quad 1409.$$

Division by hand (Euler didn't have an electronic calculator, only his brain) immediately shows that

$$F_5 = 641 \cdot 6700417.$$

Additionally, 6700417 is not a multiple of any of the primes listed, and these are the only primes $p \equiv 1 \pmod{128}$ that are less than $\sqrt{6700417}$. But a prime factor of 6700417 is a prime factor of F_5 and therefore must be 1 mod 128, so 6700417 has no prime factors less than $\sqrt{6700417}$. Therefore, 6700417 must be prime, and we have the complete factorization.

Remark. Euler proved only the congruence $p \equiv 1 \pmod{2^{n+1}}$, so he had to consider a few more primes than we did. The improvement to $p \equiv 1 \pmod{2^{n+2}}$ is due to Lucas in the 1800s.

What happens if we try Euler's idea on the Fermat numbers that actually are prime? Let's look at $F_4 = 65537$. If p is a prime factor of F_4, then $p \equiv 1 \pmod{64}$. The first two such primes are 193 and 257. But $193 \nmid 65537$ and $257 > \sqrt{65537}$, so F_4 is prime.

For $F_3 = 257$, any prime factor must be 1 mod 32. The first such prime is 97, which is greater than $\sqrt{257}$. So we find that F_3 is prime because there are no possible prime factors up to its square root. The same happens for F_2.

We now have some insight into why the first few Fermat numbers are prime: F_2, F_3 are not allowed any prime factors up to their square roots by Proposition 8.3. Moreover, F_4 is allowed only one possibility and it doesn't work. Finally, F_5 is allowed several and one of them is a factor. As n gets larger, there are many more potential divisors, so there is a higher chance that there is a factor. In fact, F_n is known to be composite for $5 \leq n \leq 32$ (but complete factorizations are known only up to $n = 11$).

8.1.2 Mersenne Numbers

If p is prime, the pth **Mersenne number** is

$$M_p = 2^p - 1.$$

The first few are

$$M_2 = 3, \quad M_3 = 7, \quad M_5 = 31, \quad M_7 = 127,$$
$$M_{11} = 2047 = 23 \times 89, \quad M_{13} = 8191.$$

As in the case of Fermat numbers, there is a restriction on the possible prime factors of Mersenne numbers.

Proposition 8.4. *Let p and q be primes and suppose that $q \mid 2^p - 1$. Then $q \equiv 1 \pmod{p}$.*

Proof. If $2^p \equiv 1 \pmod{q}$, then $\text{ord}_q(2) \mid p$. But p is prime, so $\text{ord}_q(2) = 1$ or p. If $\text{ord}_q(2) = 1$, then $2^1 \equiv 1 \pmod{q}$, which is impossible. Therefore, $\text{ord}_q(2) = p$. Corollary 8.2 says that $p \mid q - 1$, so $q \equiv 1 \pmod{p}$. $\square$

As of July 2013, there were 48 known Mersenne primes. The largest has $p = 57885161$ and is an integer with more than 17 million decimal digits. In Chapter 11, we'll discuss how to test Mersenne numbers for primality.

8.2 Primitive Roots

We know that the order of an integer mod p divides $p - 1$. If the order of g equals $p - 1$, we say that g is a **primitive root** for p.

Example. $\text{ord}_5(2) = 4$, so 2 is a primitive root for 5.

Example. $\text{ord}_7(2) = 3$, so 2 is not a primitive root for 7.

There is another way to characterize primitive roots. The powers of 2 mod 5 are

$$2^0 \equiv 1, \ 2^1 \equiv 2, \ 2^2 \equiv 4, \ 2^3 \equiv 3, \ 2^4 \equiv 1, \ 2^5 \equiv 2, \ 2^6 \equiv 4, \ \ldots.$$

So 1, 2, 3, 4 are congruent to powers of 2 mod 5. However, the powers of 2 mod 7 are

$$1, 2, 4, 1, 2, 4, \ldots,$$

so 3, 5, 6 are not congruent to powers of 2 mod 7.

Proposition 8.5. *Let p be a prime and assume $\gcd(g, p) = 1$. The following are equivalent:*
(a) g is a primitive root for p (so $\mathrm{ord}_p(g) = p - 1$)
(b) every integer that is nonzero mod p is congruent to a power of g mod p.

Proof. Let g be a primitive root for p. Look at the numbers $1, g, g^2, g^3, \ldots g^{p-2} \pmod{p}$. We claim they are distinct. Suppose, instead, that $g^i \equiv g^j \pmod{p}$, with $0 \leq i < j \leq p - 2$. Then $g^{j-i} \equiv 1 \pmod{p}$. By Theorem 8.1, $p - 1 = \mathrm{ord}_p(g)$ divides $j - i$. Since $0 < j - i \leq p - 2 < p - 1$, this is impossible. Therefore, the powers of g mod p give $p-1$ distinct nonzero congruence classes. Since there are only $p - 1$ nonzero congruence classes mod p, each is represented by a power of g. This proves that (a) implies (b).

Conversely, suppose that (b) is true for g. Let $m = \mathrm{ord}_p(g)$. The powers of g are

$$1, g, g^2, g^3, \ldots, g^{m-1} \pmod{p},$$

since $g^m \equiv 1 \pmod{p}$, and the powers of g start repeating at this point. Therefore, there are exactly m different powers of g mod p. Since (b) says that there are $p - 1$ numbers mod p that are congruent to powers of p, we must have $m = p - 1$, so g is a primitive root for p. □

Here is a property of primitive roots that will be useful in later chapters.

Proposition 8.6. *Let g be a primitive root for the prime p. Then*

$$g^{(p-1)/2} \equiv -1 \pmod{p}.$$

Proof. Let $x \equiv g^{(p-1)/2} \pmod{p}$. Then

$$x^2 \equiv g^{p-1} \equiv 1 \pmod{p}.$$

Therefore, $x \equiv \pm 1 \pmod{p}$ by Corollary 5.10. If $x \equiv 1 \pmod{p}$, then we have $g^{(p-1)/2} \equiv x \equiv 1$, which contradicts the fact that the order of g is $p - 1$. Therefore, $x \equiv -1$, which is what we wanted to prove. □

Section 8.2   Primitive Roots   205

What we just proved is the case $i = (p-1)/2$ of the next proposition (this proposition can be generalized; see Exercise 30).

Proposition 8.7. *Let p be prime and let g be a primitive root for p. Let i be an integer. Then*

$$ord_p(g^i) = \frac{p-1}{\gcd(i, p-1)}. \tag{8.1}$$

Proof. Let
$$m = ord_p(g^i).$$
Then $g^{im} \equiv 1 \pmod{p}$. By Corollary 8.2, $im \equiv 0 \pmod{p-1}$. Let
$$d = \gcd(i, p-1).$$
Then
$$\frac{i}{d}m \equiv 0 \pmod{\frac{p-1}{d}}.$$
Since $\gcd(i/d, (p-1)/d) = 1$, we can divide the congruence by i/d and get $m \equiv 0 \pmod{(p-1)/d}$. In particular, $m \geq (p-1)/d$.
Since i/d is an integer,
$$\left(g^i\right)^{(p-1)/d} \equiv \left(g^{p-1}\right)^{i/d} \equiv (1)^{i/d} \equiv 1 \pmod{p}.$$
(Why do we care that i/d is an integer? If it's not, then $1^{i/d}$ might have more than one possible value, and the whole calculation becomes ambiguous.) By Theorem 8.1, $m \mid (p-1)/d$, so $m \leq (p-1)/d$.
The two inequalities combine to yield $m = (p-1)/d$. This is the equation of the proposition. □

Example. An easy calculation shows that 2 is a primitive root for 13. We have $2^8 \equiv 9 \pmod{13}$. The proposition says that
$$ord_{13}(9) = \frac{12}{\gcd(8, 12)} = 3.$$

The proposition allows us to find all primitive roots for p as soon as we know one primitive root.

Corollary 8.8. *Let g be a primitive root for the prime p. The primitive roots for p are the numbers congruent to $g^i \pmod{p}$ for $\gcd(i, p-1) = 1$.*

Proof. Since g is a primitive root, every number that is nonzero mod p is congruent to g^i for some i. By Proposition 8.7, $\text{ord}_p(g^i) = p-1$ if and only if $\gcd(i, p-1) = 1$. □

Example. The numbers relatively prime to 12 are $1, 5, 7, 11$, so the primitive roots for 13 are

$$2, \quad 2^5 \equiv 6, \quad 2^7 \equiv 11, \quad 2^{11} \equiv 7.$$

Why did we stop at $i = 11$? Isn't $\gcd(17, 12) = 1$, for example? Fermat's theorem tells us that everything starts over at $2^{12} \equiv 1$, so

$$2^{17} \equiv 2^{12} \cdot 2^5 \equiv 2^5 \equiv 6 \pmod{13}.$$

Therefore, the exponents past $p - 1 = 12$ give us nothing new.

Theorem 8.9. *Let p be prime. There are $\phi(p-1)$ primitive roots g for p with $1 \leq g < p$.*

Proof. Let g be a primitive root. The other primitive roots are exactly $g^i \pmod{p}$ with $1 \leq i \leq p-1$ with $\gcd(i, p-1) = 1$. There are $\phi(p-1)$ such values of i, so we're done.

Wait! Are we done? We missed an important point. How do we know there are any primitive roots? We needed at least one to get the proof started. In Section 8.6, we'll fill in this gap and show that there is always at least one primitive root for each prime. □

Example. The first prime after 100000 is 100003. (What happened to 100001? Look back at the test for divisibility by 11.) The number of primitive roots for 100003 is

$$\phi(100002) = 28560,$$

so more than 1/4 of the numbers mod 100003 are primitive roots. The first ten primitive roots are

$$2, 3, 5, 7, 29, 31, 32, 34, 46, 47.$$

(Once we know that 2 is a primitive root, why do we know that 32 is also a primitive root? See Corollary 8.8.)

How do we find a primitive root? Since they are fairly abundant, we

Section 8.2   Primitive Roots

can look at 2, 3, 5, 6, 7, 8, ... until we find one. (Why did we omit 4? It's a square, and squares cannot be primitive roots for primes greater than 2. See Exercise 23.) But when we look at our list of *potential* primitive roots, how do we know that a number is *actually* a primitive root without checking successive powers?

Example. Suppose we want to show that 6 is a primitive root mod 41. Let $m = \text{ord}_{41}(6)$. Since $m \mid 40$ by Corollary 8.2, we see that m is one of the following:
$$1, 2, 4, 5, 8, 10, 20, 40.$$
A calculation shows that $6^{20} \equiv -1 \pmod{41}$, so $m \neq 20$. Without any more calculation, we can see that $m \neq 5$: If $6^5 \equiv 1 \pmod{41}$ then $6^{20} \equiv (6^5)^4 \equiv 1^4 \equiv 1$, which is not the case. Similarly, we see that m cannot be a divisor of 20. The only choices left are $m = 8$ and $m = 40$. A calculation shows that $6^8 \equiv 10 \pmod{41}$, so $m \neq 8$. We conclude that $m = 40$, so 6 is a primitive root for 41.

Let's analyze what we did. We needed only two computations:
$$6^{20} \not\equiv 1 \quad \text{and} \quad 6^8 \not\equiv 1 \pmod{41}.$$
The first removed all divisors of 20, namely, 1, 2, 4, 5, 10, 20, from consideration. The second removed all divisors of 8, namely, 1, 2, 4, 8, from consideration. The only possibility that remained was 40. What was so special about 20 and 8? They are 40/2 and 40/5. In other words, they are $(p-1)/\ell$, where ℓ runs through the prime divisors of $p-1$.

Proposition 8.10. *Let p be a prime and let $h \not\equiv 0 \pmod{p}$. The following are equivalent:*
(a) h is a primitive root for p.
(b) For each prime ℓ dividing $p-1$,
$$h^{(p-1)/\ell} \not\equiv 1 \pmod{p}.$$

Proof. If h is a primitive root, then
$$\text{ord}_p(h) = p - 1 > (p-1)/\ell > 0$$
for each ℓ, so
$$h^{(p-1)/\ell} \not\equiv 1 \pmod{p}.$$
Conversely, suppose (b) is true. Let $m = \text{ord}_p(h)$. We know that m

divides $p-1$. If $m \neq p-1$, let ℓ be a prime dividing $(p-1)/m$, so $\ell k = (p-1)/m$ for some k. Then

$$mk = (p-1)/\ell,$$

and
$$h^{(p-1)/\ell} \equiv (h^m)^k \equiv 1^k \equiv 1 \pmod{p},$$

which contradicts (b). Since the assumption that $m \neq p-1$ produced a contradiction, we must have $m = p-1$, which means that h is a primitive root for p. □

In practice, the proposition allows us to find primitive roots for a prime p fairly quickly, as long as we know the factorization of $p-1$. Unfortunately, if we cannot factor $p-1$, there does not seem to be a quick way of verifying that a number is a primitive root.

Up to now, we have been starting with a prime and looking for a primitive root. Let's reverse the procedure, starting with an integer and asking for primes for which it is a primitive root. The number $h = 2$ is a primitive root for the primes

$$3, 5, 11, 13, 19, \ldots,$$

and, in fact, it is a primitive root for 67 of the 168 primes less than 1000. Is 2 a primitive root for infinitely many primes? This type of question was perhaps first raised by Gauss, who asked whether there are infinitely many primes p where the decimal expansion of $1/p$ has period $p-1$. As we'll see in the next section, this is the same as asking whether 10 is a primitive root for infinitely many primes. In the 1920s, Emil Artin conjectured that if h is not a square or the negative of a square, then it is a primitive root for infinitely many primes. In fact, he conjectured that h is a primitive root for a positive proportion of primes. That is, he predicted that there is a constant C_h such that

$$\lim_{x \to \infty} \frac{\#\{p \mid p \leq x \text{ is prime and } h \text{ is a primitive root for } p\}}{\pi(x)} = C_h,$$

where $\pi(x)$ is the number of primes less than or equal to x. For $h = 2$, the conjectured constant is

$$C_2 = .3739558\cdots,$$

Section 8.3   Decimals

which is close to the ratio $67/168 = 0.3988095\cdots$ for primes up to 1000.

In 1967, Christopher Hooley proved Artin's conjecture on primitive roots, under the assumption of the Generalized Riemann Hypothesis (a generalization of the Riemann Hypothesis [see Section 15.3]; it is still unproved). In 1985, Roger Heath-Brown, inspired by work of Ram Murty and Rajiv Gupta, showed, without any unproved assumptions, that there are at most two primes that are not primitive roots for infinitely many primes. So, for example, at least one of 2, 3, 5 is a primitive root for infinitely many primes, but we don't know whether it's 2, 3, or 5 (but probably it's all three).

CHECK YOUR UNDERSTANDING:
2. Show that 2 is a primitive root for 19.
3. Find all primitive roots g for 19 with $1 < g < 19$.

8.3   Decimals

Look at the long division calculation of 1/7:

```
            0.142857
       7 ) 1.000000
            7
            ─
            30
            28
            ──
             20
             14
             ──
              60
              56
              ──
               40
               35
               ──
                50
                49
                ──
                 1
```

Now, look at the powers of 10 mod 7:

$$10^1 \equiv 3, \quad 10^2 \equiv 2, \quad 10^3 \equiv 6, \quad 10^4 \equiv 4, \quad 10^5 \equiv 5, \quad 10^6 \equiv 1.$$

Do you see anything in common? The remainders in the long division are 3, 2, 6, 4, 5, 1, 3,..., and these are the powers of 10 reduced mod 7. For example, the 6 that is obtained as the remainder in the third step of the long division is the remainder when 1000 is divided by 7. This is the same as $10^3 \equiv 6 \pmod{7}$. Moreover, when we get the remainder 1 after 6 steps, the decimal starts repeating. This corresponds to $10^6 \equiv 1 \pmod{7}$.

These observations lead to the following. We define the **period length** of a repeating decimal to be the length of the block that is repeated. For example, the decimal expansion of 1/7 has period length 6, and the decimal expansion of $1/3 = .3333\cdots$ has period length 1. If the repetition of the decimal begins immediately after the decimal point, the decimal is called **purely periodic**. So, $1/3 = .3333\cdots$ and $1/7 = .142857142857\cdots$ are purely periodic.

Proposition 8.11. *Let $m \geq 3$ satisfy $\gcd(m, 10) = 1$, and let $0 < b < m$ with $\gcd(b, m) = 1$. The period length for the decimal expansion of b/m is $\mathrm{ord}_m(10)$. Therefore, the period length for the decimal expansion of b/m divides $\phi(m)$.*

Proof. A periodic decimal expansion that repeats every n terms has the form

$$\frac{A}{10^n} + \frac{A}{10^{2n}} + \frac{A}{10^{3n}} + \cdots = \frac{A/10^n}{1 - 10^{-n}} = \frac{A}{10^n - 1}$$

with $1 \leq A \leq 10^n - 1$. (We have used the formula for the sum of a geometric series: $a + ar + ar^2 + \cdots = a/(1-r)$. See Appendix A.) For example,

$$\frac{3}{11} = .272727\cdots = \frac{27}{10^2} + \frac{27}{10^4} + \frac{27}{10^6} + \cdots = \frac{27}{10^2 - 1}.$$

If the decimal expansion of b/m repeats every n terms, then $b/m = A/(10^n - 1)$ for some A. This means that $(10^n - 1)b = Am$, so m divides $(10^n - 1)b$. Since $\gcd(m, b) = 1$, we must have $m \mid 10^n - 1$ by Proposition 1.14, so $10^n \equiv 1 \pmod{m}$.

Conversely, suppose that $10^n \equiv 1 \pmod{m}$. Then we can write $10^n -$

Section 8.3   Decimals   211

$1 = am$ for some integer a. Since $0 < b < m$, we have $0 < ab < am = 10^n - 1$. Moreover,

$$\frac{b}{m} = \frac{ab}{am} = \frac{ab}{10^n - 1} = \frac{ab}{10^n} + \frac{ab}{10^{2n}} + \frac{ab}{10^{3n}} + \cdots,$$

so the decimal expansion of b/m repeats every n terms.

We conclude that the decimal expansion of b/m repeats every n terms if and only if $10^n \equiv 1 \pmod{m}$. The smallest such positive n is the period of b/m and is $\text{ord}_m(10)$. The fact that $\text{ord}_m(10)$ divides $\phi(m)$ is Corollary 8.2. □

Examples.
$1/7 = 0.\overline{142857}$ has period length 6, and $\text{ord}_7(10) = 6$. (The notation $0.\overline{142857}$ means that 142857 is infinitely repeated, so $0.\overline{142857} = 0.142857142857\cdots$.)
$1/11 = 0.\overline{09}$ has period length 2, and $\text{ord}_{11}(10) = 2$.
$1/13 = 0.\overline{076923}$ has period length 6, and $\text{ord}_{13}(10) = 6$.
$5/21 = 0.\overline{238095}$ has period length 6, and $\text{ord}_{21}(10) = 6$.

The following well-known result is closely related to the preceding discussion. An **eventually periodic** decimal is one that starts repeating from some point onward. The repetition does not need to start at the beginning of the expansion. For example, $0.27272727\cdots$, $0.123757575\cdots$, and $0.123000000\cdots$ are all eventually periodic (the first is also purely periodic).

Theorem 8.12. *A decimal expansion of a real number x is eventually periodic if and only if x is rational.*

Proof. If the decimal expansion of x is periodic from the kth term after the decimal point onward, then $10^{k-1}x$ is an integer plus a purely periodic decimal. The purely periodic decimal can be evaluated using the formula for the sum of a geometric series, and this gives a rational number. Since $10^{k-1}x$ minus an integer is rational, so is x. For example,

$$10^2 \left(153.74123123123123\cdots\right)$$
$$= 15374 + .123123123 + \cdots$$
$$= 15374 + \frac{123}{999},$$

so
$$153.74123123123123\cdots = \frac{15374}{100} + \frac{123}{99900} = \frac{5119583}{33300},$$
which is rational.

Conversely, if x is rational, then for some k sufficiently large, $10^k x$ can be written in the form $t + r/s$ with t an integer and $0 < r < s$ and such that $\gcd(s, 10) = 1$ (Why? Simply multiply by a high enough power of 10 to remove 2s and 5s from the denominator.) Then r/s has a purely periodic decimal expansion (with period equal to $\text{ord}_s(10)$). Adding the integer t and dividing by 10^k yields a decimal expansion that is eventually periodic for x. This completes the proof. □

CHECK YOUR UNDERSTANDING:
4. Find the period length of the decimal expansion of 1/37.
5. Calculate $\text{ord}_{37}(10)$.

8.3.1 Midy's Theorem

Look at the following two decimal expansions:
$$\frac{3}{17} = 0.1764705882352941176470588235294\cdots$$
$$\frac{8}{13} = 0.615384615384615384\cdots.$$

Take the period 1764705882352941, which has length 16, split it in half, and add the two halves:

$$17647058$$
$$+82352941$$
$$99999999.$$

Awesome! Now try it with the other decimal:

$$615$$
$$+384$$
$$999.$$

These are examples of Midy's Theorem, which was discovered by the French mathematician E. Midy in 1836.

Section 8.3   Decimals

Theorem 8.13. *(Midy) Let $p \neq 2, 5$ be prime and let $0 < a < p$. Let*

$$\frac{a}{p} = 0.\overline{c_1 c_2 c_3 \cdots c_m}$$

be the decimal expansion of a/p, where m is the period (that is, the period is exactly m, not smaller). Suppose m is even: $m = 2n$. Then

$$c_1 c_2 \cdots c_n + c_{n+1} c_{n+2} \cdots c_{2n} = 99 \cdots 9$$

where there are n nines (this is addition of integers in base 10, so $c_1 c_2 \cdots c_n$ represents $c_1 10^{n-1} + c_2 10^{n-2} + \cdots + c_n$).

Proof. Let $A = c_1 c_2 \cdots c_n$ and $B = c_{n+1} c_{n+2} \cdots c_{2n}$. Then

$$\frac{a}{p} = \frac{10^n A + B}{10^{2n}} + \frac{10^n A + B}{10^{4n}} + \frac{10^n A + B}{10^{6n}} + \cdots$$

(this is just writing a repeating decimal as a geometric series). The formula for the sum of a geometric series (see Appendix A) implies that

$$\frac{1}{10^{2n}} + \frac{1}{10^{4n}} + \frac{1}{10^{6n}} + \cdots = \frac{10^{-2n}}{1 - 10^{-2n}} = \frac{1}{10^{2n} - 1}.$$

Therefore,

$$\frac{a}{p} = \frac{10^n A + B}{10^{2n} - 1} = \frac{10^n A + B}{(10^n - 1)(10^n + 1)}.$$

Multiply by $(10^n + 1)(10^n - 1)p$ to obtain

$$a(10^n + 1)(10^n - 1) = (10^n A + B)p.$$

Since $p \nmid a$, this implies that p divides $10^{2n} - 1 = (10^n - 1)(10^n + 1)$. Since p is prime (this is where we use this hypothesis), p divides $10^n - 1$ or $10^n + 1$.
If $p \mid 10^n - 1$, we have $\text{ord}_p(10) \leq n$, hence Proposition 8.11 implies that

$$2n = \text{ord}_p(10) \leq n,$$

which is impossible. Therefore, $p \nmid 10^n - 1$. We conclude that p divides $10^n + 1$. Therefore, $a(10^n + 1)/p$ is an integer. Since this integer equals $(10^n A + B)/(10^n - 1)$, we have

$$10^n A + B \equiv 0 \pmod{10^n - 1}.$$

But $10^n \equiv 1 \pmod{10^n - 1}$, so $A + B \equiv 0 \pmod{10^n - 1}$. Because A and B are n-digit numbers, $0 \le A, B \le 10^n - 1$. Moreover, at least one of A, B is nonzero (since $a \neq 0$), so $A + B > 0$. We cannot have $A = B = 10^n - 1 = 999\cdots 9$ since that gives $a/p = .999\cdots = 1$, contradicting $a < p$. Therefore, at least one of A, B is less than $10^n - 1$ and
$$0 < A + B < 2(10^n - 1).$$
The only multiple of $10^n - 1$ in this range is $10^n - 1$, so
$$A + B = 10^n - 1 = 999\cdots 9.$$

$\square$

8.4 Card Shuffling

Many phenomena with cards have mathematical explanations. Here we discuss what happens during a perfect riffle shuffle.

Start with a deck of $2n$ cards. For simplicity, we'll draw our diagrams with $2n = 6$ cards and call them A, B, C, D, E, F:

$$
\begin{array}{l}
A \; \rule{1cm}{0.4pt} \\
B \; \rule{1cm}{0.4pt} \\
C \; \rule{1cm}{0.4pt} \\
D \; \rule{1cm}{0.4pt} \\
E \; \rule{1cm}{0.4pt} \\
F \; \rule{1cm}{0.4pt}
\end{array}
$$

To shuffle the cards, split the pile into two piles of height n:

$$
\begin{array}{ll}
A \; \rule{1cm}{0.4pt} & D \; \rule{1cm}{0.4pt} \\
B \; \rule{1cm}{0.4pt} & E \; \rule{1cm}{0.4pt} \\
C \; \rule{1cm}{0.4pt} & F \; \rule{1cm}{0.4pt}
\end{array}
$$

Section 8.4   Card Shuffling

and then interlace them so that they alternate from the two piles, with the top card from the second pile, the next card from the first pile, etc. This yields

```
D  ─────
A  ─────
E  ─────
B  ─────
F  ─────
C  ─────
```

The process can be described mathematically. Number the positions of the cards from 1 to $2n$, with 1 being the top card and $2n$ being the bottom card. If a card starts in position j with $1 \leq j \leq n$, then it is part of the first pile. After the shuffle, it is in position $2j$. If a card starts in position j with $n+1 \leq j \leq 2n$, then it is part of the second pile. After the shuffle, the card that started in position $n+1$ ends up in position 1, the card that started in position $n+2$ ends up in position 3, and, in general, the card that started in position $n+i$ ends up in position $2i-1$. If we write $j = n+i$, then

$$2i - 1 = 2(j-n) - 1 = 2j - (2n+1).$$

The general rule for the change of position is therefore

$$j \mapsto 2j \text{ if } 1 \leq j \leq n, \text{ and } j \mapsto 2j - (2n+1) \text{ if } n+1 \leq j \leq 2n.$$

This can be stated as

$$j \mapsto 2j \pmod{2n+1}.$$

How many shuffles of this type does it take to get back to the original position? Since one shuffle is multiplying the positions by 2 mod $2n+1$, the effect of k shuffles is to multiply by 2^k mod $2n+1$. Therefore, after $\text{ord}_{2n+1}(2)$ shuffles, the cards return to their original positions.

For example, with 6 cards, we have $n = 3$ and $2n + 1 = 7$. Since $\text{ord}_7(2) = 3$, it takes three shuffles to return the deck to its original ordering. To check this, observe that a second shuffle orders the cards as B, D, F, A, C, E. One more shuffle yields A, B, C, D, E, F, as claimed.

With a deck of 52 cards, we have $2n+1 = 53$. Since $\text{ord}_{53}(2) = 52$, it takes 52 shuffles of this type to return the deck to its original order.

The situation is completely different if we modify the shuffling so that the top card is from the first pile, the next card is from the second pile, etc. After one shuffle of six cards, the order of the cards is A, D, B, E, C, F. Then the top card and the bottom cards always maintain their positions at the top and bottom. In fact, we are really doing a perfect riffle shuffle of the remaining four cards. In general, we are shuffling the interior $2n-2$ cards, so it takes $\text{ord}_{2n-1}(2)$ shuffles to return the deck to its original order. For six cards, we have $\text{ord}_5(2) = 4$, so it takes four shuffles. For 52 cards, we have $\text{ord}_{51}(2) = 8$, so after eight shuffles we have the original order.

8.5 The Discrete Log Problem

We can solve the exponential equation $2^x = 32$ by trial and error (or by recalling the powers of 2) and get that $x = 5$ is a solution. If we have a more difficult problem, say $2^x = 140$, we can still get an idea as to where to look for a solution by noticing that $2^7 = 128$ and $2^8 = 256$, so $7 < x < 8$. We can do this because $f(x) = 2^x$ is a *strictly increasing* function: as x gets larger, $f(x)$ gets larger as well. More precisely and more generally, if $b > 1$ and $x_1 > x_2$, then $b^{x_1} > b^{x_2}$.

On the other hand, if we fix b and n and look at the least positive residues of $b^x \pmod{n}$, their magnitudes are chaotic and apparently random for increasing values of x. As an example, the following table gives the least positive residue of $2^x \pmod{19}$ for $1 \leq x \leq 17$ (since $2^{18} \equiv 1 \pmod{19}$, we stop at $x = 17$).

x	1	2	3	4	5	6	7	8	9	10	11	12	13	14	15	16	17
2^x	2	4	8	16	13	7	14	9	18	17	15	11	3	6	12	5	10

Notice that once $x > 4$, so $2^x > 19$, there seems to be no pattern in the way that $2^x \pmod{19}$ changes: it increases and decreases in a haphazard way. This leads to the following:

Section 8.5   The Discrete Log Problem

Discrete Log Problem (DLP). Given a prime p, a primitive root g, and $h \not\equiv 0 \pmod{p}$, find x so that $g^x \equiv h \pmod{p}$.

The answer x is called the **discrete log** of h.

Example. If we want to solve $3^x = 1594323$ without mods, it's easy. We can check that $3^{10} = 59049$ and $3^{15} = 14348907$, so x is between 10 and 15. Quickly we find that $x = 13$ works. But now suppose we want to solve $3^x \equiv 8 \pmod{43}$. Since $3^1 = 3$ and $3^3 = 27$, we might want to guess that x is between 1 and 3 and then see if $x = 2$ works. It doesn't, since the same problem we encountered when looking at 2^x mod 19 is still present: the higher powers of 3 are more than 43, so reducing them mod 43 gives remainders that appear to be random. Soon we'll see slightly faster methods, but, for the present problem, a brute force search through possible exponents yields $x = 39$.

If you had somehow missed this solution, you'd find that $x = 81$ also works:
$$3^{81} \equiv 3^{42} \cdot 3^{39} \equiv 1 \cdot 3^{39} \equiv 8 \pmod{43}.$$

We have used Fermat's theorem to change 3^{42} to 1. In fact, $x = 39 + 42k$ works, for any integer k.

In general, Fermat's theorem tells us that if $g^x \equiv h \pmod{p}$, then $x + (p-1)k$ is also a solution for any integer k. Therefore, the discrete log can be regarded as a number mod $p-1$. We usually take the discrete log to have values from 0 to $p-2$.

The importance of the Discrete Log Problem for cryptography is the following:

The Discrete Log Problem is (believed to be) hard.

There is no proof of this statement. But people have tried to develop efficient algorithms to solve the discrete log problem, and so far no one has succeeded. Current technology allows computation of discrete logs for primes up to around 180 decimal digits.

In Chapter 9, we give some applications of discrete logs to cryptography.

8.6 Existence of Primitive Roots

This section is rather technical and can easily be skipped without impairing the understanding of the rest of the book.

In this section, we prove Theorem 8.9, which asserts that if p is a prime, then there exist primitive roots for p. If you know about fields, then you should observe that the proof we give also shows that a finite multiplicatively closed subset of a field forms a cyclic group.

A key ingredient is Proposition 8.14 below. In fact, it is really the only place where the primality of p is needed (since the use of Fermat's theorem late in the proof could be replaced by Euler's theorem).

Before getting to the proposition, we need a few remarks about polynomial congruences mod primes. Examples are $x^2 - 3x + 17 \equiv 0 \pmod{37}$, or $3x^4 - 2x^3 + x - 18 \equiv 0 \pmod{23}$, or $x^2 - 1 \equiv 0 \pmod{31}$.

An example that we saw in Corollary 5.10 is solving $x^2 - 1 \equiv 0 \pmod{p}$, where p is prime. Since $x^2 - 1 = (x+1)(x-1)$, this becomes $(x+1)(x-1) \equiv 0 \pmod{p}$. Recall that if a product of two numbers is 0 mod a prime, then one of the factors is 0 mod the prime. So we have either $x + 1 \equiv 0 \pmod{p}$ or $x - 1 \equiv 0 \pmod{p}$. This means that $x \equiv \pm 1 \pmod{p}$, so we have two solutions if p is odd, and one solution if $p = 2$ (since $+1 \equiv -1 \pmod{2}$).

Perhaps the most basic question is "How many solutions are there?" For the polynomials you have worked with in the past, namely polynomials with rational, real, or complex coefficients, the number of solutions in complex numbers is at most the degree of the polynomial. This is exactly what happened when we solved $x^2 - 1 \equiv 0 \pmod{p}$: there are at most two solutions mod p of this degree two polynomial.

The following result, due to Lagrange, says that a similar result holds for any polynomial mod a *prime*.

Proposition 8.14. *Let p be a prime and let $f(x)$ be a nonzero polynomial mod p of degree $n \geq 1$. Let $r_1, r_2, \ldots, r_s$ be distinct integers mod p such that $f(r_i) \equiv 0 \pmod{p}$. Then $s \leq n$. In other words, a polynomial of degree n has at most n roots.*

Remark. The lemma is not true for composite moduli. For example,

Section 8.6 Existence of Primitive Roots

the degree 2 polynomial congruence $x^2 - 1 \equiv 0 \pmod{8}$ has four solutions, namely $x \equiv 1,\ 3,\ 5,\ 7$.

Proof. If $s = 0$ (for example, if the polynomial has no roots mod p), then automatically $s \leq n$, so henceforth we assume that $s > 0$.

We prove the lemma by induction on n, the degree of the polynomial. For $n = 1$, the polynomial $f(x)$ is linear: $f(x) = a_1 x + a_0$ with $a_1 \not\equiv 0 \pmod{p}$. The congruence $a_1 x + a_0 \equiv 0 \pmod{p}$ has a unique solution mod p, so the lemma is true for $n = 1$.

We now assume that the lemma is true for $n = k$ and we'll use this to show that it is true for $n = k + 1$. So, suppose $f(x)$ has degree $k + 1$. Since

$$f(r_s) \equiv 0 \pmod{p},$$

we want to factor off $(x - r_s)$ and obtain a degree k polynomial, just like we do when working with real numbers or integers. But, since we're working mod p, we need to justify this step. Write

$$f(x) = a_{k+1} x^{k+1} + a_k x^k + \cdots + a_1 x + a_0.$$

Then, working mod p, we have

$$\begin{aligned} f(x) &\equiv f(x) - 0 \\ &\equiv f(x) - f(r_s) \\ &= a_{k+1} x^{k+1} + \cdots + a_1 x + a_0 - (a_{k+1} r_s^{k+1} + \cdots + a_1 r_s + a_0) \\ &= a_{k+1}(x^{k+1} - r_s^{k+1}) + a_k(x^k - r_s^k) + \cdots + a_1(x - r_s). \end{aligned}$$

We can factor $(x - r_s)$ from each term because

$$x^j - r_s^j = (x - r_s)(x^{j-1} + x^{j-2} r_s + \cdots + x r_s^{j-2} + r_s^{j-1}).$$

This yields

$$f(x) \equiv (x - r_s) g(x) \pmod{p}$$

for some polynomial $g(x)$ of degree k. (*Technical point:* A congruence between polynomials means that the coefficients of one polynomial are congruent to the corresponding coefficients of the other polynomial.)

If $s = 1$, then, since $k \geq 1$, we have $s \leq k$. Therefore, assume from now on that $s \geq 2$. Let $1 \leq i \leq s - 1$. Since we have assumed that $r_i \not\equiv r_s \pmod{p}$, and

$$0 \equiv f(r_i) \equiv (r_i - r_s) g(r_i) \pmod{p},$$

we must have
$$g(r_i) \equiv 0 \pmod{p}$$
for $1 \leq i \leq s-1$. (This is where we need p to be a prime. We're using Proposition 5.9, which says that if p is prime and $ab \equiv 0 \pmod{p}$, then $a \equiv 0 \pmod{p}$ or $b \equiv 0 \pmod{p}$.) Since $g(x)$ has degree k, the induction assumption implies that $s-1 \leq k$. Adding 1 yields $s \leq k+1$, as desired. Therefore, the lemma is true for $n = k+1$. By induction, the lemma is true for all n. □

It often happens that a polynomial congruence has no solutions. For example, $x(x-1)(x-2)+1 \equiv 0 \pmod{3}$ has no solutions, since plugging $x = 0, 1, 2$ into the polynomial yields 1 for each value of x. Therefore, Proposition 8.14 gives only an upper bound on the number of solutions. It says nothing about the existence of solutions. A similar situation arises when you work with polynomials whose coefficients are rational or real numbers. A quadratic polynomial has at most two rational or real roots, but may have none. For example, $x^2 - 2 = 0$ has no rational roots, while $x^2 + 1 = 0$ has no real roots.

We now start constructing a primitive root. We'll start by finding parts of a primitive root, one for each prime ℓ dividing $p-1$, and then put them together. Let ℓ be such a prime. Look at all numbers mod p whose order is a power of ℓ (including 1, which has order $\ell^0 = 1$). Among these numbers, let g_ℓ have order ℓ^k with $k \geq 0$ as big as possible for this ℓ. Then
$$g_\ell^i, \quad 0 \leq i \leq \ell^k - 1,$$
give ℓ^k distinct numbers mod p that are roots of the degree ℓ^k polynomial
$$x^{\ell^k} - 1 \equiv 0 \pmod{p}.$$
(Why are they distinct? See Exercise 16.) Since we have as many roots as the degree, Proposition 8.14 says that there are no other solutions to $x^{\ell^k} - 1 \equiv 0 \pmod{p}$. Let y be a number whose order mod p is a power of ℓ, say ℓ^j. Since k was chosen to be as large as possible, $j \leq k$, and ℓ^k is a multiple of ℓ^j. This means that
$$y^{\ell^k} \equiv \left(y^{\ell^j}\right)^{\ell^{k-j}} \equiv 1^{\ell^{k-j}} \equiv 1 \pmod{p}.$$
Since all solutions of $x^{\ell^k} - 1 \equiv 0 \pmod{p}$ are powers of g_ℓ, we see that

Section 8.6   Existence of Primitive Roots   221

y is a power of g_ℓ mod p. We have shown that g_ℓ generates all elements of ℓ-power order.

Now that we have a g_ℓ for each ℓ, we put them together:

$$g \equiv \prod_{\ell | p-1} g_\ell \pmod{p}$$

(the product is over all primes ℓ dividing $p - 1$). We'll prove that g is a primitive root.

Lemma 8.15. *Let $x \not\equiv 0 \pmod{p}$. Then x can be written as*

$$x \equiv \prod_{\ell | p-1} x_\ell \pmod{p},$$

where each x_ℓ has ℓ-power order.

Proof. Let

$$p - 1 = \prod_{i=1}^{r} \ell_i^{a_i}$$

be the prime factorization of $p - 1$. Let $L_i = (p-1)/\ell_i^{a_i}$. We claim that

$$\gcd(L_1, \ldots, L_r) = 1.$$

To prove the claim, suppose there is a prime q dividing the gcd. Since L_1 is a factor of $p - 1$ and q divides L_1, we see that q must be ℓ_i for some i. But ℓ_i does not divide L_i, so $q = \ell_i$ cannot divide the gcd. Therefore, no prime divides the gcd, so gcd $= 1$. By Theorem 1.13, there are integers $c_1, c_2, \ldots, c_r$ such that

$$c_1 L_1 + c_2 L_2 + \cdots + c_r L_r = 1.$$

Let

$$x_{\ell_i} \equiv x^{c_i L_i} \equiv x^{c_i(p-1)/\ell_i^{a_i}} \pmod{p}.$$

Then

$$x \equiv x^1 \equiv x^{c_1 L_1} x^{c_2 L_2} \cdots x^{c_r L_r} \equiv x_{\ell_1} x_{\ell_2} \cdots x_{\ell_r} \pmod{p}.$$

Moreover,

$$x_{\ell_i}^{\ell_i^{a_i}} \equiv x^{c_i(p-1)} \equiv 1 \pmod{p}$$

by Fermat's theorem. Therefore, the order of x_{ℓ_i} divides $\ell_i^{a_i}$, hence the order is a power of ℓ_i. This proves the lemma.   □

Let $x \not\equiv 0 \pmod{p}$. Write
$$x \equiv \prod x_\ell,$$
as in the lemma. Since x_ℓ has ℓ-power order, and every element of ℓ-power order is congruent to a power of g_ℓ,
$$g_\ell^{d_\ell} \equiv x_\ell \pmod{p}$$
for some d_ℓ. Write (with a slight change of notation from that used in the proof of Lemma 8.15)
$$p - 1 = \prod_{\ell \mid p-1} \ell^{a_\ell}.$$
Since $\operatorname{ord}_p(g_\ell)$ is a power of ℓ and divides $p - 1$, we have that
$$\operatorname{ord}_p(g_\ell) \mid \ell^{a_\ell}.$$
The Chinese Remainder Theorem says that there exists d such that
$$d \equiv d_\ell \pmod{\ell^{a_\ell}} \text{ for all } \ell \mid p - 1.$$
Therefore,
$$d \equiv d_\ell \pmod{\operatorname{ord}_p(g_\ell)},$$
so
$$g_\ell^d \equiv g_\ell^{d_\ell} \pmod{p}$$
(see Exercise 17). It follows that
$$g^d \equiv \prod_{\ell \mid p-1} g_\ell^d \equiv \prod_{\ell \mid p-1} g_\ell^{d_\ell} \equiv \prod_{\ell \mid p-1} x_\ell \equiv x \pmod{p}.$$
Therefore, x is a power of g mod p. Since x was arbitrary, every $x \not\equiv 0 \pmod{p}$ is a power of g, so g is a primitive root. $\square$

CHECK YOUR UNDERSTANDING

6. Using Proposition 5.9, but not quoting Proposition 8.14, explain why the only solutions to $(x-1)(x-3) \equiv 0 \pmod{19}$ are $x \equiv 1, 3 \pmod{19}$.

8.7 Chapter Highlights

1. $a^m \equiv 1 \pmod{n} \iff \text{ord}_n(a) \mid m$ (Theorem 8.1).
2. Primitive roots for primes exist.

8.8 Problems

8.8.1 Exercises

Section 8.1: Orders of Elements

1. Find the order of 5 mod 13.
2. Find the order of 4 mod 11.
3. Find the order of 14 mod 15.
4. Find the order of 3 mod 10.
5. Find the order of 3 mod 121.
6. (a) Show that the possible orders of elements mod 11 are 1, 2, 5, and 10.
   (b) Find the orders of 2 and of 5 mod 11.
7. (a) Show that the possible orders of elements mod 17 are 1, 2, 4, 8, and 16.
   (b) Find the order of 2 mod 17.
8. Let p be an odd prime and suppose b is an integer whose order mod p is 7. Show that $-b$ has order 14. (*Hint:* Start by showing that $(-b)^{14} \equiv 1$. Then use Theorem 8.1. Corollary 5.10 can be used to eliminate the possibility that the order is 2.)
9. Let n be a positive integer and let $\gcd(a, n) = 1$. Show that $\text{ord}_n(a) = 1$ if and only if $a \equiv 1 \pmod{n}$.
10. Let p be a prime and let q be an odd prime dividing $(3^p - 1)/2$. Show that $q \equiv 1 \pmod{p}$. (*Hint:* Look at the proof of Proposition 8.4.)

11. Let p be an odd prime and suppose that $p \mid 3^{32} + 1$.
    (a) Show that the order of 3 mod p is 64. (*Hint:* Look at the proof of Proposition 8.3.)
    (b) Show that $p \equiv 1 \pmod{64}$.

12. Let n be a positive integer and let $\gcd(b, n) = 1$. Show that $\operatorname{ord}_n(b^{-1}) = \operatorname{ord}_n(b)$.

13. If $a > 1$ and $n \geq 1$, show that $n \mid \phi(a^n - 1)$. (*Hint:* Show that the order of a mod $a^n - 1$ is n.)

14. Let n be a positive integer and suppose that $\operatorname{ord}_n(a) = m$. Let $k \mid m$. Show that there exists b such that $\operatorname{ord}_n(b) = k$.

15. Let m and n be positive integers with $\gcd(m, n) = 1$. Let $\gcd(a, mn) = 1$. Suppose $\operatorname{ord}_m(a) = b$ and $\operatorname{ord}_n(a) = c$. Show that $\operatorname{ord}_{mn}(a) = \operatorname{lcm}(b, c)$. (*Hint:* $a^j \equiv 1 \pmod{mn}$ if and only if $a^j \equiv 1 \pmod{m}$ and $a^j \equiv 1 \pmod{n}$. For the definition of "lcm," see Exercise 14 in Chapter 3.)

16. Let n be a positive integer and let $\gcd(b, n) = 1$. Let $d = \operatorname{ord}_n(b)$. Show that the numbers $1, b, b^2, \cdots, b^{d-1}$ are distinct mod n.

17. Let n be a positive integer and let $\gcd(b, n) = 1$. Show that $b^x \equiv b^y \pmod{n}$ if and only if $x \equiv y \pmod{\operatorname{ord}_n(b)}$. (*Hint:* Use Theorem 8.1 applied to $x - y$.)

Section 8.2: Primitive Roots

18. (a) Show that 3 is a primitive root mod 7.
    (b) Show that 3 is not a primitive root mod 11.

19. (a) Show that 2 is a primitive root mod 13.
    (b) Find all primitive roots mod 13.

20. (a) Find the order of 2 mod 31.
    (b) Is 2 a primitive root mod 23? Why or why not?

21. Find all primitive roots mod 17.

22. Given that 2 is a primitive root mod 101, find an element x mod 101 that has order 10.

23. Show that a square mod p cannot be a primitive root mod p when $p \geq 3$ is prime. (*Hint:* Show that a square has order dividing $(p-1)/2$.)

Section 8.8 Problems

24. Let p be a prime, and let g and k be integers. Show that if g^k is a primitive root for p, then g is a primitive root for p.

25. Let $p > 3$ be prime. Show that the product of the primitive roots g with $1 < g < p-1$ is 1 mod p. (*Hint:* See Exercise 12.)

26. Let p be an odd prime, and let g be a primitive root mod p.
    (a) Suppose that $g^j \equiv \pm 1 \pmod{p}$. Show that $j \equiv 0 \pmod{(p-1)/2}$.
    (b) Show that $\text{ord}_p(-g) = (p-1)/2$ or $p-1$.
    (c) Suppose $p \equiv 1 \pmod 4$. Show that $-g$ is a primitive root mod p.
    (d) Suppose $p \equiv 3 \pmod 4$. Show that $-g$ is not a primitive root mod p.

27. Let p be an odd prime.
    (a) Show that $0+1+2+\cdots+(p-2) \equiv -(p-1)/2 \equiv (p-1)/2 \pmod{p-1}$.
    (b) Let g be a primitive root mod p. Show that
    $$g^{0+1+2+\cdots+(p-2)} \equiv -1 \pmod p.$$
    (c) Use (b) to deduce Wilson's theorem.

28. Let p be prime, and let j be an integer with $j \not\equiv 0 \pmod{p-1}$. Let $S \equiv \sum_{a=1}^{p-1} a^j \pmod p$.
    (a) Let $b \not\equiv 0 \pmod p$. Show that $S \equiv \sum_{a=1}^{p-1}(ba)^j \pmod p$. (*Hint:* Lemma 6.4)
    (b) If b is a primitive root mod p, then $b^j - 1 \not\equiv 0 \pmod p$. Use this to prove that $S \equiv 0 \pmod p$.
    (c) Show that if $j \equiv 0 \pmod{p-1}$, then $\sum_{a=1}^{p-1} a^j \equiv -1 \pmod p$.

29. Let $p \equiv 1 \pmod 8$ be prime, and let g be a primitive root mod p. Let $y \equiv g^{(p-1)/8} \pmod p$.
    (a) Show that $y^4 \equiv -1 \pmod p$.
    (b) Let $x \equiv y + y^{-1}$. Show that $x^2 \equiv 2 \pmod p$.

30. Let n be a positive integer, and let $\gcd(b,n) = 1$. Let $d = \text{ord}_n(b)$. Let i be an integer. Then
$$\text{ord}_n(b^i) = \frac{d}{\gcd(i,d)}.$$

31. Let p be an odd prime, and let g be a primitive root for p. Let $a \not\equiv 0 \pmod{p}$, so $a \equiv g^i \pmod{p}$ for some i with $0 \leq i < p-1$.
    (a) Show that $x^2 \equiv a \pmod{p}$ has a solution if and only if i is even. (*Hint:* If a solution x exists, it is congruent to a power of g mod p.)
    (b) Show that there are exactly $(p-1)/2$ nonzero squares mod p. (This was also proved as Exercise 24 in Chapter 5.)

Section 8.3: Decimals

32. (a) Compute the decimal expansion of $1/39$.
    (b) Compute $\text{ord}_{39}(10)$.

33. (a) Compute the decimal expansion of $1/37$.
    (b) Compute $\text{ord}_{37}(10)$.

34. (a) Compute the decimal expansion of $11/21$.
    (b) Compute $\text{ord}_{21}(10)$.

35. Find three examples for Midy's theorem with primes that are different from the ones in the text.

36. Show by example that Midy's theorem can fail for $1/n$ when n satisfies $\gcd(10, n) = 1$ but n is composite.

37. Let m and n be integers ≥ 2 with $\gcd(mn, 10) = 1$ and $\gcd(m, n) = 1$. Let a be the period length of the decimal expansion of $1/m$, and let b be the period length for $1/n$. Show that the period length for $1/mn$ is $\text{lcm}(a, b)$.

Section 8.5: The Discrete Log Problem

38. Solve $2^x \equiv 3 \pmod{11}$.

39. Solve $3^x \equiv 5 \pmod{17}$.

40. (a) Suppose you know that $7^{57} \equiv 11 \pmod{101}$ and that $2^9 \equiv 7 \pmod{101}$. Find x such that $2^x \equiv 11 \pmod{101}$.
    (b) Solve $7^y \equiv 2 \pmod{101}$.

41. Let g be a primitive root for the odd prime p. Suppose $g^x \equiv h \pmod{p}$. Show that x is even if $h^{(p-1)/2} \equiv 1 \pmod{p}$, and x is odd if $h^{(p-1)/2} \equiv -1 \pmod{p}$. (This shows that it is possible to determine the parity of a discrete log without evaluating the log.)

Section 8.8   Problems   227

8.8.2 Computer Explorations

1. How many primes less than 1000 have 2 as a primitive root? How well does this compare with the estimate given at the end of Section 8.2? What if you count up to 10000? Up to 100000?

2. (a) Factor $2^{100} - 1$.
   (b) There are exactly three primes p with $\text{ord}_p(2) = 100$. Find them and show that your list is complete.

3. Find five different composite Mersenne numbers $M_p = 2^p - 1$ with p prime and factor them. Show that Proposition 8.4 holds for these examples.

8.8.3 Answers to "CHECK YOUR UNDERSTANDING"

1. Since $\text{ord}_{13}(5) \mid 12$, the possibilities are $1, 2, 3, 4, 6, 12$. Calculate $5^j \not\equiv 1 \pmod{13}$ for $j = 1, 2, 3$. However, $5^2 = 25 \equiv -1 \pmod{13}$, so squaring yields $5^4 \equiv 1 \pmod{13}$. Therefore, $\text{ord}_{13}(5) = 4$.

2. The possible orders of 2 mod 19 are divisors of $19 - 1 = 18$, namely $1, 2, 3, 6, 9, 18$. Since $2^j \not\equiv 1 \pmod{19}$ for $j = 1, 2, 3, 6, 9$, we must have $\text{ord}_{13}(2) = 18$, so 2 is a primitive root.

3. By Corollary 8.8, the primitive roots for 19 are $2^i \pmod{19}$ for $i = 1, 5, 7, 11, 13, 17$ (these are the numbers less than 18 that are relatively prime to 18). Since $2^5 \equiv 13$, $2^7 \equiv 14$, $2^{11} \equiv 15$, $2^{13} \equiv 3$, and $2^{17} \equiv 10 \pmod{19}$, the primitive roots for 19 are 2, 3, 10, 13, 14, 15.

4. $1/37 = .027027027\cdots$, so the period length is 3.

5. Since $10 \not\equiv 1 \pmod{37}$ and $10^2 \not\equiv 1 \pmod{37}$, but $10^3 \equiv 1 \pmod{37}$, we have $\text{ord}_{37}(10) = 3$, which agrees with the answer from 4.

6. If $(x-1)(x-3) \equiv 0 \pmod{19}$, then Proposition 5.9 implies that $x - 1 \equiv 0 \pmod{19}$ or $x - 3 \equiv 0 \pmod{19}$. This says that $x \equiv 1, 3 \pmod{19}$.

Chapter 9

More Cryptographic Applications

9.1 Diffie–Hellman Key Exchange

Imagine for a moment that you have the perfect cryptography system, completely impenetrable and totally unreadable by anyone who intercepts an encrypted message that you send. If you need to send classified information to agents thousands of miles away, they need the cryptographic key in order to read the message. This is a serious problem. You can't just call them up or e-mail them or text them — if you trusted the phone or Internet, you could just tell them the message and you would have no need of the cryptography. This is called the *Key Exchange Problem*, and, in the past, couriers often used a diplomatic pouch system to solve it. Alternatively, there could be a meeting with all relevant people present. Keys would then be distributed and the agents would go out into the field. The problem is that these methods are expensive and time consuming, and they require a lot of planning. In 1976, Whitfield Diffie and Martin Hellman proposed a key exchange method that can be done publicly but nevertheless maintains the secrecy of the key. (Their idea is more accurately a key establishment system. Nevertheless, since it is universally referred to as a key exchange protocol, we will continue to use this terminology.) We now describe their idea, which is based on simple properties of exponents and congruences.

Alice and Bob would like to exchange keys. All of their communications are over public channels, but we'll see that they are protected by the difficulty of computing discrete logarithms. Here is the protocol.

Diffie–Hellman Key Exchange

1. Alice and Bob agree on a prime p and a primitive root $g \bmod p$.
2. Alice chooses a secret a and then calculates $h_1 \equiv g^a \bmod p$.
3. Bob chooses a secret b and calculates $h_2 \equiv g^b \bmod p$.
4. Alice sends h_1 to Bob, and Bob sends h_2 to Alice.
5. Alice computes $k \equiv h_2^a \pmod{p}$.
6. Bob computes $k \equiv h_1^b \pmod{p}$.

Alice has computed $k \equiv h_2^a \equiv g^{ba}$, and Bob has computed $k \equiv h_1^b \equiv g^{ab}$. Since $ba = ab$, both Alice and Bob have the same number for k, which is now their shared key. You can see why this is really a key establishment protocol. There is no key that Alice or Bob initially have and can then exchange. Rather, working together, they create a key with the implicit assumption that this will be the key that they will use.

Remark: Eavesdroppers can intercept g, $g^a \pmod{p}$, and $g^b \pmod{p}$. If they can solve the Discrete Log Problem, they can use g and g^a to find a, and then raise g^b to the a power to compute $g^{ba} \pmod{p}$. It is not known whether $g^{ba} \pmod{p}$ can be computed without computing a discrete logarithm. The **Computational Diffie–Hellman Problem** asks for $g^{ba} \pmod{p}$, given p, g, $g^a \pmod{p}$, and $g^b \pmod{p}$.

Example. Alice and Bob need to create a key that they can use to send each other financial documents. They agree on the prime $p = 31981$ and a primitive root $g = 6$. Alice chooses her secret number to be 817, and Bob chooses his to be 23971.

$$\text{Alice calculates } 6^{817} \equiv 16978 \pmod{31981}$$
$$\text{and sends 16978 to Bob.}$$

$$\text{Bob calculates } 6^{23971} \equiv 28349 \pmod{31981}$$
$$\text{and sends 28349 to Alice.}$$

Alice now has 28349, which Bob has sent her, and Bob has 16978, which

Alice has sent him.

$$\text{Alice calculates} \quad 28349^{817} \equiv 7857 \pmod{31981}.$$
$$\text{Bob calculates} \quad 16978^{23971} \equiv 7857 \pmod{31981}.$$

So, 7857 is the key that they will both use.

CHECK YOUR UNDERSTANDING:
1. In the Diffie–Hellman Key Exchange, Alice and Bob agree on $p = 17$ and $g = 3$. Alice chooses $a = 5$ and Bob chooses $b = 4$. What number does Alice send to Bob, what number does Bob send to Alice, and what is the secret that they establish?

9.2 Coin Flipping over the Telephone

Bob and Alice decide to have dinner together. Bob wants to have Italian food, but Alice wants to go to a Thai restaurant. They decide to flip a coin to choose between the two. Unfortunately, they're talking about this over the phone and each one is concerned that the other will not be honest about the result of the coin flip. The idea behind the Diffie–Hellman Key Exchange can be used to solve this problem, enabling one person to do the coin flip and have the other be certain that it's being done fairly. Here's how the process works.

1. Alice and Bob agree on a prime p and share the factorization of $p - 1$. Of course, if they can't factor $p - 1$, they just choose a different prime.

2. Alice chooses two distinct primitive roots mod p, call them g_1 and g_2, and sends them to Bob.

3. Bob chooses a number n that is relatively prime to $p - 1$ and keeps it secret.

4. Bob now randomly picks either g_1 or g_2. (He can flip a real coin to help him choose.) Let's say he chooses g_i.

5. Bob sends $a \equiv g_i^n \pmod{p}$ to Alice.

6. Because Alice doesn't know Bob's choice of n, she can't know which primitive root he chose. For either choice of g_i, there is some n such that $g_i^n \equiv a \pmod{p}$, and Bob is keeping n secret. So, Alice makes a guess, randomly choosing either g_1 or g_2. She tells Bob her guess, and if she's correct she wins the coin toss. If she's wrong, Bob wins. Of course, if Alice doesn't trust Bob, how does she know he's not lying about her guess? (If that's the case, you may be wondering why she wants to have dinner with him.) To ensure that does not occur, Bob sends n to Alice. Let's say that Alice's guess was g_2. She can calculate $g_1^n \pmod{p}$. If this is a then Alice knows Bob was being honest.

Remarks. 1. If Bob claims that Alice loses, then it is a good idea for Alice to check that $\gcd(n, p-1) = 1$. See Exercise 7.

2. If Bob can find an x with $g_1^x \equiv g_2 \pmod{p}$, then he can always successfully claim to win: He chooses n and lets $g_2^n \equiv a \pmod{p}$. Then he knows that $g_1^{xn} \equiv a \pmod{n}$. After Alice makes her choice, he tells her that she guessed wrong. When she asks for the exponent, he sends n if she guessed g_1 or xn if she guessed g_2. Therefore, the prime p must be chosen large enough that Bob cannot solve the discrete log problem.

3. In the spirit of the previous remark, if Bob knows n_1 with $g_1^{n_1} \equiv a \pmod{p}$ and n_2 with $g_2^{n_2} \equiv a \pmod{p}$, then he finds all solutions to $x n_1 \equiv n_2 \pmod{p-1}$. One of these values of x (we know there is at least one value, so there are $\gcd(n_1, p-1)$ of them by Theorem 5.17) solves the discrete log problem $g_1^x \equiv g_2 \pmod{p}$. Therefore, if Bob cannot compute discrete logs, he should not be able to have two values of n, one for each of the two possible guesses by Alice.

4. Here is why Alice, not Bob, chooses the primitive roots. If Bob chooses them, he can let g_1 be any primitive root, choose x with $\gcd(x, p-1) = 1$, then let $g_2 \equiv g_1^x \pmod{p}$. The choice of x makes g_2 a primitive root, by Corollary 8.8. Bob can now use the method of the previous remark to cheat.

5. Suppose Bob decides that he wants to lose (there are many possible situations where this could be a prudent thing to do). Alice suspects this and asks for n. He could simply send a random exponent $r \neq n$, but then Alice could calculate g_1^r and g_2^r and see that neither is congruent to a mod p. Bob would then be in trouble. But if Bob sends the true

Section 9.3   Mental Poker   233

value of n, Alice can check which primitive root is correct. So, Bob is stuck. If he wins, he has to admit it.

Example. Alice and Bob agree on the prime $p = 675216457$. Alice chooses her two primitive roots,

$$g_1 = 30 \text{ and } g_2 = 39,$$

and sends them to Bob. Bob chooses 39 as his primitive root and $n = 35189$ as his number relatively prime to $p - 1 = 675216456$. He then sends Alice

$$a \equiv 39^{35819} \equiv 75914980 \pmod{675216457}.$$

Alice now has 75914980, but has no idea as to which primitive root it came from. She guesses 39 and Bob tells her she's correct, so she wins the coin flip. What would have happened if Alice had guessed 39 and Bob had told her she was wrong, that his primitive root had actually been 30? Then, as we said, Alice could ask for the n Bob had used and then she could check his honesty. If Bob sends her 35819, Alice would then calculate

$$30^{35819} \pmod{675216457}$$

and get 319436617, which is not the number that Bob had sent. Alice would then know that Bob was being dishonest.

Or, Bob could try sending her an n that was not the one he had actually used, say $n = 62137$. Then Alice would compute

$$30^{62137} \pmod{675216457}$$

and get 450455581, which is also not the number Bob had sent.

For a different method of flipping coins over the telephone, see Section 11.4.

9.3   Mental Poker

After flipping coins over the telephone for a few years, Alice and Bob decide to try something different. Bob suggests that they play poker, and Alice agrees. After Bob deals the first hand and tells Alice that she got a full house, Alice figures she's won. But rotten luck for her when Bob says that he got a royal flush. At this point, Alice realizes that they need a different way to play.

Alice goes to a local office supply store and buys 52 safes. She puts one card from a new deck of cards into each safe, locks them, puts them in a big box, and mails the box to Bob.

When Bob gets the safes, he has no idea as to what cards are in which safe. He randomly chooses five safes, puts his locks on each of them, and mails those five safes back to Alice.

When Alice receives the five safes, she takes her locks off so the only remaining locks are Bob's. So, when Alice mails these safes back to him, he can unlock each one and discover what cards he's been dealt. Bob, who still has 47 safes with Alice's locks on them, now sends five of them to Alice who can open them to discover her cards.

This is the end of the first round of dealing: Bob and Alice each have five cards, Alice has no locked safes at all, while Bob has 42 locked safes, each locked by Alice.

When Bob looks at his five cards, he decides to discard two. He puts the two cards in an empty safe, puts his lock on it and mails it to Alice. Bob now chooses two of the remaining 42 boxes he has (all with Alice's lock on them), puts his lock on them, and sends these two to Alice. Alice takes her lock off these two and sends them to Bob, who can remove his locks to get his cards.

If Alice wants to replace three cards, she puts them in a safe and mails it to Bob, who then takes three of his remaining 40 safes (with Alice's lock on them) and sends them to Alice. She can now unlock them to see her three new cards. They can then compare their cards and see who wins.

After the first game is over, Alice and Bob have spent thousands of dollars on safes and postage and realize that there has to be a cheaper way to play. They contact their local university and speak to the resident cryptographer, who suggests the following method.

Alice and Bob agree on a large prime p. Alice chooses a secret number α that is relatively prime to $p-1$ and then calculates α' with the property that $\alpha\alpha' \equiv 1 \bmod (p-1)$. Bob does the same thing, choosing a secret number β relatively prime to $p-1$ and then β' with $\beta\beta' \equiv 1 \bmod (p-1)$.

Section 9.3   Mental Poker

> **Alice's Secret:** α and α' with $\alpha\alpha' \equiv 1 \bmod (p-1)$.
> **Bob's Secret:** β and β' with $\beta\beta' \equiv 1 \bmod (p-1)$.

If we think of α as Alice's secret key, then α' is plays the same role as the decryption exponent d did in the RSA cryptosystem. (Of course, the same goes for Bob's β'.) To see this, we first note that $\alpha\alpha' \equiv 1 \pmod{p-1}$ means that $\alpha\alpha' = 1 + k(p-1)$ for some integer k. So, if M is any integer that is not a multiple of p,

$$M^{\alpha\alpha'} = M^{1+k(p-1)} = M^1(M^{p-1})^k \equiv M \cdot 1^k \equiv M \pmod{p}. \quad (9.1)$$

We've just used Fermat's theorem (again!), which says that if $\gcd(a,p) = 1$, then $a^{p-1} \equiv 1 \pmod{p}$. In our calculation, we assumed that $M \not\equiv 0 \pmod{p}$. If $M \equiv 0 \pmod{p}$, then we still have $M^{\alpha\alpha'} \equiv M \pmod{p}$, since both sides are $0 \pmod{p}$.

Alice and Bob now assign a distinct number to each of the 52 cards in a deck and share this assignation. So, they may decide that the ace of hearts is n_1, the ace of spades is n_2, ..., and the king of diamonds is n_{52}. Alice raises each of the 52 numbers to the αth power mod p so that she now has 52 numbers $c_1, c_2, ..., c_{52}$, where

$$c_i \equiv n_i^\alpha \pmod{p}.$$

This is the mathematical version of putting her locks on the 52 safes. Then, after randomly permuting these numbers, she sends them to Bob.

Bob now has 52 numbers and does not know what cards they correspond to. This is analogous to him holding, face down, a deck of 52 cards. He chooses five of the numbers (cards), $c_{i_1}, c_{i_2}, c_{i_3}, c_{i_4}, c_{i_5}$, raises each to the βth power mod p (the mathematical version of putting his locks on the safes) and sends these five numbers back to Alice:

$$\text{Alice has } c_{i_1}^{\alpha\beta}, \quad c_{i_2}^{\alpha\beta}, \quad c_{i_3}^{\alpha\beta}, \quad c_{i_4}^{\alpha\beta}, \quad c_{i_5}^{\alpha\beta}. \quad (9.2)$$

At this point it looks like things are a total mess. But, in fact, we're almost done because Alice can now raise each of her five numbers to the α' power. This will, in effect, strip away the α in the exponent, since, with c_{i_1} as an example, (9.1) tells us that

$$\left(c_{i_1}^{\alpha\beta}\right)^{\alpha'} \equiv c_{i_1}^{\alpha\beta\alpha'} \equiv c_{i_1}^{\alpha\alpha'\beta} \equiv \left(c_{i_1}^{\alpha\alpha'}\right)^\beta \equiv c_{i_1}^\beta \pmod{p}$$

(so Alice has removed her locks from the mathematical safes). After Alice does this, she has

$$c_{i_1}^\beta, \quad c_{i_2}^\beta, \quad c_{i_3}^\beta, \quad c_{i_4}^\beta, \quad c_{i_5}^\beta.$$

She now sends these five numbers to Bob, who can decrypt by raising each one to the β' power, and by (9.1) with β in place of α,

$$c_{i_1}^{\beta\beta'} \equiv c_{i_1} \pmod{p}$$

(so Bob has removed his locks). Finally, Bob has been dealt his five cards. Using a similar process, Alice gets her five cards and they can begin.

Example. Let's see how this works if Alice and Bob play a simpler game. There are only four cards (Jack, Queen, King and Ace), each player is dealt only one, and the winner is the person with the higher card. We'll first change the cards to numbers using $a = 01, b = 02, c = 03, ..., y = 25$, and $z = 26$. This gives us:

jack	queen	king	ace
10010311	1721050514	11091407	10305.

Alice and Bob agree on the prime $p = 12047353607$. Alice chooses her secret $\alpha = 11041$, and Bob chooses his secret $\beta = 70399$. (They both check that their respective secret numbers are relatively prime to $12047353607 - 1 = 12047353606$.) Using the Extended Euclidean Algorithm, Alice solves

$$11041\alpha' \equiv 1 \pmod{12047353606},$$

and Bob solves

$$70399\beta' \equiv 1 \pmod{12047353606}.$$

They get
$$\alpha' = 6467228043 \quad \text{and} \quad \beta' = 11651873071.$$

Alice now encrypts each of the four cards using their numerical values and her secret α. In this encryption, all congruences are modulo $p = 12047353607$:

Section 9.3   Mental Poker

$$10010311^\alpha \equiv 5674830805$$
$$1721050514^\alpha \equiv 4004945206$$
$$11091407^\alpha \equiv 11022499208$$
$$10305^\alpha \equiv 6975049274.$$

These cards get "shuffled" and sent to Bob, so he receives

11022499208,   4004945206,   6975049274,   and   5674830805.

Bob now chooses a card by choosing one of these four numbers (say, the first), raises it to the β power, and then sends the number to Alice.

Alice receives:
$$11022499208^{70399} \equiv 10159065536 \pmod{12047353607}.$$

In terms of our locked safe analogy, Alice now has a safe with two locks on it. She takes her lock off by raising 10159065536 to the α' power and sends the result to Bob:

Bob receives:
$$10159065536^{6467228043} \equiv 3733881457 \pmod{12047353607}.$$

Bob removes his lock by raising 3733881457 to the β' power:

Bob's card:
$$3733881457^{11651873071} \equiv 11091407 \pmod{12047353607}.$$

If we check, we see that Bob got the king.

Bob now chooses Alice's card by taking one of the three remaining numbers, say 6975049274, and sending it to Alice. Alice calculates

$$6975049274^{\alpha'} \equiv 6975049274^{6467228043}$$
$$\equiv 10305 \pmod{12047353607},$$

so her card is the ace and she wins this game.

If anyone suspects that someone lied about his or her cards, the outcomes can be checked by having Alice and Bob reveal α and β. The encrypted values of the cards can then be calculated and compared with the cards used during the game. However, there is another way to cheat; see Computer Exploration 2.

9.4 Digital Signatures

Alice applies for insurance. She calls up her broker, Bob, they negotiate, and he sends her a contract. Normally this contract is printed, put in an envelope, and mailed to Alice. She then reads it, signs it, and mails it back to Bob. This process may take more than a week and adds cost (paper, ink, and postage) to both Bob and Alice. Bob would like to send this contract via e-mail and have Alice approve it, saving them both time and money. He doesn't have to worry about encrypting the document itself, since it's a standard policy and there's nothing secretive about it. But what guarantee does he have that the person who replies, "Yes, this looks great" is really Alice? Maybe it's someone else in her household with access to her e-mail account. Or maybe it's malicious Eve who knows how to masquerade electronically as Alice. What's needed is a way for Alice to say, "Yes, this looks good. Signed, Alice" and have the recipient be certain that Alice's response really came from Alice. There are several methods that can be used to guarantee that Alice's response is as legally binding as a written signature. We'll discuss one that allows a digital signature to be attached to a message using RSA-type techniques.

Let's call the document m. In our example, m is the insurance policy that Bob has written for Alice. Just as with RSA, Alice chooses two large primes p and q, multiplies them together to get $n = pq$, and then chooses an e with $\gcd(e, \phi(n)) = 1$ and $1 < e < \phi(n)$. Next, she uses the Extended Euclidean Algorithm to calculate a d with the property that $ed \equiv 1 \pmod{\phi(n)}$. She publishes n and e, and keeps p, q, and d private.

After reading the policy, Alice creates her signature by raising m to the dth power mod n.

$$\text{Alice's Signature: } s \equiv m^d \pmod{n}.$$

Alice can now send m and s back to Bob. How does Bob know that it's really Alice who has agreed to the contract? He can go to the secure site where Alice's public information is kept and get her e and n. He can then raise the s he received to the eth power:

$$s^e \equiv (m^d)^e \equiv m^{ed} \equiv m \pmod{n}.$$

Section 9.4   Digital Signatures 239

Why is $m^{ed} \equiv m \pmod{n}$? The scheme is simply RSA encryption and decryption of m, applied in reverse order.

RSA Signatures

Alice's Public Information: n and e

Alice's Private Information: $p, q,$ and d

Alice's Signature: $s = m^d \pmod{n}$

Signed Document: (m, s)

Verification: $s^e \stackrel{?}{\equiv} m \pmod{n}$

Suppose Eve wants to subvert this arrangement. She tries to draw up a different contract, say m_1, and forge Alice's signature. In order to fool Bob, she needs to find a signature s_1 with the property that $s_1^e \equiv m_1 \pmod{n}$, and this is believed to be computationally infeasible. This is the same as starting with the RSA "ciphertext" m_1 and decrypting it to find the "plaintext" s_1. Since RSA is generally expected to be difficult to decrypt, Eve should not be able to forge Alice's signature. As a result, if a document is signed with Alice's signature, Alice cannot successfully deny that she signed it.

Of course, Eve could start with s_1 and compute $m_1 \equiv s_1^e \pmod{n}$. Then s_1 is a legitimate signature for m_1. But it is very unlikely that m_1 is a meaningful message.

CHECK YOUR UNDERSTANDING:

2. Alice has public RSA modulus $n = 91 = 7 \cdot 13$ and public exponent $e = 5$. She wants to sign the document $m = 7$.
(a) What is the signed document?
(b) Show that Bob's calculation verifies that the document you produced in part (a) is valid.

9.5 Chapter Highlights

1. Diffie–Hellman Key Exchange
2. RSA digital signatures

9.6 Problems

9.6.1 Exercises

Section 9.1: Diffie–Hellman Key Exchange

The next four exercises refer to the Diffie–Hellman Key Exchange. Alice and Bob agree on the given prime p and the given primitive root g. Alice chooses her secret a and Bob his secret b. In each exercise, find their agreed-upon key.

1. $p = 7$, $g = 3$, $a = 4$, $b = 5$.
2. $p = 11$, $g = 2$, $a = 8$, $b = 4$.
3. $p = 19$, $g = 2$, $a = 9$, $b = 11$.
4. $p = 23$, $g = 5$, $a = 7$, $b = 12$.

5. Why would it be foolish for Alice or Bob to chooses $p - 1$ or 1 as the secret exponent in the Diffie–Hellman Key Exchange algorithm? (That is, explain how an eavesdropper could find the key.)

6. What happens in the Diffie–Hellman Key Exchange if g is chosen so that $\text{ord}_p(g)$ is small, rather than having g be a primitive rood mod p? Explain why an eavesdropper can compute g^{ab} easily in this case.

Section 9.2: Coin Flipping over the Telephone

7. In the protocol for flipping coins, Bob is required to choose n so that $\gcd(n, p - 1) = 1$. This exercise shows why.

(a) To start with a simple case, suppose Bob chooses n to be a multiple of $p-1$, so $\gcd(n, p-1) = p-1$. Show what a must be and why Bob can follow the protocol to "prove" he was using the primitive root that Alice didn't guess. For definiteness, assume that Bob chooses $g_1^n \equiv a \pmod{p}$ and Alice guesses that Bob used g_1.

(b) More generally, let $\gcd(n, p-1) = d$. Show that

$$a^{(p-1)/d} \equiv 1 \pmod{p}. \tag{9.3}$$

(c) Show that the solutions to (9.3) are $g_2^{di} \pmod{p}$, for $0 \le i < (p-1)/d$. Therefore, a is one of these numbers.

(d) If $(p-1)/d$ is small, how can Bob produce n' with $g_2^{n'} \equiv a \pmod{p}$ and thereby claim that Alice loses?

Section 9.3: Mental Poker

8. The mental poker protocol uses a technique similar to the following method for encrypting messages. Alice and Bob agree on a large prime p. Alice chooses secret α and α' such that $\alpha\alpha' \equiv 1 \pmod{p-1}$, and Bob chooses secret β and β' such that $\beta\beta' \equiv 1 \pmod{p-1}$. Bob wants to send m to Alice. He computes $c_1 \equiv m^\beta \pmod{p}$ and sends c_1 to Alice. Alice computes $c_2 \equiv c_1^\alpha \pmod{p}$ and sends c_2 to Bob. Bob now computes a number c_3 and sends c_3 to Alice, who uses c_3 in a computation to recover m.

(a) What computation does Bob perform to produce c_3?
(b) What computation does Alice do with c_3 to recover m?
(c) Show that this method produces m. Explain how the congruences $\alpha\alpha' \equiv 1$ and $\beta\beta' \equiv 1 \pmod{p-1}$ are used.

Section 9.4: Digital Signatures

In the next two problems, Alice uses the RSA signature scheme with public modulus n and public encryption exponent e. She sends three documents to Bob with her signature attached. Eve creates a fourth document and forges Alice's signature. Which of the four documents is the forgery? Each ordered pair is of the form (m, s), where m is the document and s is the signature. Do not factor n and do not find the private exponent used to sign.

9. Public modulus $n = 55$ and public encryption modulus $e = 7$.

Signed documents:
(6, 51), (7, 13), (8, 19), (9, 14).

10. Public modulus $n = 51$ and public encryption modulus $e = 5$.
    Signed documents:
    (2, 32), (5, 20), (7, 24), (9, 42).

11. Alice uses the RSA signature scheme with primes $p = 3, q = 11$, and $e = 7$.
    (a) What is n, the public modulus she will use?
    (b) What is her private exponent d?
    (c) What is the signature s for the message $m = 17$?

12. Alice is using the RSA signature scheme with primes $p = 5, q = 11$, and $e = 9$.
    (a) What is n, the public modulus she will use?
    (b) What is her private exponent d?
    (c) What is the signature s for the message $m = 7$?

13. Alice is using the RSA signature scheme with primes $p = 3, q = 17$, and $e = 5$.
    (a) What is n, the public modulus she will use?
    (b) What is her private exponent d?
    (c) What is the signature s for the message $m = 14$?

14. Alice is using the RSA signature scheme with primes $p = 3, q = 17$, and $e = 7$.
    (a) What is n, the public modulus she will use?
    (b) What is her private exponent d?
    (c) What is the signature s for the message $m = 10$?

15. Alice is using the RSA signature scheme with primes $p = 7, q = 13$, and $e = 5$. She decides that $s = 11$ is a nice signature, but she needs a message to sign. What message m has $s = 11$ as its valid signature?

16. Alice is using the RSA signature scheme with primes $p = 7, q = 13$, and $e = 7$. Since 9 rhymes with "sign," she decides that $s = 9$ is a good signature, but she needs a message to sign. What message m has $s = 9$ as its valid signature?

17. Alice is using the RSA signature scheme with primes $p = 3, q = 19$, and $e = 11$. She wants to sign the message $m = 987654321$ (because she wants to certify that this is the correct launch sequence for a rocket). Since m is much larger than

the modulus $n = 57$, she computes $m_0 \equiv m \pmod{11}$, with $0 \leq m_0 \leq 10$ and signs m_0. Show that her signature is $s = 47$. (*Note:* Usually, if a document m is larger than the modulus n, Alice does not sign m. Instead, she compresses m into a number less than n using what is known as a *hash function*. In this exercise, she uses a much simpler function, namely, reduction mod 11.)

9.6.2 Computer Explorations

1. The prime 12347 has 2 as a primitive root. Suppose I tell you that $2^x \equiv 8938 \pmod{12347}$ and $2^y \equiv 9620 \pmod{12347}$, but I don't tell you x and y. Is $2^{xy} \equiv 7538 \pmod{12347}$? Is $2^{xy} \equiv 7557 \pmod{12347}$?

   The **Decisional Diffie–Hellman Problem** asks the following question: Given g, g^a, g^b, and h, is $g^{ab} \equiv h \pmod{p}$? It is not known whether this can be solved without first solving the Computational Diffie–Hellman Problem. (Recall that this asks you, given g, g^a, and g^b, to compute g^{ab}.)

2. Here's how to cheat at poker when playing with the protocol given in Section 9.3. We'll use the simplified version with only four cards (see the end of Section 9.3) in order to keep the numbers small.

   (a) Let $p = 12047353607$. Compute $x^{(p-1)/2} \pmod{p}$ for each card x (so $x = 10010311$ for the jack, etc.). Notice that the jack and king yield $-1 \pmod{p}$, while the queen and ace yield $+1 \pmod{p}$.

   (b) Compute $y^{(p-1)/2} \pmod{p}$ for each encrypted card (so $y = 5674830805$ for the jack, etc.). What do you observe? Can you explain it? (*Hint:* α is odd.)

   (c) Find a new prime p such that the ace can be separated from the other cards by this procedure.

9.6.3 Answers to "CHECK YOUR UNDERSTANDING"

1. Alice computes $3^5 \equiv 5 \pmod{17}$, and Bob computes $3^4 \equiv 13 \pmod{17}$, so Alice sends 5 to Bob and Bob sends 13 to Alice.

Then Alice computes $13^5 \equiv 13 \pmod{17}$, and Bob computes $5^4 \equiv 13 \pmod{17}$. Their common secret is 13.

2. (a) Alice's secret RSA decryption exponent satisfies $5d \equiv 1 \pmod{6 \cdot 12}$. Therefore, $d = 29$ (use the Extended Euclidean Algorithm). Alice's signature is $s \equiv m^d \equiv 7^{29} \equiv 63 \pmod{91}$. The signed document is $(m, s) = (7, 63)$.

   (b) Bob verifies by computing $s^e \equiv 63^5 \equiv 7 \pmod{91}$. Since $m = 7$, the signature is valid.

Chapter 10

Quadratic Reciprocity

10.1 Squares and Square Roots Mod Primes

We say that a number a is a square mod n if $x^2 \equiv a \pmod{n}$ has a solution. For example, 2 is a square mod 7, since $3^2 \equiv 2 \pmod 7$, and -1 is a square mod 5, since $2^2 \equiv -1 \pmod 5$. However, 2 is not a square mod 3 because $x^2 \not\equiv 2 \pmod 3$ for $x = 0, 1, 2$.

If a is a square mod n, we say that a is a **quadratic residue** mod n. If not, a is a **quadratic nonresidue**.

Proposition 10.1. *Let p be an odd prime and let $a \not\equiv 0 \pmod p$. Then $a^{(p-1)/2} \equiv \pm 1 \pmod p$. Moreover,*

$$a \text{ is a square mod } p \iff a^{(p-1)/2} \equiv 1 \pmod p.$$

Proof. Let $b \equiv a^{(p-1)/2} \pmod p$. Then $b^2 \equiv a^{p-1} \equiv 1 \pmod p$ by Fermat's theorem. By Corollary 5.10, $b \equiv \pm 1 \pmod p$, so $a^{(p-1)/2} \equiv \pm 1 \pmod p$.

If a is a square mod p, we have $x^2 \equiv a$ for some x, so

$$a^{(p-1)/2} \equiv (x^2)^{(p-1)/2} \equiv x^{p-1} \equiv 1 \pmod p,$$

by Fermat's theorem. Conversely, suppose $a^{(p-1)/2} \equiv 1 \pmod p$. Let g be a primitive root mod p. Then $g^i \equiv a$ for some i, so

$$1 \equiv a^{(p-1)/2} \equiv g^{i(p-1)/2} \pmod p.$$

Since a primitive root has order $p - 1$, Theorem 8.1 says that $p - 1$ divides $i(p-1)/2$. This implies that $(p-1)k = i(p-1)/2$ for some k, so $i = 2k$. Therefore, $a \equiv g^i \equiv (g^k)^2$, so a is a square mod p. □

Definition 10.2. *Let p be an odd prime and let a be an integer with $a \not\equiv 0 \pmod{p}$. Define the **Legendre symbol***

$$\left(\frac{a}{p}\right) = \begin{cases} +1 & \text{if } x^2 \equiv a \pmod{p} \text{ has a solution} \\ -1 & \text{if } x^2 \equiv a \pmod{p} \text{ has no solution.} \end{cases}$$

Examples. The examples at the beginning of this section can be restated as

$$\left(\frac{2}{7}\right) = +1, \quad \left(\frac{-1}{5}\right) = +1, \quad \left(\frac{2}{3}\right) = -1.$$

Proposition 10.3. *Let p be an odd prime, and let $a, b \not\equiv 0 \pmod{p}$. Then*
(a) (Euler's Criterion)

$$\left(\frac{a}{p}\right) \equiv a^{(p-1)/2} \pmod{p}$$

(b)

$$\left(\frac{a}{p}\right)\left(\frac{b}{p}\right) = \left(\frac{ab}{p}\right).$$

(c) If $a \equiv b \pmod{p}$, then $\left(\frac{a}{p}\right) = \left(\frac{b}{p}\right).$

(d)

$$\left(\frac{-1}{p}\right) = \begin{cases} +1 & \text{if } p \equiv 1 \pmod{4} \\ -1 & \text{if } p \equiv 3 \pmod{4}. \end{cases}$$

Proof. If a is a square mod p, then Proposition 10.1 says that $a^{(p-1)/2} \equiv +1 = \left(\frac{a}{p}\right) \pmod{p}$. If a is not a square mod p, then Proposition 10.1 says that $a^{(p-1)/2} \equiv -1 = \left(\frac{a}{p}\right) \pmod{p}$. This proves (a).

The congruence of (a) also holds for b and for ab. Therefore,

$$\left(\frac{a}{p}\right)\left(\frac{b}{p}\right) \equiv a^{(p-1)/2} b^{(p-1)/2} = (ab)^{(p-1)/2} \equiv \left(\frac{ab}{p}\right) \pmod{p}.$$

Section 10.1   Squares and Square Roots Mod Primes   247

Both ends of the above string of congruences are ± 1. Since $-1 \not\equiv +1 \pmod{p}$ when $p \geq 3$, the two ends must be equal. This proves (b).

If $a \equiv b \pmod{p}$, then $x^2 \equiv a \pmod{p}$ has a solution if and only if $x^2 \equiv b \pmod{p}$ has a solution. This is exactly what (c) says.

For (d), note that $(p-1)/2$ is even if $p \equiv 1 \pmod{4}$ and is odd if $p \equiv 3 \pmod{4}$. Therefore,

$$\left(\frac{-1}{p}\right) \equiv (-1)^{(p-1)/2} = \begin{cases} +1 & \text{if } p \equiv 1 \pmod{4} \\ -1 & \text{if } p \equiv 3 \pmod{4}. \end{cases}$$

$\square$

A useful fact is that $\left(\frac{x^2}{p}\right) = 1$ if $p \nmid x$. This is because x^2 is automatically a square mod p. It also follows immediately from part (b) of the proposition since $\left(\frac{x^2}{p}\right) = \left(\frac{x}{p}\right)^2 = (\pm 1)^2 = 1$.

The following is the major result about the Legendre symbol.

Theorem 10.4. *(Quadratic Reciprocity) (a) Let p and q be distinct odd primes. Then*

$$\left(\frac{q}{p}\right) = (-1)^{(p-1)(q-1)/4} \left(\frac{p}{q}\right).$$

In other words,

$$\left(\frac{q}{p}\right) = \begin{cases} \left(\frac{p}{q}\right) & \text{if at least one of } p, q \text{ is } 1 \pmod{4} \\ -\left(\frac{p}{q}\right) & \text{if } p \equiv q \equiv 3 \pmod{4}. \end{cases}$$

(b) Let p be an odd prime. Then

$$\left(\frac{-1}{p}\right) = (-1)^{(p-1)/2} = \begin{cases} +1 & \text{if } p \equiv 1 \pmod{4} \\ -1 & \text{if } p \equiv 3 \pmod{4}. \end{cases}$$

(c) Let p be an odd prime. Then

$$\left(\frac{2}{p}\right) = (-1)^{(p^2-1)/8} = \begin{cases} +1 & \text{if } p \equiv 1, 7 \pmod{8} \\ -1 & \text{if } p \equiv 3, 5 \pmod{8}. \end{cases}$$

Remarks. Parts (b) and (c) are called the *Supplementary Laws*. We will prove Quadratic Reciprocity and also the Supplementary Law for 2 in Section 10.5. We proved the Supplementary Law for -1 in Proposition 10.3. The Law of Quadratic Reciprocity is one of the high points of number theory. It was conjectured by Euler (1744) and Legendre (1785). In 1796, when he was 18 years old, Gauss gave the first proof. Gauss published six proofs during his lifetime, and two more were found in his unpublished manuscripts. Attempts to generalize Quadratic Reciprocity inspired the development of algebraic number theory in the 1800s and 1900s, and greatly influenced modern areas of research such as elliptic curves and modular forms from the 1900s up to the present.

Quadratic Reciprocity is the deepest theorem that we will meet in this book. Let's consider congruences and linear Diophantine equations as being at the easy level. The results are not surprising, although they are quite interesting and useful. Fermat's theorem and Euler's theorem are at the next level and are perhaps the first nontrivial results in number theory. In modern times, they are regarded as special cases of theorems in group theory. Finally, Quadratic Reciprocity is a very non-obvious theorem. There does not seem to be any heuristic reason that q being a square mod p should be related to p being a square mod q.

Example. Is 5 a square mod 37? Let's check:

$$\left(\frac{5}{37}\right) = \left(\frac{37}{5}\right) = \left(\frac{2}{5}\right) = -1.$$

The first equality is because of Quadratic Reciprocity. Note that the sign doesn't change because $5 \equiv 1 \pmod{4}$ (and also $37 \equiv 1 \pmod 4$). The second equality is because $37 \equiv 2 \pmod 5$ (see Proposition 10.3(c)). For the third equality, it can be verified easily that 2 is not a square mod 5. Alternatively, use Theorem 10.4(c). Since $\left(\frac{5}{37}\right) = -1$, we conclude that 5 is not a square mod 37.

Example. Is 7 a square mod 43? Let's check:

$$\left(\frac{7}{43}\right) = -\left(\frac{43}{7}\right) = -\left(\frac{1}{7}\right) = -1.$$

The first equality is because of Quadratic Reciprocity. Note that the sign changes because $7 \equiv 3 \pmod 4$ and $7 \equiv 3 \pmod 4$. The second equality is because $43 \equiv 1 \pmod 5$ (see Proposition 10.3(c)). For the

Section 10.1   Squares and Square Roots Mod Primes   249

third equality, note that 1 is a square mod 7, so $\left(\frac{1}{7}\right) = 1$. Since $\left(\frac{7}{43}\right) = -1$, we conclude that 7 is a square mod 43.

Example. Is 23 a square mod 419? Let's check:

$$\begin{aligned}
\left(\frac{23}{419}\right) &= -\left(\frac{419}{23}\right) \quad \text{(since } 23 \equiv 419 \equiv 3 \pmod{4}\text{)} \\
&= -\left(\frac{5}{23}\right) \quad \text{(since } 419 \equiv 5 \pmod{23}\text{)} \\
&= -\left(\frac{23}{5}\right) \quad \text{(since } 5 \equiv 1 \pmod{4}\text{)} \\
&= -\left(\frac{3}{5}\right) \quad \text{(since } 23 \equiv 3 \pmod{5}\text{)} \\
&= -\left(\frac{5}{3}\right) \quad \text{(since } 5 \equiv 1 \pmod{4}\text{)} \\
&= -\left(\frac{2}{3}\right) \quad \text{(since } 5 \equiv 2 \pmod{3}\text{)} \\
&= -(-1) = +1 \quad \text{(by the Supplementary Law for 2).}
\end{aligned}$$

Therefore, 23 has a square root mod 419. In the next section, we'll see how to discover that $182^2 \equiv 23 \pmod{419}$, but for the moment, let's note that Quadratic Reciprocity is nonconstructive. We find out that a square root exists, without any information about how to find it.

Example. Let's see whether 246 is a square mod the prime 257. By Proposition 10.3,

$$\left(\frac{246}{257}\right) = \left(\frac{2}{257}\right)\left(\frac{3}{257}\right)\left(\frac{41}{257}\right).$$

The Supplementary Law tells us that

$$\left(\frac{2}{257}\right) = +1.$$

By Quadratic Reciprocity,

$$\left(\frac{3}{257}\right) = \left(\frac{257}{3}\right) = \left(\frac{2}{3}\right) = -1$$

and
$$\left(\frac{41}{257}\right) = \left(\frac{257}{41}\right) = \left(\frac{11}{41}\right) = \left(\frac{41}{11}\right)$$
$$= \left(\frac{8}{11}\right) = \left(\frac{2}{11}\right)^3 = (-1)^3 = -1.$$

Therefore,
$$\left(\frac{246}{257}\right) = (+1)(-1)(-1) = +1,$$

which means that 246 is a square mod 257.

Since $246 \equiv -11 \pmod{257}$, we could have done the calculation another way:
$$\left(\frac{246}{257}\right) = \left(\frac{-11}{257}\right) = \left(\frac{-1}{257}\right)\left(\frac{11}{257}\right).$$

Since $257 \equiv 1 \pmod 4$, the Supplementary Law for -1 says that
$$\left(\frac{-1}{257}\right) = +1.$$

By Quadratic Reciprocity,
$$\left(\frac{11}{257}\right) = \left(\frac{257}{11}\right) = \left(\frac{4}{11}\right) = +1,$$

since 4 is a square mod anything, in particular mod 11. Therefore,
$$\left(\frac{246}{257}\right) = (+1)(+1) = +1.$$

Example. Is 295 a square mod 401?
$$\left(\frac{295}{401}\right) = \left(\frac{5}{401}\right)\left(\frac{59}{401}\right).$$

By Quadratic Reciprocity,
$$\left(\frac{5}{401}\right) = \left(\frac{401}{5}\right) = \left(\frac{1}{5}\right) = +1$$

Section 10.1 Squares and Square Roots Mod Primes

and

$$\left(\frac{59}{401}\right) = \left(\frac{401}{59}\right) = \left(\frac{47}{59}\right) = -\left(\frac{59}{47}\right) = -\left(\frac{12}{47}\right)$$
$$= -\left(\frac{4}{47}\right)\left(\frac{3}{47}\right) = -\left(\frac{3}{47}\right) = +\left(\frac{47}{3}\right) = \left(\frac{2}{3}\right) = -1.$$

Therefore,

$$\left(\frac{295}{401}\right) = (+1)(-1) = -1,$$

so 295 is not a square mod 401.

In the preceding examples, we factored numbers in order to apply quadratic reciprocity. In Section 10.4, we'll show that factorization is not necessary. In essence, we can pretend that every odd integer is prime and use this to evaluate Legendre symbols.

Quadratic Reciprocity is a useful computational tool, as we have seen, but it has a deeper significance. Let's consider an example. For which primes p is 5 a square mod p? We can look at 5 mod p for each p and get a list of such primes. Will there be any pattern for such p? Is there an easy way to list them? Quadratic Reciprocity gives the answer:

$$\left(\frac{5}{p}\right) = \left(\frac{p}{5}\right) = \begin{cases} +1 \text{ if } p \equiv \pm 1 \pmod{5} \\ -1 \text{ if } p \equiv \pm 2 \pmod{5}. \end{cases}$$

The primes for which 5 is a quadratic residue form two congruence classes mod 5, namely, 1 mod 5 and 4 mod 5. The miracle of Quadratic Reciprocity is that it takes a problem that depends on 5 mod p and changes it to a problem that depends on congruence conditions for p mod 5.

Let's look at another example. For which primes p is 3 a square mod p? It might be surprising, but the answer depends on p mod 12, not just p mod 3:

If $p \equiv 1 \pmod{12}$, we have

$$\left(\frac{3}{p}\right) = \left(\frac{p}{3}\right) = \left(\frac{1}{3}\right) = 1.$$

The first equality is because $p \equiv 1 \pmod{12}$ implies that $p \equiv 1 \pmod 4$. The second is because also $p \equiv 1 \pmod 3$.

If $p \equiv 5 \pmod{12}$, we have

$$\left(\frac{3}{p}\right) = \left(\frac{p}{3}\right) = \left(\frac{2}{3}\right) = -1.$$

If $p \equiv 7 \pmod{12}$, we have

$$\left(\frac{3}{p}\right) = -\left(\frac{p}{3}\right) = -\left(\frac{1}{3}\right) = -1.$$

In this case, there is a "$-$" in the first equality because both 3 and p are 3 mod 4.

If $p \equiv 11 \pmod{12}$, we have

$$\left(\frac{3}{p}\right) = -\left(\frac{p}{3}\right) = -\left(\frac{2}{3}\right) = 1.$$

As we see, it is necessary to consider the congruence class of p both mod 3 and mod 4, which means that we are looking at p mod 12 (by the Chinese Remainder Theorem). This was not necessary in the previous case, where we asked for the primes p for which 5 is a square mod p. This is because 5 is 1 mod 4, while 3 is 3 mod 4. Therefore, when we apply Quadratic Reciprocity, the "$-$" never occurs for 5 (because one of the primes, namely 5, is 1 mod 4), but it can occur for 3, depending on the congruence class of p mod 4.

More generally, if a is a nonzero integer and we ask for which primes p is a a square mod p, the answer depends only on the congruence class of p mod $4a$. This is one of the key steps (Lemma 10.11) in the proof of Quadratic Reciprocity. When $a \equiv 1 \pmod{4}$, this can be simplified to a congruence on p mod a, as in the above case with $a = 5$.

If $f(x) = ax^2 + bx + c$ is a polynomial with integer coefficients, Quadratic Reciprocity implies that the set of primes p for which $f(x) \equiv 0 \pmod{p}$ has a solution can be described by congruence conditions on p. This is because $f(x) \equiv 0 \pmod{p}$ has a solution when $b^2 - 4ac$ is a square mod p (see Section 10.3), and Quadratic Reciprocity implies that the set of primes p for which this is true can be given by congruence conditions.

CHECK YOUR UNDERSTANDING:

1. Show that $\left(\frac{3}{13}\right) = 1$ by (a) finding a solution of $x^2 \equiv 3 \pmod{13}$ and (b) using Quadratic Reciprocity.

10.2 Computing Square Roots Mod p

Suppose we have used Quadratic Reciprocity and proved that a certain number is a square mod p. How do we find a square root? In this section, we treat two cases where there are simple formulas.

Proposition 10.5. *Let $p \equiv 3 \pmod 4$ be prime. Let $x \not\equiv 0 \pmod p$ be an integer. Then exactly one of x and $-x$ is a square mod p. Let*

$$y \equiv x^{(p+1)/4} \pmod p.$$

Then $y^2 \equiv \pm x \pmod p$.

Proof. Since $p \equiv 3 \pmod 4$, Proposition 10.3 says that $\left(\frac{-1}{p}\right) = -1$, so

$$\left(\frac{-x}{p}\right) = \left(\frac{-1}{p}\right)\left(\frac{x}{p}\right) = -\left(\frac{x}{p}\right).$$

Therefore, one of $\left(\frac{x}{p}\right)$ and $\left(\frac{-x}{p}\right)$ is $+1$ and the other is -1. This means that exactly one of x and $-x$ is a square mod p.

Let $y \equiv x^{(p+1)/4} \pmod p$. Then

$$y^2 \equiv \left(x^{(p+1)/4}\right)^2 \equiv x^{(p+1)/2} \equiv x^{(p-1)/2} x \equiv (\pm 1) x \pmod p,$$

since $x^{(p-1)/2} \equiv \pm 1$ by Proposition 10.1. □

Example. Let $p = 19$ and $x = 10$. Then $(p+1)/4 = 5$, so

$$y \equiv x^{(p+1)/4} \equiv 10^5 \equiv 3 \pmod{19}.$$

Then $y^2 \equiv 9 \equiv -10 \equiv -x \pmod{19}$.

If $p \equiv 1 \pmod 4$, it is still possible to find square roots mod p quickly, but the procedure is more complicated. As an example, we have the following.

Proposition 10.6. *Let $p \equiv 5 \pmod 8$ be prime and let $x \not\equiv 0 \pmod p$ be an integer. If $x \equiv y^2 \pmod p$ then*

$$y \equiv \begin{cases} \pm x^{(p+3)/8} & \text{if } x^{(p-1)/4} \equiv 1 \pmod p \\ \pm 2^{(p-1)/4} x^{(p+3)/8} & \text{if } x^{(p-1)/4} \equiv -1 \pmod p. \end{cases}$$

Proof. Note that $x^{(p-1)/4} \equiv y^{(p-1)/2} \equiv \pm 1 \pmod p$, so the two cases of the theorem are the only possibilities.

First, let's assume that $x^{(p-1)/4} \equiv 1$. Then

$$\left(x^{(p+3)/8}\right)^2 \equiv x^{(p+3)/4} \equiv x^{(p-1)/4} x \equiv 1 \cdot x \equiv y^2 \pmod p,$$

which implies that $\pm x^{(p+3)/8} \equiv y \pmod p$.

Now, suppose that $x^{(p-1)/4} \equiv -1$. Then

$$\left(2^{(p-1)/4} x^{(p+3)/8}\right)^2 \equiv 2^{(p-1)/2} x^{(p-1)/4} x$$
$$\equiv \left(\frac{2}{p}\right)(-1) y^2 \equiv y^2 \pmod p,$$

since the Supplementary Law for 2 says that $(2/p) = -1$ when $p \equiv 5 \pmod 8$. Again, this gives the formula in the proposition. $\square$

Example. We can check via Quadratic Reciprocity that 10 is a square mod 13:

$$\left(\frac{10}{13}\right) = \left(\frac{2}{13}\right)\left(\frac{5}{13}\right) = (-1)\left(\frac{13}{5}\right) = -\left(\frac{3}{5}\right) = +1.$$

Let's find a square root: $10^{(p-1)/4} \equiv 1000 \equiv -1 \pmod{13}$, so Proposition 10.6 says that a square root is given by

$$2^{(p-1)/4} 10^{(p+3)/8} \equiv 8 \cdot 100 \equiv 7 \pmod{13}.$$

CHECK YOUR UNDERSTANDING:

2. Solve $y^2 \equiv 5 \pmod{19}$.

10.3 Quadratic Equations

Recall the quadratic formula: If $ax^2 + bx + c = 0$ with $a \neq 0$, then

$$x = \frac{-b \pm \sqrt{b^2 - 4ac}}{2a}.$$

This formula is derived by completing the square, and the same method works mod p.

Let p be an odd prime. We want to solve

$$ax^2 + bx + c \equiv 0 \pmod{p},$$

where a, b, c are integers with $a \not\equiv 0 \pmod{p}$. Since $p \nmid a$, we can divide by a and get

$$x^2 + \frac{b}{a}x + \frac{c}{a} \equiv 0.$$

Since p is odd, we can divide by 2 and write

$$\left(x + \frac{b}{2a}\right)^2 + \left(\frac{c}{a} - \frac{b^2}{4a^2}\right) \equiv 0.$$

This rearranges to

$$\left(x + \frac{b}{2a}\right)^2 \equiv \frac{b^2 - 4ac}{4a^2}.$$

There is a solution exactly when $b^2 - 4ac$ is a square mod p. If it is a square mod p, we can take the square root. This yields the following.

Proposition 10.7. *Let p be an odd prime and let a, b, c be integers with $p \nmid a$. Then $ax^2 + bx + c \equiv 0 \pmod{p}$ has a solution mod p if and only if $b^2 - 4ac$ has a square root mod p. If this is the case, the solutions are*

$$x \equiv \frac{-b \pm \sqrt{b^2 - 4ac}}{2a} \pmod{p}.$$

Example. Consider

$$x^2 + 5x + 7 \equiv 0 \pmod{13}.$$

Then $b^2 - 4ac = -3 \equiv 6^2 \pmod{13}$. The solutions are
$$x \equiv \frac{-5 \pm 6}{2} \equiv \frac{1}{2}, \frac{-11}{2} \equiv 7, 1 \pmod{13}.$$
As explained in Section 5.6, we change $1/2 \pmod{13}$ to $7 \pmod{13}$ because $2 \cdot 7 \equiv 1 \pmod{13}$.

Example. Consider $x^2 + 11x + 7 \equiv 0 \pmod{1009}$ (note: 1009 is prime). We have $b^2 - 4ac = 93$. We could try to find a square root of 93 mod 1009, but we should first use Quadratic Reciprocity to see if this is even possible:

$$\left(\frac{93}{1009}\right) = \left(\frac{3}{1009}\right)\left(\frac{31}{1009}\right) = \left(\frac{1009}{3}\right)\left(\frac{1009}{31}\right) = \left(\frac{1}{3}\right)\left(\frac{17}{31}\right)$$
$$= 1 \cdot \left(\frac{31}{17}\right) = \left(\frac{-3}{17}\right) = \left(\frac{-1}{17}\right)\left(\frac{3}{17}\right) = 1 \cdot \left(\frac{17}{3}\right) = \left(\frac{2}{3}\right) = -1.$$

Fortunately, we didn't try to look for a square root since our calculation shows that there isn't one. So this quadratic equation doesn't have any solutions mod 1009.

CHECK YOUR UNDERSTANDING:

3. Solve $x^2 + 3x - 2 \equiv 0 \pmod{13}$.

10.4 The Jacobi Symbol

Let m be a positive odd integer. Write the prime factorization of m as
$$m = p_1^{a_1} p_2^{a_2} \cdots p_r^{a_r}.$$
Let $\gcd(b, m) = 1$. Define the **Jacobi symbol**
$$\left(\frac{b}{m}\right) = \left(\frac{b}{p_1}\right)^{a_1} \left(\frac{b}{p_2}\right)^{a_2} \cdots \left(\frac{b}{p_r}\right)^{a_r},$$
where $\left(\frac{b}{p_i}\right)$ is the Legendre symbol. For example,
$$\left(\frac{2}{15}\right) = \left(\frac{2}{3}\right)\left(\frac{2}{5}\right) = (-1)(-1) = +1.$$

Section 10.4   The Jacobi Symbol   257

Since 2 is not a square mod 3, it cannot be a square mod 15. Therefore, we issue the following:

Warning: $(b/m) = +1$ does not imply that b is a square mod m.

What, then, is it good for? For our purposes, it allows us to compute Legendre symbols without factoring, as we'll see shortly as a consequence of the following result.

Theorem 10.8. *(Reciprocity for Jacobi symbols) Let m, n be positive odd integers with $\gcd(m, n) = 1$.*
(a)
$$\left(\frac{m}{n}\right) = (-1)^{(m-1)(n-1)/4} \left(\frac{n}{m}\right).$$

In other words,

$$\left(\frac{m}{n}\right) = \begin{cases} \left(\dfrac{n}{m}\right) & \text{if at least one of } m, n \text{ is } 1 \pmod 4 \\ -\left(\dfrac{n}{m}\right) & \text{if } m \equiv n \equiv 3 \pmod 4. \end{cases}$$

(b) (Supplementary Law for -1)

$$\left(\frac{-1}{m}\right) = (-1)^{(m-1)/2} = \begin{cases} +1 & \text{if } m \equiv 1 \pmod 4 \\ -1 & \text{if } m \equiv 3 \pmod 4. \end{cases}$$

(c) (Supplementary Law for 2)

$$\left(\frac{2}{m}\right) = (-1)^{(m^2-1)/8} = \begin{cases} +1 & \text{if } m \equiv 1, 7 \pmod 8 \\ -1 & \text{if } m \equiv 3, 5 \pmod 8. \end{cases}$$

Proof. Let's start with (a). Write $m = m_1 m_2^2$ and $n = n_1 n_2^2$, with m_1 and n_1 squarefree. Then

$$\left(\frac{m}{n}\right) = \left(\frac{m_1}{n}\right)\left(\frac{m_2}{n}\right)^2 = \left(\frac{m_1}{n}\right)$$

$$= \left(\frac{m_1}{n_1 n_2^2}\right) = \left(\frac{m_1}{n_1}\right)\left(\frac{m_1}{n_2}\right)^2 = \left(\frac{m_1}{n_1}\right),$$

and similarly

$$\left(\frac{n}{m}\right) = \left(\frac{n_1}{m_1}\right).$$

We'll prove (a) for m_1 and n_1. Since the square of an odd number is 1 mod 8, we have $m \equiv m_1 \pmod 8$ and $n \equiv n_1 \pmod 8$. This means that
$$(m-1)(n-1)/4 \equiv (m_1-1)(n_1-1)/4 \pmod 2$$
and
$$(-1)^{(m-1)(n-1)/4} = (-1)^{(m_1-1)(n_1-1)/4},$$
so the pluses and minuses work the same for m_1 and n_1 as for m and n. Therefore, if we prove (a) for m_1 and n_1 then

$$\begin{aligned}\left(\frac{m}{n}\right) &= \left(\frac{m_1}{n_1}\right) \\ &= (-1)^{(m_1-1)(n_1-1)/4}\left(\frac{n_1}{m_1}\right) \\ &= (-1)^{(m-1)(n-1)/4}\left(\frac{n_1}{m_1}\right) \\ &= (-1)^{(m-1)(n-1)/4}\left(\frac{n}{m}\right),\end{aligned}$$

so we obtain (a) for m, n.

Let the prime factorizations of m_1 and n_1 be
$$m_1 = p_1 p_2 \cdots p_r, \qquad n_1 = q_1 q_2 \cdots q_s.$$
Then
$$\left(\frac{m_1}{n_1}\right) = \prod_{j=1}^{s}\left(\frac{m_1}{q_j}\right) = \prod_{i=1}^{r}\prod_{j=1}^{s}\left(\frac{p_i}{q_j}\right)$$
and
$$\left(\frac{n_1}{m_1}\right) = \prod_{i=1}^{r}\left(\frac{n_1}{p_i}\right) = \prod_{i=1}^{r}\prod_{j=1}^{s}\left(\frac{q_j}{p_i}\right).$$

To change from (m_1/n_1) to (n_1/m_1), we use Quadratic Reciprocity for primes to change each (p_i/q_j) to (q_j/p_i). We pick up a factor of -1 exactly when $p_i \equiv q_j \equiv 3 \pmod 4$. Therefore,
$$\left(\frac{m_1}{n_1}\right) = (-1)^t \left(\frac{n_1}{m_1}\right),$$
where t is the number of pairs (p_i, q_j) where $p_i \equiv q_j \equiv 3 \pmod 4$. The number of such pairs is $t_1 t_2$, where t_1 is the number of primes $p_i \equiv 3$

Section 10.4   The Jacobi Symbol

(mod 4) dividing m_1 and t_2 is the number of primes $q_j \equiv 3 \pmod 4$ dividing n_1.

Suppose t_1 and t_2 are odd. The product of an odd number of primes that are 3 mod 4 is 3 mod 4, so in this case, $m_1 \equiv n_1 \equiv 3 \pmod 4$. Also, $t = t_1 t_2$ is odd. Therefore,

$$\left(\frac{m_1}{n_1}\right) = -\left(\frac{n_1}{m_1}\right).$$

Now suppose that at least one of t_1 and t_2 is even. The product of an even number of primes that are 3 mod 4 is 1 mod 4, so at least one of m_1 and n_1 is 1 mod 4. Also, $t = t_1 t_2$ is even. Therefore,

$$\left(\frac{m_1}{n_1}\right) = \left(\frac{n_1}{m_1}\right).$$

We conclude that $(m_1/n_1) = (n_1/m_1)$ if at least one of m_1 and n_1 is 1 mod 4, and $(m_1/n_1) = -(n_1/m_1)$ if both are 3 mod 4. This is exactly statement (a) for m_1 and n_1. As we showed above, this yields the result for m and n.

The reasoning for (b) is similar. Write $m = m_1 m_2^2$ as in the proof of (a) and let

$$m_1 = p_1 p_2 \cdots p_r$$

be the prime factorization of m_1. Then $m \equiv m_1 \pmod 4$ and

$$\left(\frac{-1}{m}\right) = \left(\frac{-1}{m_1}\right) = \prod_{i=1}^{r}\left(\frac{-1}{p_i}\right) = (-1)^{t_1},$$

where t_1 is the number of prime factors $p_i \equiv 3 \pmod 4$. As in the proof of (a), we see that t_1 is even if and only if $m_1 \equiv 1 \pmod 4$, so

$$\left(\frac{-1}{m_1}\right) = (-1)^{(m_1-1)/2} = (-1)^{(m-1)/2},$$

which yields (b).

The proof of (c) is very similar to that of (b), but we need the following easily verified facts. A product of primes $p \equiv \pm 1 \pmod 8$ is ± 1 (mod 8). A product of an even number of primes $p \equiv \pm 3 \pmod 8$ is

$\pm 1 \pmod 8$. A product of an odd number of primes $p \equiv \pm 3 \pmod 8$ is $\pm 3 \pmod 8$. By the reasoning in (b), we have

$$\left(\frac{2}{m}\right) = \left(\frac{2}{m_1}\right) = \prod_{i=1}^{r}\left(\frac{2}{p_i}\right) = (-1)^{u_1},$$

where u_1 is the number of prime factors $p_i \equiv \pm 3 \pmod 8$. Since u_1 is odd exactly when m_1 (and therefore m) is $\pm 3 \pmod 8$, we obtain the equality of the first and third parts of statement (c). The facts that $(-1)^{(m^2-1)/8} = +1$ if $m \equiv \pm 1 \pmod 8$ and $(-1)^{(m^2-1)/8} = -1$ if $m \equiv \pm 3 \pmod 8$ are easily verified (for example, $m \equiv \pm 3 \pmod 8 \Rightarrow m^2 \equiv 9 \pmod{16} \Rightarrow (m^2-1)/8 \equiv 1 \pmod 2$). □

Example. Let's calculate $\left(\frac{246}{257}\right)$. We did this in Section 10.1, but we needed to factor 123 as 3×41. With the Jacobi symbol, this factorization is unnecessary. The only factorization involves 2:

$$\left(\frac{246}{257}\right) = \left(\frac{2}{257}\right)\left(\frac{123}{257}\right)$$

$$= (+1)\left(\frac{123}{257}\right) \quad \text{(since } 257 \equiv 1 \pmod 8\text{)}$$

$$= \left(\frac{257}{123}\right) \quad \text{(by Reciprocity for the Jacobi symbol)}$$

$$= \left(\frac{11}{123}\right) \quad \text{(since } 257 \equiv 11 \pmod{123}\text{)}$$

$$= -\left(\frac{123}{11}\right) \quad \text{(since } 123 \equiv 11 \equiv 3 \pmod 4\text{)}$$

$$= -\left(\frac{2}{11}\right) = -(-1) = +1.$$

Notice how much easier this was than when we did the calculation in Section 10.1 using only the Legendre symbol. With larger numbers, the advantage becomes even greater, since we do not need to check whether odd numbers are primes and we do not need to factor odd integers, which can be extremely difficult for large numbers. Just remember not to assign too much meaning to the results of intermediate steps. For example, our calculation shows that $(11/123) = +1$. This does not mean that 11 is a square mod 123. In fact, it is not, since 11 is not a square mod 3.

CHECK YOUR UNDERSTANDING:

4. Compute $\left(\frac{501}{1013}\right)$.
5. Show that $\left(\frac{3}{35}\right) = +1$ but $x^2 \equiv 3 \pmod{35}$ has no solutions.

10.5 Proof of Quadratic Reciprocity

The proof of Quadratic Reciprocity is rather technical and can easily be skipped without impairing the understanding of the rest of the book.

Recall the proof of Fermat's theorem. We had a prime p and an $a \not\equiv 0 \pmod{p}$. We computed $a^{p-1} \cdot (p-1)!$ and showed that this was a rearrangement of $(p-1)! \bmod p$. Dividing by $(p-1)!$ yielded $a^{p-1} \equiv 1 \pmod{p}$. In order to work with Legendre symbols, Proposition 10.3 says that we need to be able to work with $a^{(p-1)/2}$. The natural thing to do is compute $a^{(p-1)/2} \cdot ((p-1)/2)!$ and see if this is a rearrangement of $((p-1)/2)! \bmod p$. This is almost the case. Let's analyze what occurs when $p = 11$ and $a = 7$.

$$\begin{aligned} 7^5 \cdot 1 \cdot 2 \cdot 3 \cdot 4 \cdot 5 &\equiv 7 \cdot 14 \cdot 21 \cdot 28 \cdot 35 \pmod{11} \\ &\equiv 7 \cdot 3 \cdot 10 \cdot 6 \cdot 2 \\ &\equiv (-4) \cdot 3 \cdot (-1) \cdot (-5) \cdot 2 \\ &\equiv (-1)^3 \cdot 1 \cdot 2 \cdot 3 \cdot 4 \cdot 5 \pmod{11}. \end{aligned}$$

Divide by $1 \cdot 2 \cdot 3 \cdot 4 \cdot 5$ to get

$$7^5 \equiv (-1)^3 \pmod{11}.$$

By Proposition 10.3, this says that

$$\left(\frac{7}{11}\right) \equiv 7^5 \equiv (-1)^3 \equiv -1 \pmod{11}.$$

Observe that we changed the signs of the numbers in the second line that were between $11/2$ and 11, and we changed the product to $(-1)^3 \cdot 5!$, which allowed us to compute $\left(\frac{7}{11}\right)$. The following result shows that this computation is part of a general phenomenon.

Lemma 10.9. *(Gauss's Lemma) Let p be an odd prime and let $a \not\equiv 0 \pmod{p}$. Let n be the number of integers in the set*

$$\{a, 2a, 3a, \ldots, (p-1)a/2\}$$

that are congruent to an integer between $p/2$ and p. Then

$$\left(\frac{a}{p}\right) = (-1)^n.$$

For example, when $p = 11$ and $a = 7$, we have

$$\{7,\ 2 \cdot 7,\ 3 \cdot 7,\ 4 \cdot 7,\ 5 \cdot 7\}.$$

Reducing the numbers in this set modulo 11, we get

$$\{7, 3, 10, 6, 2\}.$$

Since $11/2 < k < 11$ for $k = 7, 10, 6$, we have $n = 3$.

Proof. Let $0 < i < (p-1)/2$. We claim that there exists a unique j with $0 < j < p/2$ such that

$$aj \equiv \pm i \pmod{p}.$$

Since $\gcd(a, p) = 1$, there exists j' with $0 < j' < p$ such that $aj' \equiv i \pmod{p}$. If $0 < j' < p/2$, we're done: let $j = j'$. If $p/2 < j' < p$, let $j = p - j'$. Then $0 < j < p/2$ and

$$aj \equiv a(p - j') \equiv -aj' \equiv -i \pmod{p}.$$

Therefore, j exists.

Now suppose that $aj_1 \equiv \pm aj_2 \pmod{p}$, with $0 < j_1, j_2 < p/2$. Divide by a to get

$$j_1 \equiv \pm j_2 \pmod{p}.$$

Since $0 < j_1 + j_2 < p$, we cannot have $j_1 + j_2 \equiv 0 \pmod{p}$, so $j_1 \not\equiv -j_2 \pmod{p}$. Therefore, $j_1 \equiv j_2 \pmod{p}$. Since $0 < j_1, j_2 < p$, we must have $j_1 = j_2$. Therefore, the solution of $aj \equiv \pm i \pmod{p}$ is unique. This proves the claim.

Consider the numbers $a, 2a, 3a, \ldots, (p-1)a/2$. The claim says that if

Section 10.5   Proof of Quadratic Reciprocity   263

we change the signs of n of them, we obtain numbers congruent to each of the numbers from 1 to $(p-1)/2$. Therefore,

$$a \cdot 2a \cdot 3a \cdots (p-1)a/2 \equiv (-1)^n ((p-1)/2)! \pmod{p}.$$

Separating out the factors of a yields

$$a^{(p-1)/2} \cdot ((p-1)/2)! \equiv (-1)^n ((p-1)/2)! \pmod{p}.$$

Dividing by $((p-1)/2)!$ yields $a^{(p-1)/2} \equiv (-1)^n \pmod{p}$. Proposition 10.3 implies

$$\left(\frac{a}{p}\right) \equiv a^{(p-1)/2} \equiv (-1)^n \pmod{p}.$$

This completes the proof. □

Gauss's Lemma immediately allows us to compute $(2/p)$. The numbers in the lemma are

$$2, 4, 6, \ldots, (p-1).$$

Approximately half are less than $p/2$ and the others lie between $p/2$ and p. We need to determine the middle point precisely.

If $p = 1+4k$, then $2, 4, 6, \ldots, 2k < p/2$ and $p/2 < 2k+2, 2k+4, \ldots, 4k < p$. Therefore, $n = k = (p-1)/4$. If $p \equiv 1 \pmod{8}$, then $n = (p-1)/4$ is even, and if $p \equiv 5 \pmod{8}$, then $n = (p-1)/4$ is odd, so

$$\left(\frac{2}{p}\right) = \begin{cases} +1 \text{ if } p \equiv 1 \pmod{8} \\ -1 \text{ if } p \equiv 5 \pmod{8}. \end{cases}$$

If $p = 3+4k$, then $2, 4, 6, \ldots, 2k < p/2$ and $p/2 < 2k+2, 2k+4, \ldots, 4k+2 < p$. Therefore, $n = k+1 = (p+1)/4$. If $p \equiv 3 \pmod{8}$, then $n = (p+1)/4$ is odd, and if $p \equiv 7 \pmod{8}$, then $n = (p+1)/4$ is even, so

$$\left(\frac{2}{p}\right) = \begin{cases} -1 \text{ if } p \equiv 3 \pmod{8} \\ +1 \text{ if } p \equiv 7 \pmod{8}. \end{cases}$$

This completes the proof of the Supplementary Law for 2.

We're now ready to tackle the main Quadratic Reciprocity Law. We need another expression for n.

Lemma 10.10. *Let a be a positive integer and let j_0 be the largest even integer less than or equal to a. Let n be as in Lemma 10.9. Then n is the number of integers in the union of the intervals*

$$(j-1)p/(2a) < x < jp/(2a)$$

as j runs through the even integers $2 \leq j \leq j_0$.

Example. Let $p = 11$ and $a = 7$. Then $j_0 = 6$, so j takes the values $2, 4, 6$. We have the intervals

$$11/14 < x < 22/14, \quad 33/14 < x < 44/14, \quad 55/14 < x < 66/14.$$

The integers 1, 3, 4 are in these intervals, so $n = 3$.

Proof. To evaluate n, we need to find when ai is congruent to some k with $p/2 < k < p$. If

$$(j-1)p/2 < ai < jp/2$$

for an even integer j, then $ai \equiv k \pmod{p}$ with $p/2 < k < p$. For example, if $5p/2 < ai < 6p/2$, then $p/2 < ai - 2p < p$, and $ai \equiv ai - 2p \pmod{p}$.

Note that

$$(j-1)p/2 < ai < jp/2 \iff (j-1)p/(2a) < i < jp/(2a).$$

Therefore, recalling that j must be even, we must count integers i that lie in intervals $(j-1)p/(2a) < x < jp/(2a)$ for even integers j. But which integers j do we use?

We have $ai \leq a(p-1)/2$. Recall that j_0 is the largest even integer less than or equal to a, so

$$j_0 \leq a \leq j_0 + 1$$

($a = j_0$ if a is even, and $a = j_0 + 1$ if a is odd). Then

$$a(p-1)/2 < (j_0 + 1)p/2,$$

so the last ai, namely $a(p-1)/2$, lies before the interval $(j_0 + 1)p/2 < x < (j_0 + 2)p/2$. Therefore, we can stop at $j = j_0$ or before. Since

$$a(p+1)/2 > j_0 p/2,$$

the next multiple of a past those we are counting, namely $a(p+1)/2$, lies past $pj_0/2$. The conclusion is that if we stop at the interval $(j_0-1)p/2 < x < j_0p/2$, then we count all the ai up to $a(p-1)/2$ that lie in the desired intervals, but we do not count $a(p+1)/2$ and beyond. This completes the proof of the lemma. □

The following lemma is a key step in the proof of Quadratic Reciprocity.

Lemma 10.11. *Let p and q be odd primes and let a be a positive integer with $\gcd(a, pq) = 1$. If $p \equiv \pm q \pmod{4a}$, then*

$$\left(\frac{a}{p}\right) = \left(\frac{a}{q}\right).$$

This result says something rather surprising. We know that (a/p) depends only on $a \pmod{p}$, but the lemma says that it also depends only on $p \pmod{4a}$. This is a congruence condition for the lower argument, not the usual upper argument. For example, we know that $(3/7) = -1$. Since $127 \equiv 7 \pmod{4 \cdot 3}$, the lemma immediately tells us that $(3/127) = -1$, too. Why should the fact that 3 is not a square mod 7 imply that 3 is also not a square mod 127? This is the mystery of Quadratic Reciprocity.

Proof. First, suppose that $p \equiv q \pmod{4a}$. Write $q = p + 4am$ for some integer m. Let the notation be as in the previous lemma, and let n_p be the n for a and p, as in Lemma 10.9, and let n_q be the n for a and q. Consider an interval

$$(j-1)p/(2a) < i < jp/(2a).$$

The corresponding interval for q is

$$(j-1)q/(2a) < i < jq/(2a).$$

Substituting $q = p + 4am$, we obtain

$$(j-1)p/(2a) + 2(j-1)m < i < jp/(2a) + 2jm.$$

This is just the interval for p shifted to the right by $2(j-1)m$ and then lengthened by $2m$. If there are $n_{p,j}$ integers i in the interval for p, then there are $n_{q,j} = n_{p,j} + 2m$ integers i in the interval for q. In particular,

$$n_{p,j} \equiv n_{q,j} \pmod{2}$$

for all j. Therefore,

$$n_p = n_{p,2} + n_{p,4} + \cdots + n_{p,j_0} \equiv n_{q,2} + n_{q,4} + \cdots + n_{q,j_0} = n_q \pmod{2}.$$

It follows that

$$\left(\frac{a}{p}\right) = (-1)^{n_p} = (-1)^{n_q} = \left(\frac{a}{q}\right).$$

Now assume that $p \equiv -q \pmod{4a}$. Write $p+q = 4at$ for some integer t. Let $n_{q,j}$ denote the number of integers i satisfying

$$(j-1)q/(2a) < i < jq/(2a).$$

This interval can be rewritten as

$$2t(j-1) - (j-1)p/(2a) < i < 2tj - jp/(2a).$$

We can shift this by $2tj$ and multiply by -1 to find that $n_{q,j}$ is the number of integers i satisfying

$$jp/(2a) < i < 2t + (j-1)p/(2a).$$

Since $n_{p,j}$ is the number of integers i satisfying

$$(j-1)p/(2a) < i < jp/(2a),$$

we can put these two together and find that $n_{p,j} + n_{q,j}$ is the number of integers satisfying

$$(j-1)p/(2a) < i < (j-1)p/(2a) + 2t.$$

This is an interval of length $2t$, and neither endpoint is an integer, since $(j-1)p$ is odd and $2a$ is even. Therefore, there are $2t$ integers in the interval, so $n_{p,j} + n_{q,j} = 2t$. This implies that

$$n_{p,j} \equiv n_{q,j} \pmod{2}.$$

As in the case $p \equiv q \pmod{4a}$, this implies that $n_p \equiv n_q \pmod{2}$ and therefore

$$\left(\frac{a}{p}\right) = (-1)^{n_p} = (-1)^{n_q} = \left(\frac{a}{q}\right).$$

$\square$

Section 10.5 Proof of Quadratic Reciprocity

The proof of Quadratic Reciprocity is almost done. Suppose first that $p \equiv q \pmod{4}$. Write $q = p + 4\ell$. Then

$$\left(\frac{q}{p}\right) = \left(\frac{4\ell}{p}\right) \quad \text{(since } q \equiv 4\ell \pmod{p}\text{)}$$
$$= \left(\frac{4}{p}\right)\left(\frac{\ell}{p}\right) = \left(\frac{\ell}{p}\right) \quad \text{(since 4 is a square)}$$
$$= \left(\frac{\ell}{q}\right) \quad \text{(since } p \equiv q \pmod{4\ell}\text{, so Lemma 10.11 applies)}$$
$$= \left(\frac{4\ell}{q}\right)$$
$$= \left(\frac{-p}{q}\right) \quad \text{(since } 4\ell \equiv -p \pmod{q}\text{)}.$$

If $p \equiv q \equiv 1 \pmod{4}$, then $(-1/q) = 1$, so

$$\left(\frac{q}{p}\right) = \left(\frac{-p}{q}\right) = \left(\frac{-1}{q}\right)\left(\frac{p}{q}\right) = \left(\frac{p}{q}\right).$$

If $p \equiv q \equiv 3 \pmod{4}$, then $(-1/q) = -1$, and the same calculation yields

$$\left(\frac{q}{p}\right) = -\left(\frac{p}{q}\right).$$

Now suppose $p \equiv -q \pmod{4}$. Write $p + q = 4w$. Then

$$\left(\frac{q}{p}\right) = \left(\frac{4w}{p}\right) \quad \text{(since } q \equiv 4w \pmod{p}\text{)}$$
$$= \left(\frac{4}{p}\right)\left(\frac{w}{p}\right) = \left(\frac{w}{p}\right) \quad \text{(since 4 is a square)}$$
$$= \left(\frac{w}{q}\right) \quad \text{(by Lemma 10.11)}$$
$$= \left(\frac{4w}{q}\right) \quad \text{(since 4 is a square)}$$
$$= \left(\frac{p}{q}\right) \quad \text{(since } p \equiv 4w \pmod{q}\text{)}.$$

This completes the proof of Quadratic Reciprocity. □

10.6 Chapter Highlights

1. The Legendre symbol: $\left(\frac{a}{p}\right) = +1$ if a is a square mod p and $= -1$ if a is not a square mod p.
2. Quadratic Reciprocity (Theorem 10.4).
3. The Jacobi symbol.

10.7 Problems

10.7.1 Exercises

Section 10.1: Squares and Square Roots Mod Primes

1. Show that $\left(\frac{3}{11}\right) = 1$ by finding a solution of $x^2 \equiv 3 \pmod{11}$.

2. Evaluate $\left(\frac{38}{79}\right)$.

3. Evaluate $\left(\frac{3}{17}\right)$.

4. Evaluate $\left(\frac{31}{103}\right)$.

5. Does $x^2 \equiv 19 \pmod{101}$ have any solutions? (Justify your answer; do not find the solutions if they exist.)

6. Does $x^2 \equiv 23 \pmod{79}$ have any solutions? (Justify your answer; do not find the solutions if they exist.)

7. (a) Does $x^2 \equiv 2 \pmod{209}$ have solutions? Note that $209 = 11 \cdot 19$ is not prime. Justify your answer.
   (b) Does $x^2 \equiv 47 \pmod{209}$ have solutions? Justify your answer.

8. (a) Show that $\left(\dfrac{7}{107}\right) = -1$.
   (b) Show that $7^{53} \equiv -1 \pmod{107}$. (*Hint:* $(107-1)/2 = 53$.)
   (c) Find the order of 7 mod 107.

9. Suppose $p = 2^{(2^m)} + 1$ is prime, with $m \geq 1$.
   (a) Show that $p \equiv 2 \pmod 3$.
   (b) Show that $\left(\dfrac{3}{p}\right) = -1$.
   (c) Show that the order of 3 mod p is a power of 2.
   (d) Show that the order of 3 mod p is $p-1$, so 3 is a primitive root for every Fermat prime.
   (e) Use the method of (c) and (d) to show that if p is a Fermat prime and a satisfies $\left(\dfrac{a}{p}\right) = -1$, then a is a primitive root mod p.

10. Let p and q be primes with $p \equiv q \pmod{28}$. Use Quadratic Reciprocity (not Lemma 10.11) to show that $\left(\dfrac{7}{p}\right) = \left(\dfrac{7}{q}\right)$.

11. Let p and q be odd primes with $p \equiv q \pmod 5$. Show that $\left(\dfrac{5}{p}\right) = \left(\dfrac{5}{q}\right)$.

12. (a) Let p be a prime with $p \equiv 9 \pmod{28}$. Show that $\left(\dfrac{7}{p}\right) = 1$.
    (b) More generally, let p and q be primes with $p \equiv 9 \pmod{4q}$. Show that $\left(\dfrac{q}{p}\right) = 1$.

13. Let $p \equiv 1 \pmod 3$ be prime and $b \not\equiv 0 \pmod p$. Show that b is a cube mod p if and only if $b^{(p-1)/3} \equiv 1 \pmod p$. (*Hint:* Imitate the proof of Proposition 10.1.)

14. Let $p \equiv 2 \pmod 3$ be prime and let $b \not\equiv 0 \pmod p$. Let $x \equiv b^{(2p-1)/3} \pmod p$. Show that $x^3 \equiv b \pmod p$. This shows that every number mod p is a cube when $p \equiv 2 \pmod 3$.

15. Give congruence conditions mod 20 for primes p for which -5 is a square mod p. (*Hint:* Look at the end of Section 10.1.)

16. (a) Show that if $n \geq 5$ then $1! + 2! + 3! + \cdots + n! \equiv 3 \pmod 5$.

(b) Show that $1! + 2! + 3! + \cdots + n!$ is a square for $n = 1$ and $n = 3$, but is not a square for all other values of $n \geq 1$.

17. (a) Let p be an odd prime and let $n \geq p$. Suppose that $q = n! + 1$ is prime. Show that $\left(\dfrac{p}{q}\right) = 1$, except when $p = 3$ and $q = 7$.
(b) There are 11 values of n with $1 \leq n \leq 200$ such that $n! + 1$ is prime. Find them. For each such n, except $n = 1, 2, 3$, produce p and q that give an example for part (a).

18. Let p be a prime with $p \equiv 3 \pmod 4$, and suppose that $q = 2p + 1$ is also prime.
(a) Show that 2 is a square mod q.
(b) Show that $2^p \equiv 1 \pmod q$, so $q \mid 2^p - 1$. (This explains the fact that $23 \mid 2^{11} - 1$.)
It is not yet proved that there are infinitely many primes p such that $2p + 1$ is prime. It is also not yet proved that there are infinitely many primes p for which M_p is composite.

19. Let p be an odd prime, and let q be a prime divisor of $2^p - 1$.
(a) Show that 2 is a square mod q. (*Hint:* What is $\left(\dfrac{2}{q}\right)^p$?)
(b) Show that $q \equiv \pm 1 \pmod 8$.

20. Prove that there are infinitely many primes $p \equiv 3 \pmod 8$. (*Hint:* Let $N = (p_1 p_2 \cdots p_n)^2 + 2$. Show that $N \equiv 3 \pmod 8$ and that N must have a prime factor $p \not\equiv 1 \pmod 8$.)

21. Prove that there are infinitely many primes $p \equiv 7 \pmod 8$. (*Hint:* Let $N = (p_1 p_2 \cdots p_n)^2 - 2$. Show that $N \equiv 7 \pmod 8$ and that N must have a prime factor $p \not\equiv 1 \pmod 8$.)

22. (a) Let $p \equiv 2 \pmod 3$. Show that $\left(\dfrac{-3}{p}\right) = -1$. (*Hint:* Consider the cases $p \equiv 1 \pmod 4$ and $p \equiv 3 \pmod 4$ separately.)
(b) Prove that there are infinitely many primes $p \equiv 1 \pmod 3$. (*Hint:* Let $N = (p_1 p_2 \cdots p_n)^2 + 3$.)

23. Let n be an integer, and let $p \neq 3$ be a prime dividing $n^2 + n + 1$.
(a) Show that p is odd.
(b) Show that $p \mid (2n + 1)^2 + 3$.
(c) Use Part (a) of Exercise 22 to conclude that $p \equiv 1 \pmod 3$.

24. Let p be an odd prime, and let g be a primitive root mod p. Let
$$S = \sum_{j=1}^{p-1} \left(\frac{j}{p}\right).$$

(a) Show that
$$S = \sum_{j=1}^{p-1} \left(\frac{gj}{p}\right).$$

(b) Show that $S = \left(\frac{g}{p}\right) S$.

(c) Show that $\left(\frac{g}{p}\right) = -1$.

(d) Show that $S = 0$.

(e) Show that the number of nonzero squares mod p equals the number of nonzero nonsquares mod p. (This was also proved in Exercise 24 of Chapter 5 and Exercise 31 of Chapter 8.)

Section 10.2: Computing Square Roots Mod p

25. Use the method of Section 10.2 to solve $y^2 \equiv 2 \pmod{23}$.
26. Use the method of Section 10.2 to solve $y^2 \equiv 11 \pmod{19}$.
27. Use the method of Section 10.2 to solve $y^2 \equiv 7 \pmod{29}$.
28. Use the method of Section 10.2 to solve $y^2 \equiv 5 \pmod{29}$.

Section 10.3: Quadratic Equations

29. Does $x^2 + 7x - 10 \equiv 0 \pmod{37}$ have any solutions?
30. Does $x^2 + 5x + 7 \equiv 0 \pmod{23}$ have any solutions?
31. Find the solutions to $x^2 + 16x + 9 \pmod{23}$.
32. Find the solutions to $x^2 - x + 5 \equiv 0 \pmod{11}$.

Section 10.4: The Jacobi Symbol

33. Compute $\left(\frac{35}{223}\right)$.
34. Compute $\left(\frac{35}{73}\right)$.

35. Compute $\left(\dfrac{203}{3511}\right)$.

36. Compute $\left(\dfrac{55}{401}\right)$.

37. (a) Compute $\left(\dfrac{35}{143}\right)$.

    (b) Use the result of (a) to show that 35 is not a square mod 143.

    (c) Show that 35 is not a square mod 11, and conclude (without using (b)) that 35 is not a square mod 143 ($= 11 \cdot 13$).

38. (a) Compute $\left(\dfrac{5}{77}\right)$.

    (b) Use the result of (a) to show that 5 is not a square mod 77.

    (c) Show that 5 is not a square mod 7, and conclude (without using (b)) that 5 is not a square mod 77.

39. (a) Compute $\left(\dfrac{3}{35}\right)$.

    (b) Does the result of (a) show that 3 is a square mod 35?

    (c) Show that 3 is not a square mod 5, and conclude that 3 is not a square mod 35.

40. (a) Compute $\left(\dfrac{13}{55}\right)$.

    (b) Does the result of (a) show that 13 is not a square mod 55?

    (c) Show that 13 is not a square mod 5, and conclude that 13 is not a square mod 55.

Section 10.5: Proof of Quadratic Reciprocity

41. Use Gauss's Lemma 10.9 to calculate $\left(\dfrac{3}{7}\right)$ and $\left(\dfrac{3}{13}\right)$. Check that you get the same answers as you get using Quadratic Reciprocity.

42. Use Gauss's Lemma to show that if p is an odd prime, then $\left(\dfrac{1}{p}\right) = 1$.

43. Use Gauss's Lemma to evaluate $\left(\dfrac{-1}{p}\right)$ for odd primes p.

10.7.2 Answers to "CHECK YOUR UNDERSTANDING"

1. (a) $4^2 \equiv 3 \pmod{13}$, so $\left(\frac{3}{13}\right) = 1$.
   (b) $\left(\frac{3}{13}\right) = \left(\frac{13}{3}\right) = \left(\frac{1}{3}\right) = 1$.
2. Since $19 \equiv 3 \pmod 4$, we compute $5^{(p+1)/2} \equiv 5^5 \equiv 3125 \equiv 9 \pmod{19}$. The solutions are $y \equiv \pm 9 \pmod{19}$.
3. The quadratic formula says that
$$x \equiv \frac{-3 \pm \sqrt{3^2 + 8}}{2} \equiv \frac{-3 \pm \sqrt{4}}{2}$$
$$\equiv -1/2, -5/2 \equiv 6, 4 \pmod{13}.$$

4.
$$\left(\frac{501}{1013}\right) = \left(\frac{1013}{501}\right) = \left(\frac{11}{501}\right) = \left(\frac{501}{11}\right)$$
$$= \left(\frac{6}{11}\right) = \left(\frac{2}{11}\right)\left(\frac{3}{11}\right) = (-1)(-1)\left(\frac{11}{3}\right)$$
$$= \left(\frac{2}{3}\right) = -1.$$

5. $\left(\frac{3}{35}\right) = -\left(\frac{35}{3}\right) = -\left(\frac{2}{3}\right) = 1$. However, $\left(\frac{3}{5}\right) = -1$, which means that $x^2 \equiv 3 \pmod 5$ has no solution, and therefore $x^2 \equiv 3 \pmod{35}$ has no solution.

Chapter 11

Primality and Factorization

Two fundamental computational problems in number theory are finding large primes and factoring composite numbers. Although these problems are related, they have intrinsic differences. It is relatively easy to prove that a number is composite but, once this has been done, it is much harder to find its factors. With the popularity of cryptographic algorithms that use primes, primality testing and factoring have become very important. In this chapter, we show how to decide whether a number is prime or composite. Then we discuss some factorization methods that are used in practice.

11.1 Trial Division and Fermat Factorization

The easiest method for primality testing and factorization is trial division. If we want to factor an integer $n > 1$, we divide n by all the primes $p \leq \sqrt{n}$ (by Proposition 1.8) until we find a prime factor. If we don't find a factor, n is prime. If we find a factor p_1, then we continue to try to factor $n_1 = n/p_1$.

To factor n_1, we could start at the beginning, dividing 2, 3, 5, $\cdots$ into n_1. But we have already taken care of the primes less than p_1, so we can start at p_1 and continue until either we get to $\sqrt{n_1}$ or we find another prime factor p_2. Then we try to factor $n_2 = n_1/p_2$. This process eventually terminates and yields the factorization of n.

Example. Let $n = 43771$. It is not divisible by 2, 3, and 5, but it is

divisible by 7:
$$43771 = 7 \times 6253.$$
Now we divide 7, 11, 13 into 6253 and find that
$$6253 = 13 \times 481.$$
Next, we try to factor 481, starting with 13. In fact,
$$481 = 13 \times 37.$$
Finally, we try to factor 37. Again, we start at 13, but now we are past $\sqrt{37} \approx 6.08$, so we are done. The factorization is
$$6253 = 7 \times 13^2 \times 37.$$

Example. Let $n = 401$. Dividing 401 by the primes 2, 3, 5, 7, 11, 13, 17, 19, we find that 401 has no prime factors less than or equal to $\sqrt{401} \approx 20.02$, so 401 is prime.

Doing these divisions is rather time-consuming and this method is impractical for large values of n. However, there are a few tricks that make the calculations faster. For example, suppose we want to check that $19 \nmid 401$. If you notice that $401 = 420 - 19$, you can use Corollary 1.5 to see that
$$19 \nmid 401 \iff 19 \nmid 420 \iff 19 \nmid 42 \times 10.$$
But 19 is prime and $19 \nmid 10$, so we have to check only that $19 \nmid 42$, which is easy to do. Therefore, $19 \nmid 401$.

A slight extension of this technique handles other situations. For example, to check whether 7 divides 401, we observe that $401 - 21 = 380$, so
$$7 \nmid 401 \iff 7 \nmid 380 \iff 7 \nmid 38 \cdot 10 \iff 7 \nmid 38,$$
and this last fact is easily checked. With a little practice, this type of technique lets you check for factors of 3-digit and 4-digit numbers in your head. If you get bored while driving, this lets you check the license plate number ahead of you for primality.

Trial division is good for searching for small prime factors, but cryptographic applications often use products of large primes. Suppose that

Section 11.1   Trial Division and Fermat Factorization   277

you want to factor n, where $n = pq$ and both p and q are primes of 100 digits. The Prime Number Theorem (see the Introduction) says that there are approximately 4×10^{97} primes of 100 digits, so trial division might take a long time. For example, if you have a computer that can do 10^{10} divisions per seconds, 10^{97} trial divisions take 10^{87} seconds, which is approximately 3×10^{79} years. Therefore, better factorization methods are needed. Some of these will be discussed in later sections.

A factorization method that does not use trial division is due to Fermat and uses the identity

$$x^2 - y^2 = (x+y)(x-y).$$

As an example, suppose you want to factor $n = 247$. Observe that $247 = 256 - 9 = 16^2 - 3^2$, so

$$247 = 16^2 - 3^2 = (16+3)(16-3) = 19 \times 13.$$

There are two ways to use this method in general to factor an integer n, and it is instructive to compare them. One way is to look at $n+1$, $n+4$, $n+9$, etc., until you find a square:

$$n + k^2 = s^2.$$

Then $n = (s+k)(s-k)$. The method works well when there is a factor of n that is very close to $\sqrt{n}$. But it is less effective in other cases. For example, if we want to factor 43771, it takes 45 steps to get to

$$43771 + 45^2 = 45796 = 214^2,$$

which yields the factorization

$$43771 = 259 \times 169$$

(this is not the complete factorization since $43771 = 7 \times 37 \times 13^2$). This is slower than the trial division used above; however, trial division is slower on $n = 247$, so the speed depends on the nature of the prime factors.

There is a better way, and this is what we'll call the **Fermat Factorization Method**. When writing $n = x^2 - y^2$, instead of running through values of y, use x. For example, if we want to write $247 = x^2 - y^2$, then we need $x \geq \sqrt{247} \approx 15.7$, so the first possible x is 16. We check that

$$16^2 - 247 = 9 = 3^2,$$

so we have $247 = 16^2 - 3^2 = 19 \times 13$. In general, we can start with $x = x_0 = \lceil \sqrt{n} \rceil$, namely the largest integer greater than or equal to $\sqrt{n}$. We then check $x_1 = x_0 + 1, x_2 = x_0 + 2, \ldots$ until we find an x_j such that $x_j^2 - n$ is a square.

Let's try this speed-up on $n = 43771$. Since $\sqrt{43771} \approx 209.2$, we start with $x_0 = 210$. Since $210^2 - 43771 = 329$, which is not a square, we move to $x_1 = 211$ and compute $211^2 - 43771 = 750$, which is not a square. Continuing, we soon get to $214^2 - 43771 = 2025 = 45^2$. This yields

$$43771 = 214^2 - 45^2 = 259 \times 169.$$

This improved method took 5 steps, compared to the 45 steps in the previous version. The reason is that the squares larger than 43771 are much farther apart than the small squares. Therefore, going from one value of x to the next bypasses several values of y. However, even this better method is too slow to be useful for large numbers.

When the RSA Algorithm was invented, it was suggested that the two primes, p and q, should be chosen of slightly different sizes, say, 100 digits and 104 digits. The idea was to keep Fermat Factorization from factoring $n = pq$ and breaking the system. However, even if p and q are of the same length, it is extremely unlikely that Fermat Factorization factors n. For example, suppose p and q are each 100 digits long and that they have the same initial 20 digits (very unlikely, unless your random number generator is malfunctioning), but the last 80 digits differ. Then the Fermat Factorization would take around 10^{60} steps (Exercise 8). In other words, Fermat Factorization has limited applicability with large numbers.

CHECK YOUR UNDERSTANDING:

1. Factor 851 by the Fermat Factorization Method by considering (a) $851 + y^2$ for $y = 1, 2, 3, \ldots$ and (b) $x^2 - 851$ for $x > \sqrt{851}$.

11.2 Primality Testing

Suppose that you need a 100-digit prime. The Prime Number Theorem (see the Introduction) says that the density of primes near a large number x is approximately $1/\ln x$, so the chance that a random 100-digit number is prime is around

$$\frac{1}{\ln(10^{100})} \approx \frac{1}{230}.$$

If we look only at odd numbers, then the chances are 1 in 115 that a number is prime. A good strategy is to look at the odd numbers, starting at a random starting point, and test each one. Since we expect to encounter many more composites than primes, we need a fast way to recognize composites. Fermat's theorem (Theorem 6.1) comes to the rescue in the form of Proposition 11.1.

11.2.1 Pseudoprimes

Proposition 11.1. *(Fermat Primality Test) Let $n > 1$ be an odd integer. Choose an integer b with $1 < b < n$. (Often, we choose $b = 2$.) If $b^{n-1} \not\equiv 1 \pmod{n}$, then n is composite.*

Proof. If n is prime, then Fermat's theorem says that $b^{n-1} \equiv 1 \pmod{n}$, which we have assumed does not happen. Therefore, n cannot be prime. □

Example. Let's show that 209 is not prime. We start by noting that $2^4 \equiv 16 \pmod{209}$. By successive squaring, we have

$$\begin{aligned}
2^8 &\equiv 16^2 \equiv 47 &\pmod{209} \\
2^{16} &\equiv 47^2 \equiv 119 &\pmod{209} \\
2^{32} &\equiv 119^2 \equiv 158 &\pmod{209} \\
2^{64} &\equiv 158^2 \equiv 93 &\pmod{209} \\
2^{128} &\equiv 93^2 \equiv 80 &\pmod{209}.
\end{aligned}$$

Therefore,

$$2^{208} \equiv 2^{128} \cdot 2^{64} \cdot 2^{16} \equiv 80 \cdot 93 \cdot 119 \equiv 36 \not\equiv 1 \pmod{209}.$$

By Proposition 11.1, we see that 209 is composite.

The fact that we showed 209 is composite without factoring it or finding any prime divisors is the key to why this method is useful. Generally, for large numbers, factorization is much slower than modular exponentiation. Let's use the Fermat Primality Test to find the first prime after 10^{22}. We'll do so by using Proposition 11.1 with $b = 2$ to test $n = 10^{22} + k$ for primality for $k = 1, 3, 7,$ and 9. (If $k = 2, 4,$ or 8, then n is even, and if $k = 5$, then n is divisible by 5.) We calculate (with a computer)

$$2^{10^{22}} \equiv 3852143514621151784591 \pmod{10^{22} + 1}$$
$$2^{10^{22}+2} \equiv 2467293414617665173233 \pmod{10^{22} + 3}$$
$$2^{10^{22}+6} \equiv 7533933929676167258456 \pmod{10^{22} + 7}$$
$$2^{10^{22}+8} \equiv 1 \pmod{10^{22} + 9}.$$

We now know that $10^{22} + k$ is composite for $k = 1, 2, 3, 4, 5, 6, 7, 8,$ and we suspect that $10^{22} + 9$ is prime. In fact, it is prime; however, actually proving this takes a lot more work.

The Fermat Primality Test is quite accurate. If it says a number n is composite, then n is guaranteed to be composite. In practice, if it does not declare that n is composite, then it is very likely that n is prime. However, if we choose a value of b, there are n where $b^{n-1} \equiv 1 \pmod{n}$ and n is composite. The first for $b = 2$ is $n = 341$, which was discovered in 1819 by Sarrus:

$$2^{340} \equiv 1 \pmod{341}, \quad \text{but} \quad 341 = 11 \times 31.$$

There are only 2057 composite n with $3 \leq n \leq 100,000,000$ such that $2^{n-1} \equiv 1 \pmod{n}$.

Definition 11.2. *A composite integer $n > 1$ is called a **b-pseudoprime** if $b^{n-1} \equiv 1 \pmod{n}$.*

For example, 341 is a 2-pseudoprime. The number 91 is a 3-pseudoprime because $3^{90} \equiv 1 \pmod{91}$ but $91 = 7 \times 13$. However, we can use the Fermat Primality Test to check that 341 and 91 are composite, without factoring, by computing

$$3^{340} \equiv 56 \pmod{341} \quad \text{and} \quad 2^{90} \equiv 64 \pmod{91}.$$

Section 11.2   Primality Testing   281

The question now arises, if n is composite, is there some b such that $b^{n-1} \not\equiv 1 \pmod{n}$? The easy answer is "yes," since we can take b to be a prime divisor of n and then b^{n-1} is 0 mod b, so it cannot be 1 mod n. We therefore ask the question with the extra condition that $\gcd(b, n) = 1$. Of course, if we happen to have a value of b such that $\gcd(b, n) \neq 1$, then we have a factor of n, so n is not prime. But we are mostly worrying about whether, when n is composite, we can pick a few values of b with the expectation of finding one such that $b^{n-1} \not\equiv 1 \pmod{n}$.

Definition 11.3. *A composite $n > 1$ is called a* **Carmichael number** *if $b^{n-1} \equiv 1 \pmod{n}$ for all integers b with $\gcd(b, n) = 1$.*

Example. 561 and 1729 are Carmichael numbers. Let's prove this for $561 = 3 \times 11 \times 17$. Fermat's theorem says that if $b \not\equiv 0 \pmod{17}$, then $b^{16} \equiv 1 \pmod{17}$. Therefore,
$$b^{560} \equiv \left(b^{16}\right)^{35} \equiv (1)^{35} \equiv 1 \pmod{17}.$$
Similarly, when $b \not\equiv 0 \pmod{11}$, Fermat's theorem tells us that $b^{10} \equiv 1 \pmod{11}$, so
$$b^{560} \equiv \left(b^{10}\right)^{56} \equiv (1)^{56} \equiv 1 \pmod{11}.$$
Finally, when $b \not\equiv 0 \pmod 3$, Fermat tells us that $b^2 \equiv 1 \pmod 3$, which implies that $b^{560} \equiv 1 \pmod 3$. We now have, whenever $\gcd(b, 561) = 1$, that $b^{560} - 1$ is a multiple of 3, 11, and 17. Therefore, it is a multiple of $561 = 3 \times 11 \times 17$. This is exactly the statement that
$$b^{560} \equiv 1 \pmod{561}$$
for all integers b with $\gcd(b, 561) = 1$, so 561 is a Carmichael number. The proof for $1729 = 7 \times 13 \times 19$ is similar.

In 1994, Alford, Granville, and Pomerance proved that there are infinitely many Carmichael numbers. However, these numbers are fairly rare. Pinch calculated that there are 20138200 Carmichael numbers less than 10^{21}. That seems like a lot, but it means that the chances are only 1 in 5×10^{13} that a randomly chosen integer of this size is a Carmichael number.

There is an improvement of the Fermat Primality Test, and it comes with no extra effort. If a number n is a Carmichael number, the test often can be used to verify that n is not a prime.

Proposition 11.4. *(Strong Fermat Primality Test) Let $n > 1$ be an odd integer. Write $n - 1 = 2^k s$ with s odd and $k \geq 1$. Choose an integer b with $1 < b < n$. (Again, we often use $b = 2$.) Define*

$$b_0 \equiv b^s \pmod{n}$$
$$b_1 \equiv b_0^2 \pmod{n}$$
$$b_2 \equiv b_1^2 \pmod{n}$$
$$\vdots \quad \vdots$$
$$b_k \equiv b_{k-1}^2 \pmod{n}.$$

Suppose n is prime. Then $b_k \equiv 1 \pmod{n}$, and if j is the smallest index for which $b_j \equiv 1 \pmod{n}$, then either $j = 0$ or $b_{j-1} \equiv -1 \pmod{n}$.

Proof. Since n is prime, Fermat's theorem says that $b^{n-1} \equiv 1 \pmod{n}$. Since

$$b_k \equiv (b^s)^{2^k} \equiv b^{n-1} \pmod{n},$$

we have $b_k \equiv 1 \pmod{n}$. If $j > 0$, then, by the choice of j, $b_{j-1}^2 \equiv 1 \pmod{n}$. Since n is prime, the only solutions to $x^2 \equiv 1 \pmod{n}$ are $x \equiv \pm 1$, by Corollary 5.10. Since b_j is the first that is 1 mod n, we must have $b_{j-1} \equiv -1 \pmod{n}$. □

Remark: As in the Fermat Primality Test, this test does not *prove* primality, but it can be used to prove that a number is composite. However, if n is such that, in the notation of Proposition 11.4, we have $b_k \equiv 1 \pmod{n}$ and either $j = 0$ or $b_{j-1} \equiv -1 \pmod{n}$, then we guess that n is probably prime.

Example. Let $n = 41$ and $b = 2$. Then $n - 1 = 2^3 \cdot 5$, so $s = 5$ and $k = 3$. We compute

$$b_0 = 32, \quad b_1 \equiv 32^2 \equiv 40, \quad b_2 \equiv 40^2 \equiv 1, \quad b_3 \equiv 1^2 \equiv 1 \pmod{41}.$$

Note that $b_k = b_3 = 1$. The smallest index with $b_i = 1$ is $i = 2$. Then $b_{i-1} = b_1 = 40 \equiv -1 \pmod{41}$. We conclude that 41 is probably prime.

Section 11.2   Primality Testing   283

Example. Let $n = 91$ and $b = 3$. Then $90 = 2^1 \cdot 45$, so $s = 45$ and $k = 1$. We have
$$b_0 \equiv 3^{45} \equiv 27, \quad b_1 \equiv 27^2 \equiv 1 \pmod{91}.$$

The fact that $b_1 = 1$ should not be surprising: We saw earlier that 91 is a 3-pseudoprime. However, note that $b_{j-1} = b_0 = 27 \not\equiv -1 \pmod{91}$. Therefore, the Strong Fermat Primality Test says that 91 cannot be prime, since a prime would require $b_0 \equiv -1$ in this situation.

As we see in the second example, the Strong Fermat Primality Test sometimes finds composites when the Fermat Primality Test does not. The calculation of $b_0, b_1, b_2, \ldots$ can be done as part of the successive squaring used to calculate $b^{n-1} \pmod{n}$. Moreover, as soon as we find some $b_j = 1$, we can stop and look at the previous step to see if $b_{j-1} \equiv -1$, so we might not need to do the last few squarings. However, in practice, when n is composite we find that $b_k \ne 1$, so the simple Fermat Primality Test suffices. The advantage of the improved algorithm is that it lets very few composites slip through as possible primes. Moreover, as we'll see in Subsection 11.3.2, a b-pseudoprime that fails the Strong Fermat Primality Test can easily be factored.

Definition 11.5. *A composite integer $n > 1$ is a* **strong b-pseudoprime** *if n passes the Strong Fermat Primality Test for base b. That is, in the notation of Proposition 11.4, $b_k \equiv 1 \pmod{n}$, and if j is the smallest index for which $b_j \equiv 1 \pmod{n}$, then either $j = 0$ or $b_{j-1} \equiv -1 \pmod{n}$.*

Example. Let's show that $4033 = 37 \times 109$ is a strong 2-pseudoprime. Write $4032 = 2^6 \cdot 63$, so $k = 6$ and $s = 63$. Then
$$b_0 \equiv 2^{63} \equiv 3521 \pmod{4033}$$
$$b_1 \equiv 3521^2 \equiv -1 \pmod{4033}$$
$$b_2 \equiv (-1)^2 \equiv 1 \pmod{4033}.$$

Clearly, b_3, b_4, b_5, b_6 are also 1 mod 4033. Since $b_2 \equiv 1$ and the previous $b_1 \equiv -1$, we see that 4033 is a strong 2-pseudoprime.

Example. Let's show that $121 = 11^2$ is a strong 3-pseudoprime. Write $120 = 2^3 \cdot 15$, so $k = 3$ and $s = 15$. Then
$$b_0 \equiv 3^{15} \equiv 1 \pmod{121}.$$

We stop here, since $b_i \equiv 1 \pmod{121}$ for all i, so $b_3 \equiv 1$. In the notation of the definition, we have $j = 0$, which is one of the allowed possibilities for a strong pseudoprime.

Strong pseudoprimes are rare. If you look at integers less than 10^8, then 488 of them are strong 2-pseudoprimes, 582 are strong 3-pseudoprimes, and 475 are strong 5-pseudoprimes. Of the 488 strong 2-pseudoprimes, 21 are *also* strong 3-pseudoprimes, and of these there is only one strong 5-pseudoprime: 25326001. We can quickly check that 25326001 is *not* a strong 7-pseudoprime. In fact, computations of Pomerance, Selfridge, and Wagstaff show that if $n < 25 \times 10^9$ is composite, it cannot be a strong pseudoprime for all five of 2, 3, 5, 7, 11. This leads to a simple primality test: Given an integer $n < 25 \times 10^9$, check that it passes the Strong Fermat Primality Test for $b = 2, 3, 5, 7, 11$. If so, then it is prime.

In practice, if n is composite, it is easy to find a b such that n is not a strong b-pseudoprime. Gary Miller proved under the assumption of the Generalized Riemann Hypothesis (an extension of the Riemann Hypothesis; see Section 15.3), that if n is composite, then there is a $b < 2(\ln n)^2$ such that n is not a strong b-pseudoprime (the constant "2" in the formula is due to Eric Bach). Michael Rabin showed that there are at least $3(n-1)/4$ values of $b < n$ such that n is not a strong b-pseudoprime. This leads to a *probabilistic* primality test, known as the **Miller-Rabin Primality Test**: Choose a value of m. The size of m determines the accuracy of the test. Test n using m randomly chosen values of $b < n$. If n is composite, the probability that it is a strong pseudoprime for one b is at most $1/4$, so the probability that it passes all m tests is at most $1/4^m$. Therefore, if n passes all m tests, then n is prime with probability at least $1 - (1/4)^m$ (in practice, the probability is much closer to 1). We say that n is *probably prime*. This is often how primes are chosen for cryptographic algorithms.

11.2.2 Mersenne Numbers

Recall that when p is prime, the pth Mersenne number is

$$M_p = 2^p - 1.$$

This sequence of numbers has yielded many large primes. In fact, at most times in history, the largest known prime has been a Mersenne prime, partly because there is a fast test for primality of these numbers.

Proposition 11.6. *(Lucas-Lehmer Test) Let p be prime. Define a sequence by*
$$s_0 = 4, \quad s_{i+1} = s_i^2 - 2.$$
Then
$$M_p \text{ is prime} \iff s_{p-2} \equiv 0 \pmod{M_p}.$$

We omit the proof of this theorem. The Great Internet Mersenne Prime Search (see Section 1.10) uses the Lucas-Lehmer method to search for very large Mersenne primes.

Example. Let $p = 7$. Then $M_p = 2^7 - 1 = 127$. We calculate

$s_0 = 4$, $s_1 = 14$, $s_2 = 194 \equiv 67$, $s_3 \equiv 42$, $s_4 \equiv 111$, $s_5 \equiv 0 \pmod{127}$.

Therefore, M_7 is prime.

Example. Let $p = 11$. Then $M_{11} = 2047$. A calculation yields

$$s_9 \equiv 1736 \pmod{2047}.$$

Therefore, $2^{11} - 1$ is not prime. In fact, $2047 = 23 \times 89$.

11.3 Factorization

Suppose we have an integer n and we know, because of some primality test, that n is composite. How do we find the factors? In general, this is quite hard. In this section we give an introduction to methods that are used in practice.

Of course, before trying more complicated methods, it's good to do trial division by several small primes. If n is not divisible by any small primes, we turn to more sophisticated techniques.

11.3.1 $x^2 \equiv y^2$

Fermat factorization relies on writing n as a difference of two squares. A basic idea that underlies many factoring techniques is to generalize Fermat's idea and write a multiple of n as a difference of two squares.

In other words, find an x and y with $x^2 \equiv y^2 \pmod{n}$. Of course this is easy to do in a trivial way: just take $x \equiv \pm y \pmod{n}$. But, as the next result shows, if we can obtain $x^2 \equiv y^2 \pmod{n}$ nontrivially, then we can factor n.

Proposition 11.7. *(Basic Factorization Principle) Let n be a positive integer and let x and y be integers with*

$$x^2 \equiv y^2 \pmod{n} \quad \text{but} \quad x \not\equiv \pm y \pmod{n}.$$

Then $\gcd(x - y, n)$ is a nontrivial factor of n.

Proof. Of course, $\gcd(x - y, n)$ is a factor of n. We need to show that the gcd is not 1 or n. If $\gcd(x - y, n) = n$, then $n \mid x - y$, which means that $x \equiv y \pmod{n}$, contrary to assumption. Now suppose that $\gcd(x - y, n) = 1$. Since $x^2 \equiv y^2 \pmod{n}$,

$$(x - y)(x + y) = x^2 - y^2 \equiv 0 \pmod{n}.$$

But $\gcd(x - y, n) = 1$, so we can divide the congruence by $x - y$ and obtain $x + y \equiv 0 \pmod{n}$, which is the same as $x \equiv -y \pmod{n}$. This contradicts the assumptions. Therefore, the gcd is not 1 or n, so it is a nontrivial factor of n. □

Example. We have $9^2 \equiv 2^2 \pmod{77}$, but $9 \not\equiv \pm 2 \pmod{77}$. Therefore,

$$\gcd(9 - 2, 77) = 7$$

is a nontrivial factor of 77.

Finding x and y as in Proposition 11.7 yields a factorization of n. Conversely, suppose we can factor n. For example, suppose $n = pq$ is a product of two distinct odd primes, and we know the factors p and q. Choose any number y. The Chinese Remainder Theorem says that we can solve the simultaneous congruences

$$x \equiv +y \pmod{p} \quad \text{and} \quad x \equiv -y \pmod{q}.$$

Then $x \not\equiv \pm y \pmod{n}$. However, $x^2 \equiv y^2 \pmod{p}$ and $\pmod{q}$, so

$$x^2 \equiv y^2 \pmod{pq}.$$

Section 11.3 Factorization   287

This means that if we know p and q then we can produce x and y as in the proposition. For example, if $n = 7 \times 11$, then we can solve
$$x \equiv 1 \pmod{7} \quad \text{and} \quad x \equiv -1 \pmod{11}$$
and obtain $x = 43$. Then $43^2 \equiv 1^2 \pmod{77}$. We can use this to recover the factorization: $\gcd(43 - 1, 77) = 7$.

Now that we have Proposition 11.7, if we don't know the factorization of n, how do we find x and y? We could compute a lot of squares mod n until we get a match, but this might take a long time. Instead, we try to find combinations that yield squares.

Example. Let's factor $n = 3837523$. Suppose we have discovered, by some means, that
$$17847^2 \equiv 2^3 \cdot 5^3 \pmod{n}$$
$$22111^2 \equiv 2^2 \cdot 3^7 \cdot 5^2 \cdot 7$$
$$32041^2 \equiv 2^{13} \cdot 5 \cdot 7^2$$
$$44153^2 \equiv 3 \cdot 5^2 \cdot 7^3.$$

If we multiply the first and third lines, we find
$$(17847 \cdot 32041)^2 \equiv (2^8 \cdot 5^2 \cdot 7)^2 \pmod{n}.$$
However,
$$17847 \cdot 32041 \equiv 2^8 \cdot 5^2 \cdot 7 \pmod{n},$$
so this does not help us. If we multiply the second and fourth lines, we obtain
$$(22111 \cdot 44153)^2 \equiv (2 \cdot 3^4 \cdot 5^2 \cdot 7^2)^2 \pmod{n}.$$
In this case, we're luckier:
$$22111 \cdot 44153 \not\equiv \pm 2 \cdot 3^4 \cdot 5^2 \cdot 7^2 \pmod{n}$$
and
$$\gcd(22111 \cdot 44153 - 2 \cdot 3^4 \cdot 5^2 \cdot 7^2, n) = 3511,$$
so we have a nontrivial factor of n. The other factor is $n/3511 = 1093$. We can express what we have done in matrix form:
$$\begin{matrix} 17847 \\ 22111 \\ 32041 \\ 44153 \end{matrix} \begin{bmatrix} 3 & 0 & 3 & 0 \\ 2 & 7 & 2 & 1 \\ 13 & 0 & 1 & 2 \\ 0 & 1 & 2 & 3 \end{bmatrix}.$$

The entries in the columns are the exponents of the primes 2, 3, 5, 7 in the factorizations coming from the squares of the numbers labeling the rows. We want a sum of rows to have all of its entries even. For example, the sum of the first and third rows is [16, 0, 4, 2], and the sum of the second and fourth rows is [2, 8, 4, 4]. These correspond to the relations $x^2 \equiv y^2 \pmod{n}$ that we found.

This gives a general strategy for factoring an integer n. First, choose a set B of small primes. This is called a **factor base**. In the example, $B = \{2, 3, 5, 7\}$. (Factoring 200-digit integers uses factor bases with millions of primes.) Then look at squares mod n and save those that can be written as products of primes in the factor base. So, if we can find a method to produce squares congruent to products of small primes, we should be able to factor quickly. In 1931, D. H. Lehmer and R. E. Powers developed a method based on continued fractions that produces such squares. This continued fraction method was used to great advantage by Brillhart and Morrison in 1975 in their factorization of the seventh Fermat number. By using factor bases and linear algebra, they laid the groundwork for the quadratic sieve factorization method, which was the leading factorization method around 1990. Moreover, theirs was one of the first factorization projects to use the massive computing power of modern computers. A technique called the *Number Field Sieve* is currently the best method for producing squares that are congruent to products of small primes. It has been used to factor numbers as large as 230 decimal digits.

CHECK YOUR UNDERSTANDING:

2. Suppose you know that

$$97^2 \equiv 5^2 \pmod{391}.$$

Use this information to factor 391.

3. Suppose you know that

$$119^2 \equiv 50 \pmod{14111} \text{ and } 168^2 \equiv 2 \pmod{14111}.$$

Use this information to factor 14111.

11.3.2 Factoring Pseudoprimes and Factoring Using RSA Exponents

Suppose n is a b-pseudoprime, so $b^{n-1} \equiv 1 \pmod{n}$. Or, more generally, we might have an even exponent u such that

$$b^u \equiv 1 \pmod{n}.$$

Often, this can be used to factor n. The method is very similar to the Strong Fermat Primality Test.

Algorithm 11.8. *Suppose that we have an even exponent u such that $b^u \equiv 1 \pmod{n}$. Write $u = 2^k s$ with s odd and $k \geq 1$. Define*

$$b_0 \equiv b^s \pmod{n}$$
$$b_1 \equiv b_0^2 \pmod{n}$$
$$b_2 \equiv b_1^2 \pmod{n}$$
$$\vdots \quad \vdots$$
$$b_k \equiv b_{k-1}^2 \pmod{n}.$$

(Note that $b_k \equiv b^{2^k s} \equiv b^u \equiv 1 \pmod{n}$ by assumption.) Let $b_i \equiv 1$, with i as small as possible. If $i = 0$, stop. If $i \geq 1$, this means that $b_{i-1} \not\equiv 1 \pmod{n}$ and

$$b_{i-1}^2 \equiv 1^2 \pmod{n}.$$

If $b_{i-1} \equiv -1 \pmod{n}$, stop. If $b_{i-1} \not\equiv -1 \pmod{n}$, compute $\gcd(b_{i-1} - 1, n)$. This is a nontrivial factor of n.

Either the algorithm stops with no output, or it finds a nontrivial factorization of n. The justification for this algorithm is the Basic Factorization Principle (Proposition 11.7).

Example. Suppose we find that $2^{200} \equiv 1 \pmod{32817151}$. Write

$$200 = 2^3 \cdot 25.$$

Compute

$$b_0 \equiv 2^{25} \equiv 737281 \pmod{32817151}$$
$$b_1 \equiv b_0^2 \equiv 32800948$$
$$b_2 \equiv b_1^2 \equiv 1$$

(we don't need to compute b_3). We now have

$$\gcd(32800948 - 1, 32817151) = 4051,$$

which yields the factorization $32817151 = 4051 \times 8101$.

We can apply this idea to the RSA Algorithm. Let $n = pq$ be a product of two distinct primes. Suppose we know integers e and d such that $ed \equiv 1 \pmod{(p-1)(q-1)}$. In Algorithm 11.8, let $u = ed - 1$, and let $b \neq \pm 1, 0$ be any message. Then $b^{ed} \equiv b \pmod{n}$. If $\gcd(b, n) > 1$, we have a factor of n, so suppose $\gcd(b, n) = 1$. Then we can divide by b and obtain $b^{ed-1} \equiv 1 \pmod{n}$. Therefore, Algorithm 11.8 applies. If it fails to factor n, we can try a new b. With very high probability, we should succeed. This tells us that if we have some clever way of finding the decryption exponent d in RSA, then we can factor n. Turning this around, we can say that if factorization is hard, then finding d is hard.

Example. Suppose Alice has RSA modulus $n = 221$ and encryption exponent $e = 5$. You find out that $d = 77$. You calculate $de - 1 = 384 = 2^7 \cdot 3$. Take $b = 7$. Since $7^{de} \equiv 7 \pmod{221}$ (this is just RSA encryption and decryption of the message $m = 7$), we have $7^{384} \equiv 7^{de-1} \equiv 1 \pmod{221}$. Compute

$$b_0 \equiv 7^3 \equiv 122 \pmod{221}$$
$$b_1 \equiv 122^2 \equiv 77$$
$$b_2 \equiv 77^2 \equiv 183$$
$$b_3 \equiv 183^2 \equiv 118$$
$$b_4 \equiv 118^2 \equiv 1.$$

Since $118^2 \equiv 1$ but $118 \not\equiv \pm 1 \pmod{221}$, we compute

$$\gcd(118 - 1, 221) = 13,$$

so $221 = 13 \times 17$.

11.4 Coin Flipping over the Telephone

Now that Alice and Bob have learned some new factorization techniques, they want to try a new method for flipping coins over the telephone. Here is how it goes.

1. Alice chooses two large primes, p and q, and forms $n = pq$. She sends n to Bob.

2. Bob chooses a random integer r mod n and computes $s \equiv r^2$ (mod n). He sends s to Alice.

3. Alice knows p and q, and computing square roots mod primes is straightforward (see Section 10.2). So Alice computes a and b with $a^2 \equiv s \pmod{p}$ and $b^2 \equiv s \pmod{q}$.

4. Alice, using the Chinese Remainder Theorem, computes x_1 and x_2 with

$$x_1 \equiv a \pmod{p}, \qquad x_1 \equiv b \pmod{q}$$

and

$$x_2 \equiv -a \pmod{p}, \qquad x_2 \equiv b \pmod{q}.$$

She randomly chooses $j = 1$ or 2 and sends x_j to Bob.

5. If $x_j \not\equiv \pm r \pmod{n}$, Bob declares that Alice loses. If $x_j \equiv \pm r \pmod{n}$, Alice wins.

6. If Bob says that Alice loses, Alice asks for p and q. Bob computes

$$\gcd(x_j - r, n),$$

which equals p or q, and sends the factors to Alice.

Note that, by Proposition 5.22, there are 4 solutions to $x^2 \equiv s$ (mod pq). They are of the form $x \equiv \pm x_1, \pm x_2 \pmod{pq}$. One of these is $\pm r$, so there is a 50% probability that Alice chooses $\pm r$.

Example. Alice chooses $p = 31$ and $q = 43$, so $n = 1333$. Bob chooses $r = 75$ and computes $75^2 \equiv 293 \pmod{1333}$. He sends 293 to Alice. Alice calculates $293 \equiv 14 \pmod{31}$. Using Proposition 10.5 for finding square roots mod primes, she finds that $18^2 \equiv 14 \pmod{31}$. She similarly calculates $483 \equiv 35 \pmod{43}$ and finds that $11^2 \equiv 35 \pmod{43}$. Therefore, $a = 18$ and $b = 11$.

She solves
$$x_1 \equiv 18 \pmod{31}, \qquad x_1 \equiv 11 \pmod{43}$$
to get $x_1 \equiv 1258 \equiv -75 \pmod{1333}$, and she also solves
$$x_2 \equiv -18 \pmod{31}, \qquad x_2 \equiv 11 \pmod{41}$$
to get $x_2 = 1129$.

Alice chooses 1129 and sends this to Bob. Bob sees that $1129 \not\equiv \pm 75 \pmod{n}$, so he says he wins. Alice doesn't trust Bob, so she asks for p and q. Bob computes
$$\gcd(1129 - 75, 1333) = 31,$$
so he can factor n and produce the factors. Note that if Alice chose 1258 instead, then Bob would calculate $\gcd(1258 - 75, 1333) = 1$, and therefore he would quickly have to find a method to factor n if he wanted to cheat and claim that he wins.

Why does this work? Since $x_j \equiv a \pmod{p}$, we have
$$x_j^2 \equiv a^2 \equiv s \equiv r^2 \pmod{p}.$$
Similarly,
$$x_j^2 \equiv b^2 \equiv s \equiv r^2 \pmod{q},$$
so
$$x_j^2 \equiv r^2 \pmod{n}.$$

One of x_1 and x_2 is congruent to $\pm r \pmod{n}$ and the other is not. If $x_j \not\equiv \pm r \pmod{n}$, then $\gcd(x_j - r, n)$ gives a nontrivial factor of n, by the Basic Factorization Principle (Proposition 11.7). If $x_j \equiv r \pmod{n}$, then the gcd gives only the trivial factorization. If p and q are chosen large enough to make factorization infeasible, then Bob is unable to cheat.

Can Alice cheat? One way is for her to send a random x, rather than x_j. Then it is very likely that $\gcd(x - r, n) = 1$, since the probability is only $1/p$ that a random $x - r$ is a multiple of p, and similarly for q. This means that Bob cannot factor using the gcd. If he suspects this is happening, he should calculate $x^2 \pmod{n}$ and check that it is the same as $r^2 \pmod{n}$. If not, he has caught Alice cheating.

11.5 Chapter Highlights

1. Fermat Factorization (Section 11.1)
2. Fermat Primality Test (Proposition 11.1)
3. Basic Factorization Principle (Proposition 11.7)

11.6 Problems

11.6.1 Exercises

Section 11.1: Trial Division and Fermat Factorization

1. Factor 2880 using trial division.
2. Factor 85680 using trial division.
3. Factor 1152 using trial division.
4. Factor 899 using Fermat Factorization.
5. Factor 391 using Fermat Factorization.
6. Factor 551 using Fermat Factorization.
7. Factor 621
   (a) by using Fermat Factorization
   (b) by using trial division.
8. Let p and q be primes, with $p - q = t > 0$. Let $n = pq$.
   (a) In the Fermat Factorization method, if $n = x^2 - y^2$, with $x - y > 1$, show that $x = (p+q)/2$ and $y = (p-q)/2$.
   (b) Show that the Fermat method takes approximately $(p+q)/2 - \sqrt{pq}$ steps to factor n.
   (c) Show that
   $$\left(\frac{p+q}{2} - \sqrt{pq}\right)\left(\frac{p+q}{2} + \sqrt{pq}\right) = \frac{1}{4}t^2.$$
   (d) Assume that t is much smaller than p, so that $(p+q)/2 \approx p$

and $\sqrt{pq} \approx p$. Show that

$$\frac{p+q}{2} - \sqrt{pq} \approx \frac{t^2}{8p}.$$

(e) Suppose p and q are 100-digit primes (so $p, q \approx 10^{99}$) and $p - q \approx 10^{80}$ (so the first 19 or 20 digits of p and q are the same). Show that the Fermat method takes approximately 10^{60} steps.

(f) Suppose p and q are 100-digit primes (so $p, q \approx 10^{99}$) and $p - q \approx 10^{50}$ (so the first 49 or 50 digits of p and q are the same). Show that the Fermat method succeeds in very few steps.

Section 11.2: Primality Testing

9. Use the Fermat Primality Test with $b = 2$ to show that 15 is composite.

10. Using the facts that $2^{140} \equiv 67 \pmod{561}$ and $2^{280} \equiv 1 \pmod{561}$, show that 561 is composite (compare Exercise 20).

11. (a) Show that $2^{11} \equiv 1 \pmod{2047}$.
    (b) Use (a) to show that $2^{1023} \equiv 1 \pmod{2047}$.
    (c) Use (b) to show that 2047 is a strong 2-pseudoprime.

12. Show that 1729 is a Carmichael number.

13. Show that $41041 = 7 \cdot 11 \cdot 13 \cdot 41$ is a Carmichael number.

14. (a) Let $k \geq 1$ be such that $p = 6k + 1$, $q = 12k + 1$, and $r = 18k + 1$ are primes. Show that $n = pqr$ is a Carmichael number. (The case $k = 1$ gives $n = 1729$.)
    (b) Find six Carmichael numbers larger than 1729.

15. Suppose you use a pseudoprimality test to find primes for the RSA Algorithm. You choose a prime p and a Carmichael number r, which you thought was prime because it passed several pseudoprimality tests. Assume that $\gcd(p, r) = 1$. You form the product $n = pr$ as your RSA modulus, and then you choose e and d with $ed \equiv 1 \pmod{(p-1)(r-1)}$. Show that if $\gcd(m, n) = 1$, and $c \equiv m^e \pmod{n}$, then $m \equiv c^d \pmod{n}$.

Section 11.3: Factorization

16. Use the fact that $284^2 \equiv 123^2 \pmod{851}$ to factor 851.
17. Use the fact that $333^2 \equiv 427^2 \pmod{893}$ to factor 893.
18. Use the facts that $511^2 \equiv 300 \pmod{23711}$ and $1251^2 \equiv 75 \pmod{23711}$ to factor 23711.
19. Use the facts that $937^2 \equiv 2 \cdot 3^2 \pmod{n}$ and $1666^2 \equiv 2 \cdot 7^2 \pmod{n}$ to factor $n = 28321$.
20. Using the facts that $2^{140} \equiv 67 \pmod{561}$ and $2^{280} \equiv 1 \pmod{561}$, factor 561 (compare Exercise 10).
21. (a) Show that 15 is a 4-pseudoprime.
    (b) Show that 15 is not a strong 4-pseudoprime.
    (c) Use (a) and (b) to factor 15 (of course, there is an easier way to factor 15).
22. (a) Show that 35 is a 6-pseudoprime.
    (b) Show that 35 is not a strong 6-pseudoprime.
    (c) Use (a) and (b) to factor 35 (of course, there is an easier way to factor 35).

11.6.2 Computer Explorations

1. (a) There are seven 2-pseudoprimes less than 2000. Find them.
   (b) How many of the 2-pseudoprimes from (a) are strong 2-pseudoprimes?
2. (a) Find a prime p such that all of the prime factors of $p-1$ are less than 25.
   (b) Find a prime q such that $q-1$ is divisible by the prime 101.
   (c) Let $n = pq$. Use the $p-1$ method to factor n.
   (d) Find a prime $r \neq p$ such that all of the prime factors of $r-1$ are less than 25.
   (e) Try to factor $m = pr$ by the $p-1$ method. If the gcd is m, use Algorithm 11.8 to factor m.
3. Let $n = 4300736399$.
   (a) Factor n by the Fermat Factorization method.
   (b) Try to factor n by the $p-1$ method. If you don't succeed,

look at the factorization from part (a). Find the factorization of $p-1$ for each prime factor p of n. Explain why the $p-1$ factorization method is expected to fail for this n.
(*Note:* Don't be misled by this example. Usually, Fermat Factorization is slower than other methods.)

4. Suppose you are testing integers up to 1000000 for primality. The product of the primes less than 20 is 9699690. A quick way to check for prime factors less than 20 is to compute $\gcd(n, 9699690)$.
(a) Count how many integers $n \leq 1000000$ satisfy $\gcd(n, 9699690) = 1$.
(b) Count how many primes are less than 1000000. How does this compare with the simple trial division test from (a)?
(c) Compute the product

$$\prod_{p<20}\left(1-\frac{1}{p}\right) = \left(1-\frac{1}{2}\right)\left(1-\frac{1}{3}\right)\left(1-\frac{1}{5}\right)\cdots\left(1-\frac{1}{19}\right),$$

where the product is over the primes less than 20. How does this relate to the number from (a)?
(d) The probability that an integer is divisible by a prime p is $1/p$. Explain why the numbers in (a) and (c) are closely related.

11.6.3 Answers to "CHECK YOUR UNDERSTANDING"

1. (a) Look at $851 + 1^2$, $851 + 2^2$, $\ldots, 851 + 7^2 = 900 = 30^2$. This gives $851 = (30+7)(30-7) = 37 \cdot 23$.
(b) We have $\sqrt{851} \approx 29.2$, so we start with $x = 30$. This yields $30^2 - 851 = 49 = 7^2$, so we obtain $851 = (30+7)(30-7) = 37 \cdot 23$.

2. Since $97 \not\equiv \pm 5 \pmod{391}$, the Basic Factorization Principle tells us to compute $\gcd(97-5, 391) = 23$, which yields the factorization $391 = 23 \cdot 17$.

3. We have $(119 \cdot 168)^2 \equiv 50 \cdot 2 \equiv 10^2 \pmod{14111}$. Compute $\gcd(119 \cdot 168 - 10, 14111) = 103$. Therefore, $14111 = 103 \cdot 137$.

Chapter 12

Sums of Squares

12.1 Sums of Two Squares

Which integers can be written as the sum of two squares? As we'll see, the case of prime numbers is the key. The French mathematician Albert Girard observed that every prime congruent to 1 mod 4 is a sum of two squares. Fermat, in a letter to Mersenne in 1640, said that he had a proof, and it is generally believed that this is the case. The first published proof is due to Euler in 1747. The proof we give is based on a lemma of Axel Thue (1863–1922), a Norwegian mathematician who did fundamental work in combinatorics and also on the approximation of irrational numbers by rational numbers.

Theorem 12.1. *(Fermat) Let p be a prime with $p \equiv 1 \pmod{4}$. Then there exist integers x and y with $x^2 + y^2 = p$.*

Proof. We start with a lemma that gives a property that holds for primes $p \equiv 1 \pmod{4}$ but does not hold for primes $p \equiv 3 \pmod{4}$ (see Exercises 12 and 31 in Chapter 6). This lemma is also the Supplementary Law for -1 in Theorem 10.4.

Lemma 12.2. *Let p be a prime with $p \equiv 1 \pmod{4}$. Then there exists u with $u^2 \equiv -1 \pmod{p}$.*

Proof. Let g be a primitive root mod p, and let $u \equiv g^{(p-1)/4} \pmod{p}$. (We need $p \equiv 1 \pmod{4}$ so that the exponent is an integer.) Then

$$u^2 \equiv g^{(p-1)/2} \equiv -1 \pmod{p}$$

by Proposition 8.6. □

Next, we need the lemma of Thue.

Lemma 12.3. *(Thue's Lemma) Let $n \geq 2$, and let b be an integer. There exist integers x and y with $0 < |x| \leq \sqrt{n}$ and $0 \leq |y| \leq \sqrt{n}$ such that $bx \equiv y \pmod{n}$.*

Proof. Consider the list of numbers of the form $jb - k$, where $0 \leq j \leq \sqrt{n}$ and $0 \leq k \leq \sqrt{n}$. Because 0 is included, there are $\lfloor \sqrt{n} \rfloor + 1 > \sqrt{n}$ possible values for j and also more than $\sqrt{n}$ possible values for k. Therefore, there are more than n numbers in our list of numbers $jb - k$. Since there are only n congruence classes mod n, this forces two of them to be congruent mod n:

$$j_1 b - k_1 \equiv j_2 b - k_2 \pmod{n},$$

where $(j_1, k_1) \neq (j_2, k_2)$. Let $x = j_1 - j_2$ and $y = k_1 - k_2$. Then $|x| \leq \sqrt{n}$ and $|y| \leq \sqrt{n}$, and also $xb - y \equiv 0 \pmod{n}$. We're done, except for showing that $x \neq 0$.

If $x = 0$, then $y \equiv 0 \pmod{n}$. Since $|y| \leq \sqrt{n}$, this makes $y = 0$, so $(j_1, k_1) = (j_2, k_2)$, contrary to the choice of these pairs. Therefore, $x \neq 0$. This proves the lemma. □

Let u be as in Lemma 12.2. By Lemma 12.3, there exist x and y with $0 < |x| < \sqrt{p}$ and $0 \leq |y| < \sqrt{p}$, and also $ux \equiv y \pmod{p}$ (we must have $|x| < \sqrt{p}$ rather than just $|x| \leq \sqrt{p}$ because $\sqrt{p}$ is irrational, so $x \neq \sqrt{p}$). Then

$$0 < x^2 + y^2 < (\sqrt{p})^2 + (\sqrt{p})^2 = 2p.$$

Moreover,

$$x^2 + y^2 \equiv x^2 + (ux)^2 \equiv x^2(1 + u^2) \equiv 0 \pmod{p},$$

since $1 + u^2 \equiv 0 \pmod{p}$. The only multiple of p between 0 and $2p$ is p. Therefore, $x^2 + y^2 = p$, as desired. This proves Theorem 12.1. □

We now discuss arbitrary positive integers. Suppose that n and m are integers such that

$$m = a^2 + b^2 \text{ and } n = c^2 + d^2.$$

Then

$$mn = (ac - bd)^2 + (ad + bc)^2. \tag{12.1}$$

Section 12.1   Sums of Two Squares   299

Where does this identity come from? The absolute value of the complex number $a + bi$ is $|a + bi| = \sqrt{a^2 + b^2}$, so we have

$$m = |a + bi|^2 \text{ and } n = |c + di|^2.$$

Therefore,

$$\begin{aligned} mn &= |a + bi|^2 |c + di|^2 \\ &= |(a + bi)(c + di)|^2 \\ &= |(ac - bd) + (ad + bc)i|^2 \\ &= (ac - bd)^2 + (ad + bc)^2. \end{aligned}$$

Of course, it's also possible to verify the formula (12.1) by algebraic calculation.

We conclude that if two numbers are sums of squares, then so is their product. This leads to the main result of this section.

Theorem 12.4. *Let n be a positive integer. Then $n = x^2 + y^2$ for integers x, y if and only if n is a square times a product of primes $p \not\equiv 3 \pmod 4$. In other words, n is a sum of squares if and only if, in the prime factorization of n, every prime $p \equiv 3 \pmod 4$ has an even exponent.*

For example, $65 = 5 \times 13$ and $130 = 2 \times 5 \times 13$ and $405 = 3^4 \times 5$ are sums of squares, while $35 = 5 \times 7$ and $189 = 3^3 \times 7$ are not sums of squares. We also include $49 = 7^2 + 0^2$ as a sum of two squares.

Proof. First, suppose that $n = u^2 v$ is a square times a number v that is a product of primes $p \not\equiv 3 \pmod 4$. Each prime $p \not\equiv 3 \pmod 4$ is a sum of two squares: The odd primes are treated by Theorem 12.1, and the even prime is $2 = 1^2 + 1^2$. Therefore, Equation (12.1) implies that v, which is a product of primes that are sums of two squares, is a sum of two squares: $v = x_1^2 + y_1^2$. Therefore, $n = u^2 v = (ux_1)^2 + (uy_1)^2$ is a sum of two squares.

Conversely, suppose that the integer n is a sum of two squares: $n = x^2 + y^2$. Let $p \equiv 3 \pmod 4$. We need to show that if p^a is the power of p in the factorization of n, then a is even. Suppose p^j is the largest power of p that divides both x and y. Write $x = p^j x_1$ and $y = p^j y_1$, where at least one of x_1, y_1 is not a multiple of p. Without loss of generality, we may assume that $x_1 \not\equiv 0 \pmod{p}$. We have

$$n = x^2 + y^2 = p^{2j}(x_1^2 + y_1^2),$$

so $p^{2j} \mid n$. Write $n = p^{2j} n_1$, with $n_1 = x_1^2 + y_1^2$. If $p \nmid n_1$, then we conclude that p^{2j} is the power of p in the factorization of n, and the exponent $2j$ is even. On the other hand, if $p \mid n_1$, then

$$x_1^2 + y_1^2 \equiv 0 \pmod{p}.$$

Since $x_1 \not\equiv 0 \pmod{p}$, we can write $(y_1/x_1)^2 \equiv -1 \pmod{p}$. As we saw in Exercise 12 in Chapter 6, this is impossible when $p \equiv 3 \pmod{4}$. Therefore, $p \nmid n_1$, so we conclude that p occurs with an even exponent in the factorization of n. This is true for every $p \equiv 3 \pmod{4}$, so n has the form claimed in the theorem. □

12.1.1 Algorithm for Writing $p \equiv 1 \pmod{4}$ as a Sum of Two Squares

The proof of Theorem 12.1 does not readily yield the values of x and y such that $x^2 + y^2 = p$. Of course, we can simply try all values of $x < \sqrt{p}$ until $p - x^2$ is a square, but there is a faster way.

1. Find a solution to $u^2 \equiv -1 \pmod{p}$. This may be done by computing $x^{(p-1)/4} \pmod{p}$ for random values of x. If x is a quadratic nonresidue mod p, then this yields the desired u. We may assume that $0 < u < p$. If $u > p/2$, change u to $p - u$. Therefore, we may assume that $0 < u < p/2$.

2. Apply the Euclidean algorithm to p and u. Let r_k and r_{k+1} be the first two remainders less than $\sqrt{p}$.

3. $p = r_k^2 + r_{k+1}^2$.

For example, let $p = 557$. We have $2^{139} \equiv 118 \pmod{557}$, and $118^2 \equiv -1 \pmod{557}$, so we let $u = 118$. The Euclidean algorithm is

$$557 = 4 \cdot 118 + 85$$
$$118 = 1 \cdot 85 + 33$$
$$85 = 2 \cdot 33 + 19$$
$$33 = 1 \cdot 19 + 14$$
$$19 = 1 \cdot 14 + 5$$
$$14 = 2 \cdot 5 + 4$$
$$5 = 1 \cdot 4 + 1$$
$$4 = 4 \cdot 1 + 0.$$

The first two remainders less than $\sqrt{557}$ are 19 and 14. We have $19^2 + 14^2 = 557$.

This algorithm is essentially due to Charles Hermite and Joseph Serret (independently) in 1848. Improvements were made by Henry Smith and John Brillhart. For the proof that the algorithm works, see F. W. Clarke, W. N. Everitt, L. L. Littlejohn, and S. J. R. Vorster, "H. J. S. Smith and the Fermat two squares theorem," *Amer. Math. Monthly* 106 (1999), no. 7, 652–665.

CHECK YOUR UNDERSTANDING:

1. Using the fact that $22^2 \equiv -1 \pmod{97}$, write 97 as a sum of two squares.

12.2 Sums of Four Squares

The amazing fact that every positive integer is a sum of four squares was probably known to Diophantus, but the first proof is due to Lagrange in 1770. As in the case of sums of two squares, the key case is that of primes.

Theorem 12.5. *(Lagrange) Every positive integer is a sum of four squares.*

Proof. To start, we prove the theorem when $n = p$ is prime. We need an analogue of Lemma 12.2.

Lemma 12.6. *Let p be prime. Then there are integers u, v such that $u^2 + v^2 + 1 \equiv 0 \pmod{p}$.*

Proof. The case $p = 2$ is trivial (let $u = 1$ and $v = 0$), so assume p is odd. There are $(p-1)/2$ nonzero squares mod p by Exercise 24 in Chapter 5, or by Exercise 31 in Chapter 8. Since 0 is also a square, there are $1 + (p-1)/2 = (p+1)/2$ squares mod p. If we take each square and subtract it from -1, we get $(p+1)/2$ numbers of the form $-1 - v^2 \pmod{p}$. We have two sets: the squares with $(p+1)/2$ elements, and the numbers $-1 - v^2$ with $(p+1)/2$ elements. There are only p congruence classes mod p, and $(p+1)/2 + (p+1)/2 > p$. Therefore, the two sets must overlap. This means that $u^2 \equiv -1 - v^2 \pmod{p}$ for some u, v, which says that $u^2 + v^2 + 1 \equiv 0 \pmod{p}$. □

Example. As an example of the lemma, let $p = 11$. The squares mod 11 are
$$\{0, 1, 4, 9, 5, 3\},$$
and the numbers of the form $-1 - v^2$ are
$$\{-1, -2, -5, -10, -6, -4\} \equiv \{10, 9, 6, 1, 5, 7\} \pmod{11}.$$

Note that 1 is in both sets, which means that $1^2 \equiv -1 - 3^2$, or $1^2 + 3^2 + 1 \equiv 0 \pmod{11}$.

We now use a two-dimensional analogue of Thue's Lemma.

Lemma 12.7. *Let $n \geq 2$, and let a, b, c, d be integers. There exist w, x, y, z, with at least one of y, z nonzero, such that*
$$0 \leq |w| \leq \sqrt{n}, \quad 0 \leq |x| \leq \sqrt{n}, \quad 0 \leq |y| \leq \sqrt{n}, \quad 0 \leq |z| \leq \sqrt{n}$$
and
$$w \equiv ay + bz \pmod{n}, \quad x \equiv cy + dz \pmod{n}.$$

Proof. Make a list of all pairs $(i - ak - b\ell, j - ck - d\ell)$ with $0 \leq i, j, k, \ell \leq \sqrt{n}$. Since we are including 0, there are $\lfloor \sqrt{n} \rfloor + 1 > \sqrt{n}$ values of each of i, j, k, ℓ. Therefore, there are more than $(\sqrt{n})^4 = n^2$ pairs. There are

Section 12.2   Sums of Four Squares   303

only n^2 pairs of congruence classes mod n, so we must have two pairs that are congruent mod n:

$$i_1 - ak_1 - b\ell_1 \equiv i_2 - ak_2 - b\ell_2 \pmod{n},$$
$$j_1 - ck_1 - d\ell_1 \equiv j_2 - ck_2 - d\ell_2 \pmod{n},$$

where $(i_1, j_1, k_1, \ell_1) \neq (i_2, j_2, k_2, \ell_2)$. Let

$$w = i_1 - i_2, \quad x = j_1 - j_2, \quad y = k_1 - k_2, \quad z = \ell_1 - \ell_2.$$

Then w, x, y, z satisfy the inequalities and the congruences of the lemma. It remains to show that at least one of y, z is nonzero.

If $y = z = 0$, then $w \equiv x \equiv 0 \pmod{n}$. Since $|w|, |x| \leq \sqrt{n}$, we must have $x = w = 0$. But $w = x = y = z = 0$ implies that $(i_1, j_1, k_1, \ell_1) = (i_2, j_2, k_2, \ell_2)$, contrary to the choice of these 4-tuples. This completes the proof of the lemma. □

We now return to the proof of Lagrange's theorem. We start by showing that if p is a prime number, then p is a sum of four squares. Let u, v be as in Lemma 12.6. By Lemma 12.7, there exist w, x, y, z, with at least one nonzero and with all less than $\sqrt{p}$ (they cannot equal $\sqrt{p}$ because $\sqrt{p}$ is not an integer) such that

$$w \equiv uy + vz \pmod{p}, \quad x \equiv vy - uz \pmod{p}.$$

Then

$$w^2 + x^2 + y^2 + z^2 \equiv (uy + vz)^2 + (vy - uz)^2 + y^2 + z^2$$
$$\equiv (u^2 + v^2 + 1)y^2 + (v^2 + u^2 + 1)z^2$$
$$\equiv 0 + 0 \equiv 0 \pmod{p}$$

(the cross terms with yz cancel). Moreover,

$$0 < w^2 + x^2 + y^2 + z^2 < (\sqrt{p})^2 + (\sqrt{p})^2 + (\sqrt{p})^2 + (\sqrt{p})^2 = 4p.$$

The only multiples of p between 0 and $4p$ are $p, 2p$, and $3p$, so

$$w^2 + x^2 + y^2 + z^2 = p, 2p, \text{ or } 3p.$$

If $w^2 + x^2 + y^2 + z^2 = p$, we're done. We need to treat the cases where the sum equals $2p$ or $3p$.

Suppose that $w^2 + x^2 + y^2 + z^2 = 2p$. Then either all of w, x, y, z are odd, or two of them are odd, or all are even (if one of them is odd or three of them are odd, then $w^2 + x^2 + y^2 + z^2$ is odd, so does not equal $2p$). By rearranging the order, we can assume that

$$w \equiv x \pmod{2}, \quad y \equiv z \pmod{2}.$$

Then $(w \pm x)/2$ and $(y \pm z)/2$ are integers, and

$$\left(\frac{w+x}{2}\right)^2 + \left(\frac{w-x}{2}\right)^2 + \left(\frac{y+z}{2}\right)^2 + \left(\frac{y-z}{2}\right)^2$$
$$= \frac{w^2 + x^2 + y^2 + z^2}{2} = p$$

(expand it out; a lot of cross terms cancel), so we obtain p as the sum of four squares.

Now suppose that $w^2 + x^2 + y^2 + z^2 = 3p$. As in the case $2p$, we want to manipulate this to get p as a sum of four squares. If $p = 3$, then $3 = 1^2 + 1^2 + 1^2 + 0^2$, so we know 3 is a sum of four squares. Therefore, we assume that $p \neq 3$. If $w \equiv x \equiv y \equiv z \equiv 0 \pmod{3}$, then $w^2 + x^2 + y^2 + z^2 \equiv 0 \pmod 9$, while $3p \not\equiv 0 \pmod 9$, contradiction. Therefore, at least one of w, x, y, z is nonzero mod 3. Since $0^2 \equiv 0 \pmod{3}$ and $(\pm 1)^2 \equiv 1 \pmod{3}$, it is easy to see that the only way for $w^2 + x^2 + y^2 + z^2$ to add up to $3p$ is for three of w, x, y, z to be $\pm 1 \pmod{3}$ and one of them to be $0 \pmod{3}$. By rearranging the order, if necessary, we can assume that $z \equiv 0 \pmod{3}$. By changing the signs of w, x, y, if necessary, we may assume that $w \equiv x \equiv y \equiv 1 \pmod{3}$. Let

$$x_1 = (w + x + y)/3, \quad x_2 = (w - x + z)/3,$$
$$x_3 = (-w + y - z)/3, \quad x_4 = (x - y + z)/3.$$

Then x_1, x_2, x_3, x_4 are integers. A straightforward but lengthy calculation shows that

$$x_1^2 + x_2^2 + x_3^2 + x_4^2 = (w^2 + x^2 + y^2 + z^2)/3 = p.$$

Therefore, p is a sum of four squares in all cases.

To prove that every positive integer is a sum of four squares, we need

Section 12.2   Sums of Four Squares   305

to extend Equation(12.1) to four squares. Here's the relevant identity:

$$(a^2 + b^2 + c^2 + d^2)(e^2 + f^2 + g^2 + h^2)$$
$$= (ae - bf - cg - dh)^2 + (af + be + ch - dg)^2$$
$$+ (ag + ce + df - bh)^2 + (ah + de + bg - cf)^2. \quad (12.2)$$

If you don't believe this, multiply it out and check it. To see how it can be deduced using quaternions, see Exercise 11. However, this identity was discovered before quaternions were discovered!

Every prime is a sum of four squares, and Equation (12.2) says that products of sums of four squares are sums of four squares. Since every integer $n \geq 2$ is a product of primes, each such n is a sum of four squares. Finally, $1 = 1^2 + 0^2 + 0^2 + 0^2$, so 1 is also a sum of four squares. This completes the proof. □

In 1834, Carl Gustav Jacobi gave a formula for the number of ways of writing n as a sum of four squares. Let $S(n)$ be the sum of all the divisors d of n that are not multiples of 4. For example, when $n = 20$ we have

$$S(20) = 1 + 2 + 5 + 10 = 18.$$

Then there are exactly $8S(n)$ ways of writing n as a sum of four squares, where different orders of the summands are regarded as different, and the squares can be squares of positive or negative numbers. So, for example, we can write

$$1 = 0^2 + 0^2 + 0^2 + 1^2$$
$$= 0^2 + 0^2 + 0^2 + (-1)^2$$
$$= 0^2 + 0^2 + 1^2 + 0^2$$
$$= 0^2 + 0^2 + (-1)^2 + 0^2$$
$$= 0^2 + 1^2 + 0^2 + 0^2$$
$$= 0^2 + (-1)^2 + 0^2 + 0^2$$
$$= 1^2 + 0^2 + 0^2 + 0^2$$
$$= (-1)^2 + 0^2 + 0^2 + 0^2.$$

Note that $8S(1) = 8$, which is the number of ways we've written 1 as a sum of four squares.

CHECK YOUR UNDERSTANDING

2. Use trial and error to write 71 as a sum of four squares (there are two completely distinct ways of doing this).

12.3 Other Sums of Powers

Since we now know which integers are sums of two squares, and we know that all positive integers are sums of four squares, you might be wondering about sums of three squares. In 1796, Gauss proved that every integer that is not of the form $4^j(8k+7)$, with $j, k \geq 0$, is a sum of three squares. For example, 42 is not of this form, and $42 = 5^2 + 4^2 + 1^2$. Moreover, numbers of the form $4^j(8k+7)$ are not sums of three squares (it is not difficult to prove this by congruences mod powers of 2). For example, 15 has this form (with $j = 0$ and $k = 1$), and it is easy to check that 15 is not a sum of three squares. The proof of Gauss's theorem is quite difficult and is beyond the scope of this book.

What about higher powers? Edward Waring conjectured in 1770 that, for each positive integer k, there exists a number $g(k)$ such that every positive integer is a sum of $g(k)$ nonnegative kth powers. For example, Lagrange's theorem says that for $k = 2$, every integer is a sum of $g(2) = 4$ squares. Waring's conjecture was proved by Hilbert in 1909. For $k = 3$, Waring's problem asks us to write an integer as a sum of nonnegative cubes. The numbers 23 and 239 each require nine nonnegative cubes, but all other positive integers require at most eight cubes. The number 1290740 is a sum of six cubes. It is conjectured that every number larger than 1290740 is a sum of five nonnegative cubes, but so far this has not been proved.

Instead of looking at sums of powers, we could consider differences. In 1844, Eugène Charles Catalan conjectured that

$$3^2 - 2^3 = 1$$

is the only example of two perfect powers differing by 1. In other words, he predicted that if m and n are integers with $m, n > 1$, then the only

solution to
$$x^m - y^n = 1$$
in positive integers x, y is $x = 3, y = 2, m = 2, n = 3$. Catalan's conjecture was proved in 2002 by Preda Mihăilescu.

12.4 Chapter Highlights

1. Every prime $p \equiv 1 \pmod 4$ is a sum of two squares.
2. Every positive integer is a sum of four squares.

12.5 Problems

12.5.1 Exercises

1. Write 41 as a sum of two squares.
2. Write 97 as a sum of two squares.
3. Determine which of the following numbers are sums of two squares: (a) $221 = 17 \times 13$, (b) $259 = 7 \times 37$, (c) $5040 = 2 \times 3 \times 4 \times 5 \times 6 \times 7$.
4. Write 65 as a sum of two squares in two different ways.
5. Use the facts that $13 = 2^2 + 3^2$ and $17 = 4^2 + 1^2$, plus Equation 12.1, to write $221 = 13 \cdot 17$ as a sum of two squares.
6. Use the facts that $29 = 2^2 + 5^2$ and $37 = 6^2 + 1^2$, plus Equation 12.1, to write $1073 = 29 \cdot 37$ as a sum of two squares.
7. Use the algorithm in Subsection 12.1.1, plus the fact that $546^2 \equiv -1 \pmod{1237}$ to write the prime 1237 as a sum of two squares.
8. Use the algorithm in Section 12.1.1, plus the fact that $17682^2 \equiv -1 \pmod{100049}$ to write the prime 100049 as a sum of two squares.

9. Write 123 as a sum of four squares.

10. Write 23 as a sum of four squares.

11. The *quaternions* are expressions of the form $z = a+bi+cj+dk$, where a, b, c, d are real numbers and i, j, k satisfy the relations

$$i^2 = j^2 = k^2 = -1, \quad ij = k = -ji,$$
$$jk = i = -kj, \quad ki = j = -ik.$$

(a) Show that

$$(a_1+b_1i + c_1j + d_1k)(a_2 + b_2i + c_2j + d_2k)$$
$$= (a_1a_2 - b_1b_2 - c_1c_2 - d_1d_2)$$
$$+ (a_1b_2 + a_2b_1 + c_1d_2 - c_2d_1)i$$
$$+ (a_1c_2 + a_2c_1 - b_1d_2 + b_2d_1)j$$
$$+ (a_1d_2 + a_2d_1 + b_1c_2 - b_2c_1)k.$$

(b) Define

$$\overline{a + bi + cj + dk} = a - bi - cj - dk$$

and

$$N(a + bi + cj + dk) = (a + bi + cj + dk)(\overline{a + bi + cj + dk}).$$

Show that

$$N(a + bi + cj + dk) = a^2 + b^2 + c^2 + d^2.$$

(c) It can be shown that if z_1 and z_2 are two quaternions, then $N(z_1 z_2) = N(z_1)N(z_2)$. Show that this relation is the same as Equation (12.2).

12. Show that if $n \equiv 7 \pmod{8}$, then n is not a sum of 3 squares. (*Hint:* Lemma 4.7.)

13. Show that every cube is congruent to 0, 1, or -1 mod 9. Deduce that there are infinitely many positive integers that are not sums of three nonnegative cubes.

14. (a) Express 23 and 239 as sums of 9 nonnegative cubes.
    (b) Show that 23 and 239 are not sums of 8 nonnegative cubes.

Section 12.5   Problems   309

(It has been shown that these are the only two numbers requiring 9 cubes.)

(c) Using a computer, express 1290740 as a sum of six cubes. (It is conjectured that every number larger than 1290740 is a sum of five nonnegative cubes.)

15. There is an integer $n \leq 100$ that is a sum of 19 fourth powers but is not a sum of 18 fourth powers. Find it.

16. Let n be a sum of two squares in two ways, so we have $n = a^2 + b^2 = c^2 + d^2$, with $a \geq b \geq 0$ and $c \geq d \geq 0$. Also, we may assume that $a > c$. Let $x = ad + bc$ and $y = ad - bc$. It can be shown that $\gcd(x, n)$ and $\gcd(y, n)$ are nontrivial factors of n. Use the fact that $1000009 = 1000^2 + 3^2 = 972^2 + 235^2$ to factor 1000009.

Answers to "CHECK YOUR UNDERSTANDING"

1. The Euclidean algorithm for $\gcd(97, 22)$ is

$$97 = 4 \cdot 22 + 9$$
$$22 = 2 \cdot 9 + 4$$
$$9 = 2 \cdot 4 + 1.$$

The first two remainders less than $\sqrt{97}$ are 9 and 4. We have $97 = 9^2 + 4^2$.

2. The two ways are $71 = 7^2 + 3^2 + 3^2 + 2^2$ and $71 = 6^2 + 5^2 + 3^2 + 1^2$. Note that the "greedy algorithm" of choosing the largest square ($8^2 = 64$) less than 71 does not work in this case.

Chapter 13

Arithmetic Functions

13.1 Perfect Numbers

Suppose that you are playing with sums of divisors of numbers (yes, mathematicians do things like this) and you discover the following two examples:

$$1 + 2 + 3 = 6$$
$$1 + 2 + 4 + 7 + 14 = 28.$$

What is so special about these? 1, 2, 4, 7, 14 are the positive divisors of 28 that are less than 28. Similarly, 1, 2, 3 are the divisors of 6 that are less than 6. Are there any other such examples? In fact, here is another:

$$1 + 2 + 4 + 8 + 16 + 31 + 62 + 124 + 248 = 496.$$

Definition 13.1. *A positive integer n is called* **perfect** *if n is the sum of the positive divisors of n that are less than n. It is called* **deficient** *if the sum of these divisors is less than n, and it is* **abundant** *if the sum of these divisors is greater than n.*

Examples. 6, 28, and 496 are perfect. However, 15 is deficient because $1 + 3 + 5 < 15$, and 12 is abundant because $1 + 2 + 3 + 4 + 6 > 12$.

Definition 13.2. *When n is a positive integer, $\sigma(n)$ is the sum of all positive divisors of n (including n).*

Examples. $\sigma(6) = 1 + 2 + 3 + 6 = 12$, $\quad \sigma(15) = 1 + 3 + 5 + 15 = 24$

Proposition 13.3. *Let p be prime and let $k \geq 1$. Then*

$$\sigma(p^k) = 1 + p + p^2 + \cdots + p^k = \frac{p^{k+1} - 1}{p - 1}.$$

Proof. The divisors of p^k are $1, p, p^2, \ldots, p^k$, so the sum of these is $\sigma(p^k)$. The second equality follows from letting $x = p$ in the identity

$$x^{k+1} - 1 = (x - 1)(x^k + x^{k-1} + \cdots + x^2 + x + 1)$$

(if we divide both sides by $x - 1$ we have the formula for a geometric sum; see Appendix A). □

The advantage of using $\sigma(n)$ rather than the sum of the divisors less than n is the following.

Proposition 13.4. *If m and n are positive integers with $\gcd(m, n) = 1$, then*

$$\sigma(mn) = \sigma(m) \cdot \sigma(n).$$

This property is called **multiplicativity** and will be proved in the next section (Corollary 13.11).

Example. $\sigma(3) = 1 + 3 = 4$ and $\sigma(5) = 1 + 5 = 6$. The proposition says that $\sigma(15) = \sigma(3)\sigma(5) = 4 \cdot 6 = 24$, which agrees with our previous calculation.

Since we are including n in $\sigma(n)$, we have

$$n \text{ is perfect} \iff \sigma(n) = 2n.$$

We can use the multiplicativity of $\sigma(n)$ to produce all even perfect numbers.

Theorem 13.5. *Let n be an even perfect number. Then there is a prime p such that $2^p - 1$ is prime and*

$$n = 2^{p-1}(2^p - 1).$$

Conversely, every n of this form, with p and $2^p - 1$ prime, is perfect.

Remark. The formula that gives perfect numbers is in Euclid's *Elements* (Book IX, Proposition 36). The fact that all even perfect numbers have this form was proved by Euler.

Section 13.1   Perfect Numbers

Proof. Let's show that the formula yields perfect numbers. Since $2^p - 1$ is prime,
$$\sigma(2^p - 1) = 1 + (2^p - 1) = 2^p.$$
Also,
$$\sigma(2^{p-1}) = 1 + 2 + 4 + \cdots + 2^{p-1} = 2^p - 1$$
by Proposition 13.3. The multiplicativity of σ (Proposition 13.4) says that
$$\sigma(n) = \sigma(2^{p-1}(2^p - 1)) = \sigma(2^{p-1}) \cdot \sigma(2^p - 1) = (2^p - 1)(2^p) = 2n,$$
so n is perfect.

Now suppose n is perfect and even. We want to show that n has the desired form, namely, $n = 2^{p-1}(2^p - 1)$.

Write $n = 2^k m$, with m odd and $k \geq 1$. Then
$$\sigma(n) = \sigma(2^k) \cdot \sigma(m).$$
As before, we have $\sigma(2^k) = 1 + 2 + 4 + \cdots + 2^k = 2^{k+1} - 1$. Although we don't know much about m, let's see what we can say. If $m = 1$, then $n = 2^k$ and $\sigma(n) = 2^{k+1} - 1 \neq 2n$. Therefore, $m > 1$, and we have at least two distinct divisors of m, namely, 1 and m. Write
$$\sigma(m) = 1 + m + s,$$
where $s \geq 0$ is the sum of the other divisors of m. The key step will be to show that $s = 0$.

Since n is assumed to be perfect,
$$2^{k+1} m = 2n = \sigma(n) = \sigma(2^k) \cdot \sigma(m) = (2^{k+1} - 1)(1 + m + s).$$

When this is expanded and rearranged, we get
$$m = (2^{k+1} - 1)(s + 1). \tag{13.1}$$

Therefore, $s + 1$ is a divisor of m. Since n is even, we have $k \geq 1$ and $2^{k+1} - 1 > 1$. Equation (13.1) implies that $m > s + 1$. If $s \neq 0$, then $s + 1$ cannot be 1, so $m > s + 1 > 1$ and we have found another divisor of m, namely, $s + 1$. This means that
$$\sigma(m) \geq 1 + m + (s + 1),$$

But $\sigma(m) = 1 + m + s$, so $1 + m + s \geq 1 + m + (s+1)$, which is impossible. Therefore, we are forced to have $s = 0$. Since s is the sum of the divisors of m other than 1 and m, we conclude that m has no other divisors. The only numbers with this property are 1 and primes. Since we already showed that $m > 1$, we must have m prime. Also, Equation (13.1) implies that

$$m = (2^{k+1} - 1)(s+1) = 2^{k+1} - 1.$$

Therefore,
$$n = 2^k m = 2^k(2^{k+1} - 1),$$

which shows that n has the desired form. Since we showed that m is prime, $2^{k+1} - 1$ is prime. As shown in Section 1.10, this means that $k+1$ is prime. This completes the proof. □

Recall that primes of the form $M_p = 2^p - 1$ are called **Mersenne primes**. The first few Mersenne primes are

$$M_2 = 3, \quad M_3 = 7, \quad M_5 = 31, \quad M_7 = 127, \quad M_{13} = 8191.$$

The corresponding perfect numbers are

$$6, \quad 28, \quad 496, \quad 8128, \quad 33550336.$$

It is generally expected (but not yet proved) that there are infinitely many Mersenne primes, and therefore that there are infinitely many even perfect numbers. At present, the largest known Mersenne prime is $M_{57885161}$, and the corresponding perfect number is $2^{57885160}(2^{57885161} - 1)$.

What about odd perfect numbers? No odd perfect numbers are known, and it is conjectured that there are none. It has been proved that an odd perfect number must have at least 9 distinct prime factors. Results such as this lead most people to think that it is unlikely that odd perfect numbers exist. However, perhaps we should be slightly cautious. Most numbers have at least 9 prime factors, but we tend to meet only those with a few prime factors, so it's hard to have good intuition about such numbers.

CHECK YOUR UNDERSTANDING:
1. Show that 18 is an abundant number.

13.2 Multiplicative Functions

The relation $\sigma(mn) = \sigma(m) \cdot \sigma(n)$ when $\gcd(m, n) = 1$, which was very useful in the previous section, holds for several other functions. In this section, we develop the theory of such functions.

Definition 13.6. *A real-valued (or complex-valued) function $f(x)$ on the positive integers is called a* **multiplicative function** *if $f(mn) = f(m) \cdot f(n)$ for all pairs of positive integers m, n with $\gcd(m, n) = 1$.*

The condition that $f(mn) = f(m)f(n)$ when $\gcd(m, n) = 1$ might seem strange. Why don't we require the multiplicativity for all m, n? In several number theoretic functions, complete multiplicativity (that is, no restriction on the gcd) does not occur. For example, Euler's ϕ function is multiplicative, but $\phi(4 \cdot 6) = 8 \neq 4 = \phi(4)\phi(6)$.

Examples. You've already seen some of these examples, while some others are new.

1. The Euler ϕ function. For example,

    $$\phi(5) = 4, \quad \phi(7) = 6, \quad \phi(35) = 24 = \phi(5) \cdot \phi(7).$$

2. The function $f(n) = n$ is multiplicative, since $f(mn) = mn = f(m) \cdot f(n)$. We do not need the condition that $\gcd(m, n) = 1$ in this case.

3. The function $f(n) = 1$ for all n is multiplicative. Again, we do not need the condition that $\gcd(m, n) = 1$ in this case.

4. The sum of divisors function, $\sigma(n)$, is multiplicative. This property was used in the previous section and will be proved in this section.

5. Let n be a positive integer, and let $\tau(n)$ be the number of positive divisors of n (that is, the number of integers d such that $1 \leq d \leq n$ and $d \mid n$). For example,

    $$\tau(4) = 3, \quad \tau(75) = 6, \quad \tau(300) = 18.$$

    We'll show later (see Corollary 13.10) that τ is multiplicative.

6. Define the Möbius function:

$$\mu(n) = \begin{cases} (-1)^r & \text{if } n \text{ is a product of } r \text{ distinct primes.} \\ 0 & \text{otherwise.} \end{cases}$$

For example,

$$\mu(15) = (-1)^2 = +1, \quad \mu(12) = 0,$$
$$\mu(2) = -1, \quad \mu(1) = (-1)^0 = +1.$$

Then μ is multiplicative. This is Exercise 20.

Here is a very useful property of multiplicative functions.

Proposition 13.7. *If f and g are multiplicative functions and $f(p^j) = g(p^j)$ for all primes p and all $j \geq 0$, then $f(n) = g(n)$ for all positive integers n.*

Proof. Let $n = p_1^{a_1} p_2^{a_2} \cdots p_r^{a_r}$ be the prime factorization of n. Since $\gcd(p_i^{a_i}, p_j^{a_j}) = 1$ if $i \neq j$, the multiplicativity of f implies that

$$f(n) = f(p_1^{a_1}) \cdot f(p_2^{a_2} \cdots p_r^{a_r})$$
$$= f(p_1^{a_1}) \cdot f(p_2^{a_2}) \cdot f(p_3^{a_3} \cdots p_r^{a_r})$$
$$= \cdots$$
$$= f(p_1^{a_1}) \cdot f(p_2^{a_2}) \cdots f(p_r^{a_r})$$
$$= g(p_1^{a_1}) \cdot g(p_2^{a_2}) \cdots g(p_r^{a_r})$$
$$= g(p_1^{a_1} \cdot p_2^{a_2} \cdots p_r^{a_r})$$
$$= g(n). \qquad \square$$

If we have one multiplicative function, we can produce another as follows.

Proposition 13.8. *Let f be a multiplicative function and let*

$$g(n) = \sum_{d \mid n} f(d)$$

(the sum is over the positive divisors of n). Then g is multiplicative.

Proof. We start with a lemma.

Section 13.2   Multiplicative Functions   317

Lemma 13.9. *Let m and n be positive integers with $\gcd(m,n) = 1$. Let d be a positive divisor of mn. Then d has a unique decomposition $d = d_1 d_2$ with $d_1 \mid m$ and $d_2 \mid n$ (where d_1, d_2 are positive integers).*

Example. Let $m = 56 = 2^3 \cdot 7$ and $n = 75 = 3 \cdot 5^2$. Let $d = 70$. Then
$$70 \mid 4200 = 56 \cdot 75.$$
The lemma says that we can write $70 = 14 \cdot 5$, where $d_1 = 14 \mid 56$ and $d_2 = 5 \mid 75$.

Proof. Let $m = p_1^{a_1} p_2^{a_2} \cdots p_r^{a_r}$ and $n = q_1^{b_1} q_2^{b_2} \cdots q_s^{b_s}$ be the prime factorizations of m and n. A divisor d of mn has the form
$$d = p_1^{a_1'} p_2^{a_2'} \cdots p_r^{a_r'} q_1^{b_1'} q_2^{b_2'} \cdots q_s^{b_s'},$$
with $0 \leq a_i' \leq a_i$ for all i and $0 \leq b_j' \leq b_j$ for all j. Since we require that $d_1 \mid m$, and since $\gcd(m,n) = 1$, we have $\gcd(d_1, n) = 1$, so the primes q_j cannot appear in d_1. Similarly, the primes p_i cannot appear in d_2. Therefore, we have no choice and we must take
$$d_1 = p_1^{a_1'} p_2^{a_2'} \cdots p_r^{a_r'}, \qquad d_2 = q_1^{b_1'} q_2^{b_2'} \cdots q_s^{b_s'}.$$
Then $d = d_1 d_2$, with $d_1 \mid m$ and $d_2 \mid n$, as required. $\square$

We now can prove Proposition 13.8. Let g be as defined in the proposition, and let $\gcd(m,n) = 1$. Then
$$g(mn) = \sum_{d \mid mn} f(d) = \sum_{d_1 \mid m,\, d_2 \mid n} f(d_1 d_2).$$
Since $d_1 \mid m$ and $d_2 \mid n$, and since $\gcd(m,n) = 1$, we have $\gcd(d_1, d_2) = 1$. The function f is assumed to be multiplicative, so $f(d_1 d_2) = f(d_1) f(d_2)$. The sum becomes
$$\sum_{d_1 \mid m} \sum_{d_2 \mid n} f(d_1) f(d_2) = \left(\sum_{d_1 \mid m} f(d_1) \right) \left(\sum_{d_2 \mid n} f(d_2) \right) = g(m) g(n).$$
Therefore, g is multiplicative. $\square$

Corollary 13.10. *$\tau(n)$ (the number of divisors of n) is multiplicative.*

Proof. In Proposition 13.7, let $f(n) = 1$ for all n. Then f is certainly multiplicative. The function $g(n) = \sum_{d|n} 1$ is the sum of a set of 1s, one for each divisor of n. In other words, $g(n)$ is the number of divisors of n, which is $\tau(n)$. The proposition implies that $\tau(n)$ is multiplicative. $\square$

When $n = p^k$, it's easy to evaluate $\tau(p^k)$. The divisors of p^k are $1, p, p^2, \ldots, p^k$, so $\tau(p^k) = k+1$. The multiplicativity of τ implies that

$$\tau(p_1^{a_1} p_2^{a_2} \cdots p_r^{a_r}) = (a_1+1)(a_2+1)\cdots(a_r+1). \tag{13.2}$$

For example, $\tau(12) = \tau(2^2 \cdot 3) = (2+1)(1+1) = 6$.

Corollary 13.11. *$\sigma(n)$ (the sum of the divisors of n) is multiplicative.*

Proof. In Proposition 13.7, let $f(n) = n$, so

$$g(n) = \sum_{d|n} d = \sigma(n).$$

The proposition implies that $\sigma(n)$ is multiplicative. $\square$

The proof of the next corollary shows how Proposition 13.7 can be used to show that two multiplicative functions are equal.

Corollary 13.12. *Let $n \geq 1$. Then*

$$\sum_{d|n} \phi(d) = n,$$

where the sum is over all positive divisors of n.

Proof. Let $f(n) = \phi(n)$, which is multiplicative by Proposition 6.6. Then $g(n) = \sum_{d|n} \phi(d)$ is multiplicative. Inspired by Proposition 13.7, we compute $g(p^j)$. The divisors of p^j are $1, p, p^2, \ldots, p^j$. Therefore,

$$\begin{aligned} g(p^j) &= \phi(1) + \phi(p) + \phi(p^2) + \cdots + \phi(p^j) \\ &= 1 + (p-1) + (p^2 - p) + \cdots + (p^j - p^{j-1}) \\ &= p^j. \end{aligned}$$

Let $h(n) = n$, which is multiplicative. Then $g(p^j) = h(p^j)$ for all primes p and all $j \geq 0$. By Proposition 13.7, $g(n) = h(n)$ for all $n \geq 1$. This is the statement of the corollary. $\square$

Section 13.2   Multiplicative Functions   319

Example. Let $n = 12$. Then $\phi(1) + \phi(2) + \phi(3) + \phi(4) + \phi(6) + \phi(12) = 1 + 1 + 2 + 2 + 2 + 4 = 12$.

Corollary 13.13. *Let $n \geq 1$. Then*

$$\sum_{d|n} \mu(d) = \begin{cases} 1 & \text{if } n = 1 \\ 0 & \text{if } n > 1, \end{cases}$$

where μ is the Möbius function defined in the examples preceding Proposition 13.7.

Proof. Let $h(n) = 1$ if $n = 1$ and $h(n) = 0$ if $n > 1$. Then h is multiplicative: $h(mn) = 0$ unless $m = n = 1$, and $h(m) \cdot h(n) = 0$ unless $m = n = 1$. Also, $h(1)h(1) = h(1 \cdot 1)$. In Proposition 13.8, let

$$f(n) = \mu(n) \text{ and } g(n) = \sum_{d|n} \mu(d).$$

When $n = p^j$ and $j \geq 1$, we have

$$g(p^j) = \mu(1) + \mu(p) + \mu(p^2) + \cdots + \mu(p^j)$$
$$= 1 - 1 + 0 + \cdots + 0$$
$$= 0 = h(p^j).$$

Since $h(p^0) = 1$ and $g(p^0) = g(1) = 1$, we have proved that $g(p^j) = h(p^j)$ for all primes p and all $j \geq 0$. Proposition 13.7 implies that $g(n) = h(n)$ for all $n \geq 1$. This is the statement that we were trying to prove. □

Corollary 13.13 allows us to solve for f in terms of g in Proposition 13.8.

Proposition 13.14. *(Möbius inversion) Suppose f is any function (not necessarily multiplicative) and*

$$g(n) = \sum_{d|n} f(n).$$

Then

$$f(n) = \sum_{d|n} \mu(d) g(\frac{n}{d}).$$

Proof. The definition of g says that

$$g(\frac{n}{d}) = \sum_{e|(n/d)} f(e),$$

which allows us to write

$$\sum_{d|n} \mu(d)g(\frac{n}{d}) = \sum_{d|n} \mu(d) \sum_{e|(n/d)} f(e).$$

In the right-hand side, we have pairs of integers d and e such that $d \mid n$ and $e \mid (n/d)$. This simply means that $de \mid n$. Notice that

$$d \mid n \text{ and } e \mid (n/d) \iff e \mid n \text{ and } d \mid (n/e).$$

The sum becomes

$$\sum_{e|n} \sum_{d|(n/e)} \mu(d)f(e) = \sum_{e|n} \left(\sum_{d|(n/e)} \mu(d)\right) f(e) = f(n)$$

because Corollary 13.13 says that

$$\sum_{d|(n/e)} \mu(d) = \begin{cases} 1 \text{ if } n = e \\ 0 \text{ if } n > e, \end{cases}$$

which makes all the terms in the inner sum 0 except for the one with $n = e$. This completes the proof. □

This proposition gives us some interesting formulas. Here is one for the τ function. Since

$$\tau(n) = \sum_{d|n} 1,$$

Proposition 13.14 tells us that

$$1 = \sum_{d|n} \mu(d)\tau(\frac{n}{d}).$$

For example, let $n = 12$. Then

$$\mu(1)\tau(12) + \mu(2)\tau(6) + \mu(3)\tau(4) + \mu(4)\tau(3) + \mu(6)\tau(2) + \mu(12)\tau(1)$$
$$= 6 - 4 - 3 + 0 + 2 + 0 = 1.$$

Section 13.3 Chapter Highlights

If we apply Proposition 13.14 to the formula in Corollary 13.12, we recover a formula for $\phi(n)$. First, we have

$$\phi(n) = \sum_{d|n} \mu(d)\frac{n}{d}.$$

When $n = p^k$, this says that

$$\phi(p^k) = \mu(1)\frac{p^k}{1} + \mu(p)\frac{p^k}{p} + 0 + 0 + \cdots = p^k - p^{k-1} = p^k\left(1 - \frac{1}{p}\right).$$

The multiplicativity of ϕ yields the well-known formula

$$\phi(n) = \prod_{p|n} p^k\left(1 - \frac{1}{p}\right) = n\prod_{p|n}\left(1 - \frac{1}{p}\right).$$

CHECK YOUR UNDERSTANDING:
2. How many divisors does 60 have?
3. Evaluate $\sigma(40)$.

13.3 Chapter Highlights

1. ϕ, σ, τ, and μ are multiplicative.
2. Even perfect numbers have the form $2^{p-1}(2^p - 1)$, where $2^p - 1$ is prime.

13.4 Problems

13.4.1 Exercises

1. (a) List the divisors of 2, of 4, of 15, of 30, and of 60.
   (b) Compute $\sigma(n)$ for $n = 2, 4, 15, 30$, and 60.
   (c) Show that $\sigma(4)\sigma(15) = \sigma(60)$ but $\sigma(2)\sigma(30) \neq \sigma(60)$.

2. How many divisors does 1000 have?
3. What is the sum of the divisors of 20?
4. Find a value of n that has exactly 32 divisors.
5. Show that $\phi(n) = n - 1$ if and only if n is prime.
6. Show that $\sigma(n) = n + 1$ if and only if n is prime.
7. Show that if $n = 2^p - 1$ is prime, then $\sigma(\sigma(n)) = 2n + 1$.
8. Show that $\sigma(n) = 5$ has no solution.
9. Find four pairs (m, n) of integers with $5 \leq m < n$ such that $\phi(m) = \phi(n)$.
10. We know that $\phi(n) = \phi(n+1)$ when $n = 3$. Find a value of $n \geq 4$ with $\phi(n) = \phi(n+1)$.
11. Show that there is no integer n with $\phi(n) = 14$.
12. The first few perfect numbers are 6, 28, 496, 8128.
    (a) Show that the last digit of an even perfect number is always 6 or 8.
    (b) Show by example that the last digits of the perfect numbers do not always alternate between 6 and 8.
13. Show that n is perfect if and only if $2 = \sum_{d|n} \frac{1}{d}$ (for example, $2 = 1 + \frac{1}{2} + \frac{1}{3} + \frac{1}{6}$).
14. (a) Use the multiplicativity of $\tau(n)$ (= number of divisors) to show that $\tau(n)$ is odd if and only if n is a square.
    (b) Pair a divisor d of n with n/d. Note that a divisor d is paired with itself if and only if $n = d^2$. Deduce the result of part (a) directly from this observation.
15. There is a long row of doors labeled 1 through 10000. Originally, they are all closed. The first person opens all of them. The second person closes the 2nd, the 4th, the 6th, etc. The third person changes the status (that is, opens a closed door and closes an open door) of the 3rd, the 6th, the 9th, etc. This continues, with the nth person changing the status of the doors that are multiples of n until the 10000th person changes the status of the 10000th door. When this is finished, which doors are open and which are closed?
16. Descartes thought he had found that $n = 3^2 \cdot 7^2 \cdot 11^2 \cdot 13^2 \cdot 22021$

is an odd perfect number. It appears that he thought 22021 is prime and calculated

$$\sigma(3^2)\sigma(7^2)\sigma(11^2)\sigma(13^2)(1+22021) = 2n.$$

Using the facts $\sigma(7^2) = 3 \cdot 19$ and $\sigma(13^2) = 183 = 3 \cdot 61$, factor 22021.

17. Let $\sigma(n) = \sum_{d|n} d$ be the sum of the divisors of n, and let $\phi(n)$ be the Euler phi function. Let $n = pq$ be the product of two distinct primes. Show that $\sigma(n) + \phi(n) = 2n + 2$.

18. Let m and n be positive integers.
    (a) Show that
    $$\phi(m)\phi(n) = \phi(mn) \prod_{p|\gcd(m,n)} \left(1 - \frac{1}{p}\right).$$
    (b) Show that $\phi(m)\phi(n) \leq \phi(mn)$, and that there is equality if and only if $\gcd(m,n) = 1$.

19. Let $\sigma(n) = \sum_{d|n} d$ be the sum of the divisors of n.
    (a) Let p be an odd prime. Show that $\sigma(p^a)$ is odd if and only if a is even.
    (b) Show that $\sigma(2^a)$ is always odd.
    (c) Show that $\sigma(n)$ is odd if and only if n is a square or twice a square.
    (d) Show that a square cannot be a perfect number (as Lenstra remarked, this says that there are no "perfect squares").

20. Show that the Möbius function $\mu(n)$ is multiplicative.

21. (a) Let μ be the Möbius function. Let $g(n) = \sum_{d|n} \mu(d)^2$. Show that $g(n)$ is multiplicative.
    (b) Let $\omega(n)$ be the number of distinct prime factors dividing n (for example, $\omega(60) = 3$ since 2, 3, 5 divide 60). Show that $2^{\omega(n)}$ is multiplicative.
    (c) Show that
    $$\sum_{d|n} \mu(d)^2 = 2^{\omega(n)}.$$

22. Assume that m is perfect or abundant. Let $n \geq 2$ be an integer with $\gcd(m,n) = 1$. Show that mn is abundant.

23. We saw in the text that

$$1 = \sum_{d|n} \mu(d)\tau(\frac{n}{d}).$$

Suppose that $n = p_1 p_2 \cdots p_k$ is the product of k distinct primes. Show that this formula is the same as

$$1 = (2-1)^k = 2^k - \binom{k}{1}2^{k-1} + \binom{k}{2}2^{k-2} - \cdots + (-1)^k.$$

24. (a) Show that if $n > 1$, then

$$\frac{\sigma(n)}{n} < \prod_{p|n} \frac{p}{p-1},$$

where the product is over the distinct primes dividing n.
(b) Show that if n is an odd perfect number, then n has at least three distinct prime factors. (*Hint:* If there are two prime factors $p > q \geq 3$, then $\frac{p}{p-1}\frac{q}{q-1} \leq \frac{5}{4}\frac{3}{2} < 2$.)

25. Show that if $\sigma(m)$ is prime, then $\tau(m)$ is prime. (*Hint:* If $(p^n - 1)/(p-1)$ is prime, then n is prime.)

26. Two distinct positive integers a and b are called an **amicable pair** if b equals the sum of the divisors of a that are less than a, and a equals the sum of the divisors of b that are less than b. Show that 220 and 284 form an amicable pair.

13.4.2 Computer Explorations

1. Rosser and Schoenfeld showed that if $n \geq 3$, then

$$n/\phi(n) < 1.7811 \cdot \ln(\ln n) + \frac{5}{2\ln(\ln n)}$$

except for one value of n. Find that n. (*Hint:* Look at products of distinct small primes.)

2. Carmichael conjectured that if x and m are positive integers with $\phi(x) = m$, then there is an integer $y \neq x$ with $\phi(y) = m$. In 1998, Ford showed that this statement is true for all $m < 10^{10^{10}}$. Verify that it is true for all $m \leq 100$. (*Note:* For

some values of m, there might not be an x with $\phi(x) = m$. The conjecture says that if there is one, then there is at least one more. The inequality of Computer Exploration 1 can be used to bound the largest x with $\phi(x) \leq 100$.)

3. Mertens conjectured that $|\sum_{n \leq x} \mu(n)| \leq \sqrt{x}$ for all $x \geq 1$. Odlyzko and te Riele showed in 1985 that there are (very large, non-explicit) values of x such that the conjecture is false. Show that it is true for all $x \leq 1000$.

4. A positive integer n is called **highly composite** if n has more divisors than any smaller number. That is, $\tau(n) > \tau(m)$ for all m with $1 \leq m < n$. The first few highly composite numbers are 1, 2, 4, 6, 12, 24.
(a) There are 20 highly composite numbers less than 10000. Can you find them?
(b) Notice that the numbers $k!$ for $0 \leq k \leq 7$ are highly composite. Show that 8! and 9! are not highly composite.

5. A positive integer n is called **highly abundant** if the sum of the divisors of n is larger than the sum of the divisors of each smaller positive integer. That is, $\sigma(n) > \sigma(m)$ for all m with $1 \leq m < n$. The first few highly abundant numbers are 1, 2, 3, 4, 6, 8, 10.
(a) There are eight highly abundant numbers between 500 and 1000. Can you find them?
(b) $n = 9! = 362880$ is not highly abundant. Find $m < n$ with $\sigma(m) > \sigma(n)$.

13.4.3 Answers to "CHECK YOUR UNDERSTANDING"

1. The divisors of 18 that are less than 18 are 1, 2, 3, 6, 9. Their sum is 21, which is larger than 18. Therefore, 18 is abundant.

2. By Equation (13.2), $\tau(60) = \tau(2^2 \cdot 3 \cdot 5) = (2+1)(1+1)(1+1) = 12$.

3. $\sigma(40) = \sigma(2^3)\sigma(5) = (1+2+4+8)(1+5) = 90$.

Chapter 14

Continued Fractions

Christiaan Huygens faced a challenge. In 1680, the Dutch astronomer was designing a mechanical planetarium for l'Académie Royale des Sciences in France. He calculated that in one year, the Earth goes through an angle of $359°45'40''31'''$ of arc, where $31'''$ denotes $31/60$ seconds. This is

$$(359 \cdot 60^3 + 45 \cdot 60^2 + 40 \cdot 60 + 31)/60 = 77708431/60$$

seconds of arc (note that this is slightly less than $360°$, which is why we need leap years). Saturn, which goes slower, goes through $2640858/60$ seconds of arc in a year. Therefore, in his planetarium, he needed gears in the ratio of

$$\frac{77708431}{2640858} \approx 29.42545$$

in order to make the relative motions of Earth and Saturn accurate. Since it was impossible to make a gear with 77708431 teeth, he approximated the ratio as

$$\frac{77708431}{2640858} \approx \frac{206}{7} \approx 29.42857.$$

Why did he choose this approximation, and how did he find it? Huygens wrote

$$\frac{77708431}{2640858} = 29 + \cfrac{1}{2 + \cfrac{1}{2 + \cfrac{1}{1 + \cfrac{1}{5 + \cfrac{1}{1 + \cdots}}}}}$$

327

(the fraction continues for 11 more terms). Expressions such as this are called **continued fractions**. Stopping before the 5 yields

$$\frac{206}{7} = 29 + \cfrac{1}{2 + \cfrac{1}{2 + \cfrac{1}{1}}}$$

which is the approximation that Huygens used.

In this chapter, we develop the theory of continued fractions, show how they give rational numbers approximating real numbers, and use them to find solutions of Pell's equation. The proofs of some of the results are rather technical, so we give only a description of the method and omit the proofs.

14.1 Rational Approximations

Suppose we have a real number, for example, $\pi = 3.14159265\cdots$, and we want to find a good approximation using rational numbers. We could use the rational number
$$\frac{314159}{100000},$$
but we can obtain a better approximation with
$$\frac{355}{113} = 3.14159292\cdots,$$
which also uses a much smaller denominator. Of course, we could use a denominator of 10000000 and obtain a better approximation, but we want to look at the size of the denominator when we consider how good an approximation is. In Exercise 18 (see also Theorem 14.1), we'll show that if x is a real number, then there are integers p and q such that

$$\left| x - \frac{p}{q} \right| < \frac{1}{q^2}. \tag{14.1}$$

For example,
$$\left| \pi - \frac{22}{7} \right| \approx .00126 < \frac{1}{7^2}, \quad \text{and} \quad \left| \pi - \frac{355}{113} \right| \approx .00000027 < \frac{1}{113^2}.$$

Section 14.1   Rational Approximations   329

However,
$$\left|\pi - \frac{314159}{100000}\right| \approx .00000265 > \frac{1}{100000^2},$$

so this is not as good an approximation.

But how do we find good approximations? The answer is continued fractions. Let's use π as an example. The easiest approximation is $\pi \approx 3$. To refine this, we look at $\pi - 3 = .14159265\cdots$. We need to approximate $.14159265\cdots$. Since it is less than 1, we invert it to get

$$\frac{1}{.14159265\cdots} = 7.0625133\cdots = 7 + .0625133\cdots$$

(calculations such as this one are easily done on a calculator). The integer part of this is 7, so we use

$$3 + \frac{1}{7} = \frac{22}{7}$$

as the next approximation. To continue the process, we need to find an approximation for $.0625133$, so we invert it and continue:

$$\frac{1}{.0625133\cdots} = 15.99659\cdots = 15 + .99659\cdots.$$

We now have

$$\pi = 3 + \cfrac{1}{7 + \cfrac{1}{15.99659\cdots}}$$

and the approximation

$$\pi \approx 3 + \cfrac{1}{7 + \cfrac{1}{15}} = \frac{333}{106}.$$

Then we approximate $.99659\cdots$ by inverting it to get

$$\frac{1}{.99659\cdots} = 1.0034\cdots.$$

We now have

$$\pi = 3 + \cfrac{1}{7 + \cfrac{1}{15 + \cfrac{1}{1.0034\cdots}}}$$

and the approximation

$$\pi \approx 3 + \cfrac{1}{7 + \cfrac{1}{15 + \cfrac{1}{1}}} = \frac{355}{113}.$$

We could continue:

$$\frac{1}{.0034\cdots} = 292.63\cdots.$$

The fact that 292 is relatively large is an indication that the previous step yields a very good approximation, namely 355/113. This will be explained in Theorem 14.1.

You might wonder why we didn't approximate 15.99659 by 16 rather than by 15. If we had done so, we would have obtained the *nearest integer continued fraction*. The theory for these is very similar to the theory we'll be describing for ordinary continued fractions. However, note that the algorithm we're using corrects itself: since 16 is better than 15, the next fraction we get is $15 + \frac{1}{1}$, which is 16. The nearest integer method skips a step here.

This process can be done for any real number x. Recall that $\lfloor x \rfloor$ denotes the largest integer less than or equal to x. Let $a_0 = \lfloor x \rfloor$ and $x_1 = 1/(x - a_0)$. For $k \geq 1$, let

$$a_k = \lfloor x_k \rfloor \quad \text{and} \quad x_{k+1} = 1/(x_k - a_k).$$

If $a_k = x_k$, which means that x_k is an integer, we stop. In general, we form the continued fraction

$$[a_0, a_1, a_2, a_3, \ldots] \stackrel{\text{def}}{=} a_0 + \cfrac{1}{a_1 + \cfrac{1}{a_2 + \cfrac{1}{a_3 + \cdots}}}$$

(the notation $[a_0, a_1, a_2, a_3, \ldots]$ saves a lot of paper). As we'll see later, when we stop after any a_n we obtain rational approximations that always satisfy Equation (14.1). Note that for each k we have

$$x = [a_0, a_1, \ldots, a_{k-1}, x_k]. \tag{14.2}$$

Section 14.1   Rational Approximations   331

Technical point. Suppose we compute the continued fraction of $x = -5.4321$. Then $a_0 = \lfloor x \rfloor = -6$, and we have $x = -6 + .5679 = -6 + 1/1.76087\cdots$. Therefore, after the initial step, we are computing using only positive numbers. Therefore, throughout this chapter, we work only with continued fraction expansions of positive real numbers.

If we allow arbitrarily large denominators, we can approximate a real number arbitrarily closely. But, as was the case with Huygens in the introduction to this chapter, it is often desirable to approximate a real number using small denominators. The following quantifies the accuracy of the approximation obtained from a continued fraction, where the estimate is given in terms of the size of the denominator of the rational number.

Theorem 14.1. *If the continued fraction for x is $[a_0, a_1, a_2, \ldots]$, then*

$$\left| x - \frac{p_k}{q_k} \right| < \frac{1}{a_{k+1} q_k^2}.$$

We omit the proof. The theorem explains why we obtain good approximations by stopping just before large values of a_n. Since $a_{k+1} \geq 1$, we always have

$$\left| x - \frac{p_k}{q_k} \right| < \frac{1}{q_k^2}.$$

But if a_k is large, then the right-hand side in the theorem is small, so the continued fraction gives a better than expected approximation. For example, 22/7 is a good approximation for π because the next a_n is 15. The approximation 355/113 is very good because the next a_n is 292, which is quite large.

CHECK YOUR UNDERSTANDING:

1. Compute the first three terms of the continued fraction of 2.357; that is, write $2.357 = [a_0, a_1, a_2, x_3]$.

14.2 Evaluating Continued Fractions

If you tried to evaluate

$$3 + \cfrac{1}{7 + \cfrac{1}{15 + \cfrac{1}{1}}} = 3 + \cfrac{1}{7 + \cfrac{1}{16}} = 3 + \cfrac{1}{\frac{113}{16}} = \frac{355}{113},$$

you probably found it rather annoying. And if you wanted to include one more term, namely the 292, you'd have to start all over. Fortunately, there is a faster way.

Proposition 14.2. *Let $a_0, a_1, a_2, \ldots$ be positive real numbers. Let*

$$p_{-2} = 0 \qquad p_{-1} = 1$$
$$q_{-2} = 1 \qquad q_{-1} = 0.$$

For $k \geq 0$, define

$$p_k = a_k p_{k-1} + p_{k-2}$$
$$q_k = a_k q_{k-1} + q_{k-2}.$$
(14.3)

Then

$$\frac{p_n}{q_n} = [a_0, a_1, a_2, \ldots, a_n]$$

for each $n \geq 0$.

The fraction p_k/q_k is called the **kth convergent** of the continued fraction. Although we're mostly interested in the situation where the a_k are integers, we allow them to be real numbers here since this is sometimes useful. We could also allow the a_k to be negative, as long as we don't end up dividing by 0, but we won't need this.

Rather than sorting out the mess of indices in the proposition, it's easier to calculate the numbers p_k and q_k as follows. Make a table with the a_k in the top row and $p_{-2}, p_{-1}, q_{-2}, q_{-1}$ at the beginning, then fill in the p_k and q_k:

Section 14.2 Evaluating Continued Fractions

		a_0	a_1	a_2	a_3	a_4
0	1	p_0	p_1	p_2	p_3	p_4
1	0	q_0	q_1	q_2	q_3	q_4

For example, for $[3, 7, 15, 1]$, we start with

		3	7	15	1
0	1				
1	0				

Then we fill in the values in the table:

		3	7	15	1
0	1	3	22	333	355
1	0	1	7	106	113

For example, to get the entry 333 that is below 15, multiply 15 times the entry 22 in the previous column and add to it the entry 3 from the column before that:
$$15 \times 22 + 3 = 333.$$
To get 355, multiply 1×333 and add 22:
$$1 \times 333 + 22 = 355.$$

Similarly, to obtain $q_k = 113$ in the bottom row, multiply 1 (the a_k at the top of the column) times 106 (which is q_{k-1}) and add 7 (which is q_{k-2}).

Note that we obtain all the intermediate approximations along the way. If we decide to include one more a_k, we can simply continue, rather than start over.

CHECK YOUR UNDERSTANDING:

2. Evaluate $[1, 3, 5, 7, 9]$ as a rational number.

14.3 Pell's Equation

Now let's turn our attention to Pell's equation. We want to find integers x and y that are solutions to

$$x^2 - dy^2 = 1.$$

If we rewrite this as

$$\frac{x^2}{y^2} - d = \frac{1}{y^2},$$

we see that x/y is approximately $\sqrt{d}$. With this in mind, let's compute the continued fraction for $\sqrt{d}$.

For example, if we let $d = 19$ and use the recursive definition of a_k, a computation yields the continued fraction

$$[4, 2, 1, 3, 1, 2, 8, 2, 1, 3, 1, 2, 8, 2, 1, 3, 1, 2, 8, \ldots]$$

for $\sqrt{19}$. Note that, after the initial 4, it has a repeating pattern. This is a general phenomenon of continued fractions of square roots of integers. To indicate the repetition, we write

$$[4, 2, 1, 3, 1, 2, 8, 2, 1, 3, 1, 2, 8, 2, 1, 3, 1, 2, 8, \ldots] = [4, \overline{2, 1, 3, 1, 2, 8}].$$

Let's compute the numbers p_k and q_k:

		4	2	1	3	1	2	8	2	1	3	1	2
0	1	4	9	13	48	61	170	1421	3012	4433	16311	20744	57799
1	0	1	2	3	11	14	39	326	691	1017	3742	4759	13260

Since $a_6 = 8$ is relatively large, stopping with the preceding a_k should give a good approximation (as with the 292 in the expansion for π, this is made precise in Theorem 14.1). Therefore, we look at 170/39 and compute

$$170^2 - 19 \cdot 39^2 = 1.$$

If we go to the next 8 (that is, the first column that is not computed in the table) and look at the preceding convergent, we have

$$57799^2 - 19 \cdot 13260^2 = 1.$$

This is not a coincidence.

Section 14.3   Pell's Equation

Theorem 14.3. *Let d be a positive integer, not a square. The continued fraction for $\sqrt{d}$ has the form*

$$[a_0, \overline{a_1, a_2, \ldots, a_{n-1}, a_n}],$$

where $a_n = 2a_0$. We have $a_k \leq a_0$ for $1 \leq k \leq n-1$. Let $m \geq 1$, and let $x = p_{mn-1}$ and $y = q_{mn-1}$. Then

$$x^2 - dy^2 = \pm 1.$$

We now have an easy algorithm for solving $x^2 - dy^2 = 1$.

Solving $x^2 - dy^2 = 1$

1. Compute $a_0 = \lfloor \sqrt{d} \rfloor$.
2. Compute the continued fraction $[a_0, a_1, a_2, \ldots]$ for $\sqrt{d}$. Stop when $a_n = 2a_0$.
3. Let $x_1 = p_{n-1}$ and $y_1 = q_{n-1}$.
4. If $x_1^2 - dy_1^2 = 1$, let $x = x_1$ and $y = y_1$ and stop.
5. If $x_1^2 - dy_1^2 = -1$, compute $x + y\sqrt{d} = (x_1 + y_1\sqrt{d})^2$. Then $x^2 - dy^2 = 1$.

Example. Let $d = 13$. Using a calculator, we find that the continued fraction expansion for $\sqrt{13} = 3.60555\cdots$ is

$$[3, 1, 1, 1, 1, 6, \ldots].$$

Since 6 is twice the initial 3, we compute

$$[3, 1, 1, 1, 1] = \frac{18}{5}$$

and $18^2 - 13 \cdot 5^2 = -1$. Since this is -1, we compute

$$(18 + 5\sqrt{13})^2 = 18^2 + 2 \cdot 18 \cdot 5\sqrt{13} + 5^2 \cdot 13 = 649 + 180\sqrt{13}.$$

Then

$$649^2 - 13 \cdot 180^2 = 1.$$

The reason that squaring changes the -1 to $+1$ is that

$$\begin{aligned}x^2 - dy^2 &= (x+y\sqrt{d})(x-y\sqrt{d}) = (x_1+y_1\sqrt{d})^2(x_1-y_1\sqrt{d})^2 \\ &= \left((x_1+y_1\sqrt{d})(x_1-y_1\sqrt{d})\right)^2 \\ &= (x_1^2 - dy_1^2)^2 = (-1)^2 = 1.\end{aligned}$$

The reason that we can stop when we get $a_n = 2a_0$ is that all the other a_k are less than $2a_0$, so there is no danger of stopping at the wrong place. It is true that if we go a full period extra, we again get $a_{2n} = 2a_0$, and we get a solution of Pell's equation by stopping there. In fact, if we get $x_1^2 - dy_1^2 = -1$, then we can also obtain $x^2 - dy^2 = +1$ by continuing the continued fraction calculation. Then $x = p_{2n-1}$ and $y = q_{2n-1}$ give the same solution to Pell's equation as we get by squaring. But squaring is usually faster.

Occasionally, we stop as soon as we get started. For example, consider $d = 17$. The continued fraction is $[4, 8, 8, 8, \ldots]$. Then we use only a_0, so $p_0 = 4$ and $q_0 = 1$, and

$$4^2 - 17 \cdot 1^2 = -1.$$

Squaring yields $x = 33$ and $y = 8$, and $33^2 - 17 \cdot 8^2 = 1$. A similar situation occurs for all d of the form $n^2 + 1$.

CHECK YOUR UNDERSTANDING:

3. Use continued fractions to find a nontrivial solution of $x^2 - 11y^2 = 1$.

14.4   Chapter Highlights

1. Continued fractions give good rational approximations of real numbers (Theorem 14.1).

2. Continued fractions yield solutions of Pell's equation (Theorem 14.3).

Section 14.5 Problems

3. The continued fraction of the square root of a positive non-square integer has the form

$$[a_0, \overline{a_1, a_2, \ldots, a_{n-1}, a_n}],$$

where $a_n = 2a_0$ (Theorem 14.3).

14.5 Problems
14.5.1 Exercises

1. Find the convergents p_0/q_0, p_1/q_1, p_2/q_2, and p_3/q_3 of the continued fraction expansion of 1.234567 and verify that

$$\left| 1.234567 - \frac{p_n}{q_n} \right| < \frac{1}{q_n q_{n+1}}$$

for $n = 0, 1, 2, 3$.

2. Find a, b less than 100 such that

$$|.123456789 - (a/b)| < 10^{-8}.$$

3. Find a and b less than 20 such that

$$|.2357111317 - (a/b)| < .001.$$

4. Find a and b with $0 < b < 20$ such that

$$\left| \frac{3141}{5926} - \frac{a}{b} \right| < .001.$$

5. Find a and b with $0 < b < 20$ such that

$$\left| \frac{1357}{2468} - \frac{a}{b} \right| < .005.$$

6. Evaluate $[1, 2, 3, 4]$ as a rational number.
7. Evaluate $[4, 3, 2, 1]$ as a rational number.

8. For each denominator b with $1 \leq b \leq 6$, find the fraction a/b that is closest to π. Show that none of these give as close an approximation to π as $22/7$.

9. (a) Find the continued fraction for $\sqrt{7}$.
   (b) Find a nontrivial solution of $x^2 - 7y^2 = 1$.

10. (a) Find the continued fraction for $\sqrt{11}$.
    (b) Find a nontrivial solution of $x^2 - 11y^2 = 1$.

11. (a) Find the continued fraction for $\sqrt{15}$.
    (b) Find a nontrivial solution of $x^2 - 15y^2 = 1$.

12. (a) Find the continued fraction for $\sqrt{n^2 + 1}$, where $n \geq 1$ is an integer.
    (b) Find a nontrivial solution of $x^2 - (n^2 + 1)y^2 = 1$.

13. (a) Find the continued fraction for $\sqrt{n^2 - 1}$, where $n \geq 2$ is an integer.
    (b) Find a nontrivial solution of $x^2 - (n^2 - 1)y^2 = 1$.

14. (a) Find the continued fraction for $\sqrt{n^2 - 2}$, where $n \geq 3$ is an integer.
    (b) Find a nontrivial solution of $x^2 - (n^2 - 2)y^2 = 1$.

15. Show that there are infinitely many Pythagorean triples a, b, c such that $a + 1 = b$. (*Hint:* Change $a^2 + b^2 = c^2$ to an equation in two variables, multiply by 2, and complete the square to obtain a Pell equation.)

16. Show that there are infinitely many n such that both n and $2n + 1$ are squares.

17. (a) Evaluate $[1, 1], [1, 1, 1], [1, 1, 1, 1]$ as rational numbers.
    (b) Let
    $$\frac{p_n}{q_n} = [1, 1, 1, \cdots, 1],$$
    where there are $n + 1$ ones (since the first 1 is a_0, the $(n+1)$st is a_n). Use the recursion relations for p_n and for q_n to show that
    $$p_n = F_{n+2}, \qquad q_n = F_{n+1}.$$

18. Let x be a real number. Continued fractions can be used to produce rational numbers a/b with $|x - (a/b)| < 1/b^2$. Here we use a different method. For a real number y, let $\{y\} = y - \lfloor y \rfloor$

be the fractional part of y. For example, $\{2.75\} = .75$ and $\{-3.2\} = .8$.

(a) Let N be a positive integer. Consider the numbers

$$\{0\}, \{x\}, \{2x\}, \{3x\}, \ldots, \{Nx\}.$$

Show that there exist i and j with $0 \le i < j \le N$ such that

$$|\{jx\} - \{ix\}| < 1/N.$$

(*Hint:* Break up the interval from 0 to 1 into N subintervals of length $1/N$ and explain why there must be some subinterval containing two numbers.)

(b) Use (a) to show that there exists an integer b with $1 \le b \le N$, and an integer a such that $|bx - a| < 1/N$.

(c) Let a and b be as in part (b). Show that $|x - (a/b)| < 1/b^2$.

14.5.2 Computer Explorations

1. Find a nontrivial solution of $x^2 - 61y^2 = 1$.
2. Find a nontrivial solution of $x^2 - 109y^2 = 1$.
3. Calculate several terms of the continued fractions of e, $(e+1)/(e-1)$, and $(e^2+1)/(e^2-1)$ until you find a pattern. Calculate a few more terms and see if the pattern persists (if you have the correct pattern it should keep going).

14.5.3 Answers to "CHECK YOUR UNDERSTANDING"

1. We have $\lfloor 2.357 \rfloor = 2$, so $a_0 = 2$ and $x_1 = 1/.357 = 2.8011\cdots$. Since $\lfloor 2.8011 \rfloor = 2$, we have $a_1 = 2$ and $x_2 = 1/.8011 = 1.248\cdots$. Next, $a_3 = 1$ and $x_3 = 1/.248 = 4.027\cdots$. Therefore, $2.357 = [2, 2, 1, 4.027]$.

2. Write the table

		1	3	5	7
0	1	1	4	21	151
1	0	1	3	16	115

.

This yields the answer 151/115.

3. Calculate $\sqrt{11} = 3.31662\cdots$. The continued fraction starts $[3, 3, 6, \ldots]$. Since $a_0 = 3$ and $6 = 2a_0$, we stop before the 6 and evaluate $[3, 3] = 3 + 1/3 = 10/3$. Then $10^2 - 11 \cdot 3^2 = 1$, so $(x, y) = (10, 3)$ is a solution.

Chapter 15

Recent Developments

The development of number theory has always been driven by problems. People play with numbers and find patterns. Do these patterns persist? Can we prove them? In this chapter, we discuss some of the more prominent examples that have influenced number theory over the years.

15.1 Goldbach's Conjecture and the Twin Prime Problem

In 1742, Christian Goldbach wrote a letter to Euler conjecturing that every even integer from 4 onward can be written as a sum of two primes. For example, $100 = 29 + 61$ and $32 = 13 + 19$. This has become known as *Goldbach's Conjecture*. No one had any idea how to prove it, and there was really no progress until 1920, when Viggo Brun showed that every even integer is a sum of two integers, each of which has at most 9 prime factors. This led to the development of *sieve theory*, which is a vast generalization of the Sieve of Eratosthenes. The techniques were continually refined, and in 1966, Jingrun Chen showed that every sufficiently large even integer is a sum of a prime and an integer that is either prime or the product of two primes. The related question for odd integers was studied in the 1930s by Vinogradov, who showed that every sufficiently large odd integer is a sum of three primes. The problem was completed in 2013 by Helfgott, who showed that every odd integer greater than or equal to 7 is a sum of three primes.

Exercise. *(a) Show that every even integer n with $4 \leq n \leq 200$ is a sum of two primes.*
(b) There is a wide variation in the number of ways that a number n can be written as a sum of two primes. Show that 128 is a sum of two primes in only 3 ways, while 180 is a sum of two primes in 14 ways. If you have a computer, count the number of ways that 990 is a sum of two primes (answer: 52).

The **Twin Prime Problem** asks whether there are infinitely many twin primes; that is, are there infinitely many primes p such that $p+2$ is also prime? In this case, $(p, p+2)$ is called a twin prime pair. For example, $(11, 13)$ is a twin prime pair. There are 8 pairs up to 100, there are 205 up to 1000, and there are 808,675,888,577,436 pairs less than 10^{18}. It is reasonable to conjecture that there are infinitely many twin prime pairs, but no one has yet proved this.

More generally, let k be a positive even integer. It was conjectured by de Polignac in 1849 that there are infinitely many primes p such that $p+k$ is also prime. For example, when $k = 10$, there are the pairs $(31, 41)$ and $(37, 47)$. There has been spectacular recent progress on this conjecture. In 2013, Yitan Zhang proved that there is some $k \leq 70000000$ such that de Polignac's conjecture is true for k. Work of Maynard, Tao, and others has reduced this to $k \leq 246$. So now we know that there is some $k \leq 246$ such that $p+k$ is prime for infinitely many primes p. Therefore, de Polignac's conjecture is true for at least one k. Unfortunately, we don't know a specific value of k for which it works, just that there is some k (and $k \leq 264$).

Exercise. *Find the 8 twin prime pairs up to 100. Observe that for each prime pair $(p, p+2)$ on your list, $p+1$ is a multiple of 6. Can you prove this is always true for twin prime pairs (except for the pair $(3, 5)$)?*

15.2 Fermat's Last Theorem

In the late 1630s, Pierre de Fermat was reading Diophantus's *Arithmetica*, which had recently received a new translation from Greek into Latin by Bachet. *Arithmetica* consists of hundreds of algebraic problems covering linear, quadratic, and higher-degree equations and uses

Section 15.2 Fermat's Last Theorem

algebraic techniques to find their solutions. Fermat must have been intrigued by Problem 8 from Book II, which asks how "to divide a square number into two squares." Diophantus starts with a square, for example 16, and tries to write it in the form $x^2 + y^2$ with nonzero rational numbers x and y. To do so, he starts with an "obvious" solution, namely $(x, y) = (4, 0)$, and uses it to find a nontrivial solution. Take, for example, a line with slope -2 through $(4, 0)$, described parametrically by

$$x = 4 - t, \quad y = 2t$$

for a parameter t. Substitute this into $x^2 + y^2 = 16$:

$$16 = (4 - t)^2 + (2t)^2 = 16 - 8t + 5t^2.$$

This says that $8t = 5t^2$. The solution $t = 0$ corresponds to the point $(4, 0)$ that we already have. The solution $t = 8/5$ yields $(x, y) = (12/5, 16/5)$:

$$16 = \left(\frac{16}{5}\right)^2 + \left(\frac{12}{5}\right)^2.$$

Other values of the slope yield other solutions to the problem, and it is clear that any square can be decomposed into a sum of two squares in this way, since there is always an "obvious" solution to use as a starting point.

Fermat wrote the following in the margin of his copy of *Arithmetica*:

"On the other hand, it is impossible to separate a cube into two cubes, or a biquadrate into two biquadrates, or generally any power except a square into two powers with the same exponent. I have discovered a truly marvelous proof of this, which however the margin is not large enough to contain."

Fermat is asserting that the equation

$$x^n + y^n = z^n \tag{15.1}$$

has no solutions if x, y, and z are nonzero integers and $n > 2$.

Fermat never provided a proof of (15.1), except for the case $n = 4$. Near the end of his life, he made a list of his major accomplishments in number theory and mentioned only the cases $n = 3$ and $n = 4$ of (15.1).

After Fermat's death, many of his notes and letters were published by his son, Samuel. Eventually, all of Fermat's assertions and claims were resolved except (15.1), so this became known as Fermat's Last Theorem, even though it was neither the last thing he did nor was it a theorem. Throughout the 18th and 19th centuries, many mathematicians tried to prove Fermat's Last Theorem. Although none succeeded, progress was made. For example, Euler gave a proof for $n = 3$, and Dirichlet and Legendre showed in 1825 that (15.1) has no positive integer solutions for $n = 5$.

There is a slight simplification of the problem. Suppose that you want to prove that the sum of two tenth powers cannot be a tenth power. Since a tenth power is automatically a fifth power, it suffices to prove that the sum of two fifth powers cannot be a fifth power. Because every integer $n \geq 3$ is a multiple either of 4 or of an odd prime, it suffices to prove Fermat's Last Theorem for $n = 4$ and for odd primes. Since Fermat already proved it for $n = 4$, only odd prime exponents need to be considered.

In the mid-nineteenth century, the great German mathematician Ernst Kummer began the development of modern algebraic number theory. Part of his work gave criteria that guarantee the validity of Fermat's Last Theorem for a given exponent n. (If Kummer's criteria are not satisfied, it doesn't mean that Fermat's Last Theorem is false for that exponent, only that his method does not apply in that situation.) Using these ideas, Kummer proved Fermat's Last Theorem for all exponents less than 100. With extensions of these techniques and the use of computers, the list of exponents was expanded. By 1993 it was known that Fermat's Last Theorem is true for exponents up to 4000000.

In 1985, Gerhard Frey suggested a new approach based on the theory of elliptic curves. The graph of an equation of the form $y^2 = x(x - A)(x - B)$, where A and B are integers, is an example of an elliptic curve. See Figure 15.1.

A cubic polynomial $(x - r_1)(x - r_2)(x - r_3))$ has roots r_1, r_2, r_3, and a fundamental quantity attached to this polynomial is its *discriminant*

$$\Delta = (r_1 - r_2)^2 (r_1 - r_3)^2 (r_2 - r_3)^2.$$

Now suppose $a^n + b^n = c^n$ with nonzero integers a, b, c, and an integer $n \geq 3$. Form the equation $y^2 = x(x - a^n)(x + b^n)$. The discriminant of

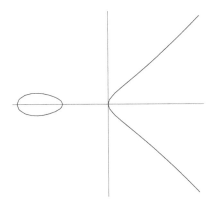

FIGURE 15.1: An elliptic curve.

the cubic polynomial, which has roots $0, a^n, -b^n$, is
$$\Delta = (0-a^n)^2(0+b^n)^2(a^n+b^n)^2 = a^{2n}b^{2n}c^{2n}.$$
Frey conjectured that if an elliptic curve has a discriminant that is an nth power and has a certain property, called *modularity*, then this puts such severe restriction on the elliptic curve that it cannot exist. In 1986, Ribet proved Frey's conjecture, and in 1994, Andrew Wiles, with the assistance of Richard Taylor, proved that Frey's elliptic curves must have the modularity property. Combined with Ribet's theorem, this says that Frey's elliptic curves cannot exist. Since a nonzero solution to the Fermat equation yields a Frey elliptic curve, the conclusion is that the Fermat equation has no nonzero solutions. This completed the proof of Fermat's Last Theorem.

15.3 The Riemann Hypothesis

Let $s > 1$ be a real number. The **Riemann zeta function** is defined as
$$\zeta(s) = \sum_{n=1}^{\infty} \frac{1}{n^s}.$$
It is a standard application of the integral test to show that this infinite series converges for each $s > 1$. In the 1700s, Euler discovered a striking

connection between the zeta function and prime numbers that gives an analytic expression of the fact that every positive integer is uniquely a product of primes.

Theorem 15.1. *Let $s > 1$. Then*
$$\zeta(s) = \prod_p (1 - p^{-s})^{-1},$$
where the product is over all prime numbers.

Proof. We ignore technicalities about convergence except to say that the absolute convergence (not just conditional convergence) of the sum and product can be used to justify our manipulations. Recall the formula for the sum of a geometric series:
$$1 + r + r^2 + r^3 + \cdots = \frac{1}{1-r} = (1-r)^{-1}$$
when $|r| < 1$. Letting $r = p^{-s}$, we obtain
$$1 + \frac{1}{p^s} + \frac{1}{p^{2s}} + \frac{1}{p^{3s}} + \cdots = (1 - p^{-s})^{-1}.$$

Consider the product
$$(1 - 2^{-s})^{-1}(1 - 3^{-s})^{-1}$$
$$= \left(1 + \frac{1}{2^s} + \frac{1}{4^s} + \frac{1}{8^s} + \cdots\right)\left(1 + \frac{1}{3^s} + \frac{1}{9^s} + \cdots\right).$$

When we expand this, we obtain
$$1 + \frac{1}{2^s} + \frac{1}{4^s} + \frac{1}{8^s} + \cdots + \frac{1}{3^s} + \frac{1}{2^s 3^s}$$
$$+ \frac{1}{4^s 3^s} + \cdots + \frac{1}{9^s} + \frac{1}{2^s 9^s} + \frac{1}{4^s 9^s} + \cdots$$
$$= 1 + \frac{1}{2^s} + \frac{1}{3^s} + \frac{1}{4^s} + \frac{1}{6^s} + \frac{1}{8^s} + \frac{1}{9^s} + \frac{1}{12^s} + \cdots$$
$$= \sum_{n \in S(2,3)} \frac{1}{n^s},$$

where $S(p, q, \ldots)$ denotes those integers (including $1 = p^0 q^0 \cdots$) whose prime factorizations involve only primes in $\{p, q, \dots\}$. Note that each integer in $S(2,3)$ occurs exactly once in the sum. For example, $12 \in S(2,3)$, and the term $1/12^s$ comes from $1/4^s \times 1/3^s$. The uniqueness of factorization says that this is the only way it can occur.

When we multiply by $(1-5^{-s})^{-1}$, we obtain

$$(1-2^{-s})^{-1}(1-3^{-s})^{-1}(1-5^{-s})^{-1} = \sum_{n \in S(2,3,5)} \frac{1}{n^s}.$$

In general, if p_m is the mth prime, we have

$$(1-2^{-s})^{-1}(1-3^{-s})^{-1}(1-5^{-s})^{-1} \cdots (1-p_m^{-s})^{-1} = \sum_{n \in S(2,3,5,\dots,p_m)} \frac{1}{n^s}.$$

Let $m \to \infty$. The left side converges to the product over all primes. Since every positive integer has a prime factorization, each positive integer n lies in $S(2,3,5,\cdots,p_m)$ for a sufficiently large integer m. Therefore, the right side converges to the sum over all positive integers n. This gives the identity in the theorem. $\square$

Euler initiated the study of $\zeta(s)$ and showed, for example, that

$$\zeta(2) = \sum_{n=1}^{\infty} \frac{1}{n^2} = \frac{\pi^2}{6}, \text{ and } \zeta(4) = \frac{\pi^4}{90}.$$

In general, when $m \geq 1$ is an integer,

$$\zeta(2m) = \frac{(-1)^{m+1}(2\pi)^{2m}B_{2m}}{2 \cdot (2m)!},$$

where B_k is the kth *Bernoulli number*, defined as the coefficient of $t^k/k!$ in the Taylor series

$$\frac{t}{e^t - 1} = 1 - \frac{1}{2}t + \frac{1}{12}t^2 - \frac{1}{720}t^4 + \frac{1}{30240}t^6 + \cdots = \sum_{k=0}^{\infty} B_k \frac{t^k}{k!}.$$

This implies that $\zeta(2m)$ is π^{2m} times a rational number. On the other hand, little is known about $\zeta(2m+1)$, where $2m+1 \geq 3$ is an odd integer. Numerical experiments indicate that it is probably not a power of π times a rational number. In 1978, Roger Apéry proved that $\zeta(3)$

is irrational. In 2000, Tanguy Rivoal and Wadim Zudilin proved, for example, that at least one of $\zeta(5), \zeta(7), \zeta(9), \zeta(11)$ is irrational. (It is conjectured that all of these are irrational.)

The sum defining the Riemann zeta function $\zeta(s)$ also converges for complex values of s with $Re(s) > 1$ (that is, $s = x + iy$ with $x > 1$). There is a natural extension of $\zeta(s)$ to all complex numbers $s \neq 1$ (for those who have studied complex analysis, the technical term is "analytic continuation"). In 1859, Bernhard Riemann showed that there is a remarkable connection between the distribution of prime numbers and the location of the solutions of $\zeta(s) = 0$ with $0 \leq Re(s) \leq 1$. He conjectured the **Riemann Hypothesis**, which predicts that the only complex solutions to $\zeta(s) = 0$ with $0 \leq Re(s) \leq 1$ have $Re(s) = 1/2$. This problem is perhaps the most famous unsolved problem in mathematics. In 1900, David Hilbert listed the Riemann Hypothesis as one of 23 important unsolved problems in a list that had a lot of influence on 20th century mathematics. In 2000, the Clay Mathematics Institute included the Riemann Hypothesis as one of its seven Millennium Prize Problems and offered $1 million for its solution. We make no prediction as to whether or not it will be on 2100's list.

In 1896, Jacques Hadamard and Charles Jean de la Vallée-Poussin independently proved that $\zeta(s) \neq 0$ when $Re(s) = 1$, and they used this to deduce the Prime Number Theorem, which states that the number of primes less than x is asymptotic to $x/\ln(x)$ as $x \to \infty$. If the full Riemann Hypothesis is proved, it will yield much more precise estimates on the number of primes less than x and also supply technical steps needed in many situations in number theory.

In 2000, Jeffrey Lagarias, building on work of Guy Robin, gave an elementary problem that is equivalent to the Riemann Hypothesis: For each positive integer n, let

$$H_n = \sum_{j=1}^{n} \frac{1}{j}.$$

The Riemann Hypothesis is equivalent to the statement that

$$\sum_{d|n} d \leq H_n + e^{H_n} \ln(H_n)$$

for all $n \geq 1$, with equality only for $n = 1$ (the sum is over the divisors d of n with $1 \leq d \leq n$).

Section 15.3   The Riemann Hypothesis

For example, let $n = 6$. Then

$$H_6 = 1 + (1/2) + (1/3) + (1/4) + (1/5) + (1/6) = 49/20.$$

We have
$$H_6 + e^{H_6} \ln(H_6) = 12.834 \cdots.$$

Since
$$\sum_{d|6} d = 1 + 2 + 3 + 6 = 12 < 12.834,$$

the inequality is verified for $n = 6$. If we could verify it for all $n \geq 1$, then we would win \$1 million.

Admittedly, if you tried to tell someone that proving that $\sum_{d|n} d \leq H_n + e^{H_n} \ln(H_n)$ is the most famous problem in mathematics, then you would probably get a weird look. But it's interesting that such a famous problem has such an elementary equivalent formulation.

Appendix A

Supplementary Topics

This appendix contains some topics that are used in a few places in the main text. The sections of the appendix may be used for reference when they are needed. However, the topics covered are very basic ideas of mathematics, so the reader who is unfamiliar with them will profit from learning them regardless of their relations to the rest of the book.

A.1 Geometric Series

Certain types of expressions occur frequently in mathematics. One of the most common is a **geometric series**, which is a sum of the form $a + ar + ar^2 + \cdots + ar^{n-1}$. Such sums can be evaluated quite easily:

Proposition A.1. *Let a and r be real numbers, with $r \neq 1$, and let $n \geq 1$. Then*

$$a + ar + ar^2 + \cdots + ar^{n-1} = \frac{ar^n - a}{r - 1}.$$

Proof. Let S be the sum. Then

$$\begin{aligned} rS &= ar + ar^2 + ar^3 + \cdots + ar^n \\ &= -a + a + ar + ar^2 + \cdots + ar^{n-1} + ar^n \end{aligned}$$

(a was subtracted from and added to the previous expression)

$$= -a + S + ar^n.$$

Therefore, $(r-1)S = -a + ar^n$. Dividing by $r-1$ yields the formula. □

This formula occurs in various contexts. For example, let $a = 1$ and $r = x$. Then
$$1 + x + x^2 + \cdots + x^{n-1} = \frac{x^n - 1}{x - 1}.$$
Multiplying by $x - 1$ yields the factorization
$$x^n - 1 = (x - 1)(x^{n-1} + x^{n-2} + \cdots + x + 1).$$
For example, when $n = 2$, this says $x^2 - 1 = (x - 1)(x + 1)$.

We can also consider an infinite geometric series.

Proposition A.2. *Let a and r be real numbers with $|r| < 1$. Then*
$$a + ar + ar^2 + ar^3 + \cdots = \frac{a}{1 - r}.$$

Proof. In Proposition A.1, let $n \to \infty$. Since $|r| < 1$, we know that $r^n \to 0$. Therefore,
$$a + ar + ar^2 + ar^3 + \cdots = \lim_{n \to \infty} \left(a + ar + ar^2 + \cdots + ar^{n-1}\right)$$
$$= \lim_{n \to \infty} \frac{ar^n - a}{r - 1}$$
$$= \frac{0 - a}{r - 1}$$
$$= \frac{a}{1 - r}.$$
□

Example. An example of this proposition is given by repeating decimals:
$$0.42424242\ldots = .42 + .42(.01)^1 + .42(.01)^2 + .42(.01)^3 + \cdots$$
$$= \frac{.42}{1 - .01} = \frac{42}{99} = \frac{14}{33},$$
since $a = .42$ is the first term and $r = .01$ is the ratio between successive terms.

Another example is
$$1 + \frac{1}{2} + \frac{1}{4} + \frac{1}{8} + \cdots = \frac{1}{1 - \frac{1}{2}} = 2.$$

Section A.2   Mathematical Induction   353

CHECK YOUR UNDERSTANDING
1. Show that $1 + 2 + 4 + \cdots + 2^{n-1} = 2^n - 1$ for all $n \geq 1$.
2. Interpret $.3333\cdots$ as a geometric series and use this to show that it equals $1/3$.

A.2   Mathematical Induction

Mathematicians like to make educated guesses and then prove that their guesses are correct. For example, if we add up even integers, we get the following:
$$2 + 4 = 6 = 2 \cdot 3$$
$$2 + 4 + 6 = 12 = 3 \cdot 4$$
$$2 + 4 + 6 + 8 = 20 = 4 \cdot 5.$$

Looking at this pattern, we make an educated guess (or conjecture) that
$$2 + 4 + 6 + \cdots + 2n = n(n+1). \tag{A.1}$$

We could then check this for several more values of n. For example, for $n = 10$, we have
$$2 + 4 + 6 + 8 + 10 + 12 + 14 + 16 + 18 + 20 = 110 = 10 \cdot 11, \tag{A.2}$$

so it works in this case. Now suppose we want to check it for $n = 11$. We don't have to go back and add up $2 + 4 + \cdots + 20 + 22$. Instead, we already know that Equation (A.2) is true, so we add 22 and get
$$2 + 4 + \cdots + 20 + 22 = 110 + 22 = 132 = 11 \cdot 12.$$

This says that Equation (A.1) holds for $n = 11$. In this way, if we know the equation holds for one value of n, we can verify it for the next value of n without recomputing the whole sum. Rather, we can simply add on one more term. Let's try this in general. Suppose we know that Equation (A.1) holds for $n = k$. This means that
$$2 + 4 + 6 + \cdots + 2k = k(k+1).$$

Let's try to deduce that it holds for the next value of n, namely $k+1$. Add $2(k+1)$ to obtain

$$(2+4+6+\cdots+2k)+2(k+1) = k(k+1)+2(k+1)$$
$$= (k+1)(k+2).$$

This is Equation (A.1) for $n = k+1$. We have shown that if we know the equation is true for one value of n, then it is true for the next value of n.

All we need now is a place to start: Let $n = 1$. The sum on the left in (A.1) has one term, namely 2, and the right side is $1(1+1) = 2$, so (A.1) is true for $n = 1$. We have just shown that it is true for the next value of n, which is 2. And then it is true for the next value of n, which is 3. This keeps going, and we find that (A.1) is true for all $n \geq 1$.

What we just did is called **mathematical induction**. The general principle is the following.

If we want to prove some statement is true for all positive integers, we need to show two things:

1. The statement is true for $n = 1$.
2. If the statement is true for $n = k$ then it is true for $n = k+1$.

We then conclude that the statement is true for all positive integers n.

Mathematical induction is often compared to proving that you can climb to the top of a ladder. You prove that you can get onto the ladder. This is the $n = 1$ case. Then you show that you can always get from one rung to the next. This is the $n = k$ to $n = k+1$ case. Therefore, you can climb every rung of the ladder.

Let's look at another example. Consider sums of odd numbers:

$$1 = 1^2$$
$$1+3 = 2^2$$
$$1+3+5 = 3^2$$
$$1+3+5+7 = 4^2.$$

We guess that

$$1+3+5+\cdots+(2n-1) = n^2. \tag{A.3}$$

Section A.2 Mathematical Induction

We have already seen that (A.3) is true for $n = 1$. Now let's assume it's true for $n = k$, so

$$1 + 3 + 5 + \cdots + (2k - 1) = k^2.$$

Add $2k + 1$ to obtain

$$(1 + 3 + 5 + \cdots + (2k - 1)) + (2k + 1) = k^2 + (2k + 1)$$
$$= (k + 1)^2.$$

This is (A.3) for $n = k + 1$, so we have proved that if (A.3) is true for $n = k$ then it is also true for $n = k + 1$. Mathematical induction tells us that (A.3) is true for all positive integers n.

As another example, let's prove that

$$n + 1 \leq 2^n$$

for all $n \geq 1$. The statement is certainly true for $n = 1$. Now assume that it is true for $n = k$, so

$$k + 1 \leq 2^k.$$

Adding 1 to each side, we obtain

$$k + 2 \leq 2^k + 1$$
$$\leq 2^k + 2^k$$
$$\leq 2^{k+1},$$

which is the statement for $n = k+1$. Therefore, if the inequality is true for $n = k$ then it is true for $n = k+1$. Mathematical induction tells us that the inequality is true for all $n \geq 1$.

Remark. Sometimes the induction does not start at $n = 1$. Suppose we want to prove that $10n + 9 \leq 2^n$ for all $n \geq 7$. We then take the base case to be $n = 7$ (instead of $n = 1$) and then check that the inequality is true for this case. We then show, as in the preceding calculation, that if the inequality is true for $n = k$ with $k \geq 7$, then it is true for $n = k + 1$:

$$10k + 9 \leq 2^k \implies 10(k+1) + 9 \leq 2^k + 10 \leq 2^k + 2^k = 2^{k+1}.$$

The last inequality used the fact that $10 \leq 2^k$ when $k \geq 7$. (In fact, this step would have worked for $k \geq 4$, but showing "true for 4" $\implies$ "true for 5" is worthless when the statement is false for 4.) We conclude that $10n + 9 \leq 2^n$ for all $n \geq 7$.

There is another form of induction that is useful. **Strong induction** assumes that a statement is true for *all* $n \leq k$ and proves it for $n = k+1$. For example, suppose we want to prove that every integer $n > 1$ is a prime or a product of primes. Assume that we have proved this for all $n \leq k$. Now consider $k+1$. If $k+1$ is prime, we're done. Otherwise, $k+1$ factors as $k + 1 = ab$ with $1 < a, b < k + 1$. The induction hypothesis says that a and b are either prime or products of primes. Therefore, $k + 1 = ab$ is a prime or a product of primes.

Closely related to strong induction is the **Well-Ordering Principle**, which states that a nonempty set of positive integers has a smallest element. For example, let's prove again that every integer $n > 1$ is either prime or a product of primes. Let S be the set of integers $n > 1$ that are neither prime nor a product of primes. Suppose that S is nonempty. The Well-Ordering Principle says that S has a smallest element. Call it s. The elements of S are not prime, so s factors as $s = ab$ with $1 < a, b < s$. Since s is the smallest element of S, both a and b are prime or products of primes. Therefore, $s = ab$ is a product of primes, contradicting the fact that s is in S. This contradiction shows that S must be empty, which is exactly the statement that every integer $n > 1$ is either prime or a product of primes. As we see, this proof is very close to the one using strong induction. It is often a matter of personal choice whether to use strong induction or the Well-Ordering Principle.

The Well-Ordering Principle can also be used to give an alternate proof of the induction example we began this section with. We want to prove that Equation (A.1)

$$2 + 4 + 6 + \cdots + 2n = n(n+1).$$

is true for all n. We begin by letting S be the set of integers $n \geq 1$ for which

$$2 + 4 + 6 + \cdots + 2n \neq n(n+1).$$

Suppose that S is nonempty. Then the Well-Ordering Principle says that S has a smallest element, which we'll call m. Since we know that (A.1) is true for $n = 1$, we know that m must be larger than 1. Since $m > 1$, we know that $m - 1$ is still a positive integer. Moreover, $m - 1 <$

Section A.2   Mathematical Induction   357

m. So $m-1$ is a positive integer smaller than the smallest element of S, so (A.1) must be true for $m-1$. This means that
$$2+4+\cdots+2(m-1) = (m-1)(m-1+1) = (m-1)m.$$
Adding $2m$ to both sides of this equation, we get
$$2+4+\cdots+2(m-1)+2m = (m-1)m+2m$$
$$= m^2 - m + 2m = m^2 + m = m(m+1).$$
But this shows that (A.1) is true for m, so it is impossible for S to be nonempty. Therefore,
$$2+4+\cdots+2n = n(n+1)$$
must be true for all $n \geq 1$.

Finally, here's an interesting example of strong induction. We'll play a game. A move in this game is to divide a pile of markers into two piles, say of size a and b. Our score for that move is ab. Suppose we start with a pile of 12 markers. We make moves by choosing a pile and breaking it into two smaller piles and adding the score to our total. We continue until all the piles are size 1. The claim is that no matter how we play our total score will be 66. For example, suppose the game proceeds as follows:

$$\begin{aligned}
12 &= 5+7 \quad (\text{score} = 5\cdot 7 = 35) \\
&= 2+3+7 \quad (\text{score} = 2\cdot 3 = 6) \\
&= 2+3+5+2 \quad (\text{score} = 5\cdot 2 = 10) \\
&= 2+3+5+1+1 \quad (\text{score} = 1\cdot 1 = 1) \\
&= 1+1+3+5+1+1 \quad (\text{score} = 1\cdot 1 = 1) \\
&= 1+1+2+1+5+1+1 \quad (\text{score} = 1\cdot 2 = 2) \\
&= 1+1+1+1+1+5+1+1 \quad (\text{score} = 1\cdot 1 = 1) \\
&= 1+1+1+1+1+2+3+1+1 \quad (\text{score} = 2\cdot 3 = 6) \\
&= 1+1+1+1+1+1+1+3+1+1 \quad (\text{score} = 1\cdot 1 = 1) \\
&= 1+1+1+1+1+1+1+2+1+1+1 \quad (\text{score} = 2\cdot 1 = 2) \\
&= 1+1+1+1+1+1+1+1+1+1+1+1 \quad (\text{score} = 1\cdot 1 = 1).
\end{aligned}$$

The total score is
$$35+6+10+1+1+2+1+6+1+2+1 = 66.$$

Proving this always yields 66 by trying all cases would be painful. However, it is much easier to consider the general statement: If we start with a pile of n markers, our total score will be $\frac{1}{2}n(n-1)$. This is certainly true for $n = 1$ since we cannot make any moves, so our total score is 0. Let's assume the statement is true for all $n \leq k$ and try to prove it for $n = k+1$.

The first move splits the pile of $k+1$ markers into two piles, say of sizes a and b, with $a + b = k + 1$. This gives us ab points. After that, the game really breaks into two subgames, one for the markers in the a pile and one for the markers in the b pile. By the induction assumption, the a pile earns us $\frac{1}{2}a(a-1)$ total points, and the b pile earns us $\frac{1}{2}b(b-1)$ total points. The total number of points we earn is the ab for the first move plus these points:

$$ab + \frac{1}{2}a(a-1) + \frac{1}{2}b(b-1) = \frac{1}{2}\left(a^2 + 2ab + b^2 - a - b\right)$$
$$= \frac{1}{2}(a+b)(a+b-1)$$
$$= \frac{1}{2}(k+1)(k).$$

Therefore, the claim holds for $n = k+1$. Strong induction tells us that the claim holds for all $n \geq 1$. In particular, if we start with $n = 12$ markers, we will always earn $\frac{1}{2}(12)(11) = 66$ points.

CHECK YOUR UNDERSTANDING

3. Use induction to prove that $1 + 2 + 4 + \cdots + 2^{n-1} = 2^n - 1$ for all $n \geq 1$.

A.3 Pascal's Triangle and the Binomial Theorem

Consider the following triangular array of numbers:

Section A.3 Pascal's Triangle and the Binomial Theorem

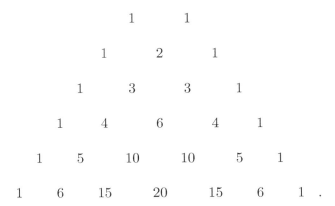

Each number is obtained by adding together the two numbers above it in the preceding row. For example, the 15 that is the third entry in the 6th row is the sum $5 + 10$ of the numbers above it in the 5th row. This array is often called **Pascal's Triangle**, after the great French mathematician and philosopher Blaise Pascal. The triangle did not originate with Pascal; it had been known for hundreds of years in many countries around the world. But Pascal organized its properties in his book *Traité du triangle arithmétique* (1653) and used them to develop probability theory and analyze gambling strategies.

In this section we define binomial coefficients and show that they are the entries in Pascal's triangle.

We all learned in high school that

$$(a+b)^2 = a^2 + 2ab + b^2,$$

and you might remember that

$$(a+b)^3 = a^3 + 3a^2b + 3ab^2 + b^3.$$

There are many times when we need to calculate $(a+b)^n$ for an arbitrary positive integer n, and the *Binomial Theorem* allows us to do just that.

In order to preview what the Binomial Theorem says, we calculate

$(a+b)^n$ for $n = 1, 2, 3, 4, 5$, and 6:

$$(a+b)^1 = a + b$$

$$(a+b)^2 = a^2 + 2ab + b^2$$

$$(a+b)^3 = a^3 + 3a^2b + 3ab^2 + b^3$$

$$(a+b)^4 = a^4 + 4a^3b + 6a^2b^2 + 4ab^3 + b^4$$

$$(a+b)^5 = a^5 + 5a^4b + 10a^3b^2 + 10a^2b^3 + 5ab^4 + b^5$$

$$(a+b)^6 = a^6 + 6a^5b + 15a^4b^2 + 20a^3b^3 + 15a^2b^4 + 10ab^5 + b^6.$$

The first thing to notice is that when calculating $(a+b)^n$, each term is of the form a^ib^j with $i+j=n$. The coefficient of a^ib^j is the jth entry in the nth row of Pascal's Triangle (where the initial 1 is counted as the 0th element of the row).

Although this is nice, we still want a general theorem that gives a formula for the value of each coefficient in the expansion of $(a+b)^n$. For this we need the binomial coefficients:

Definition A.3. *Let m and r be integers, with $0 \le r \le m$. Then*

$$\binom{m}{r} = \frac{m!}{r!(m-r)!}$$

is called a **binomial coefficient**.

If you have a set with m elements, $\binom{m}{r}$ is the number of subsets with r elements. In other words, it is the number of ways to choose a subset of r marbles from a set of m marbles. For this reason, $\binom{m}{r}$ is often read as "m choose r."

If we recall that $0! = 1$, we see that

$$\binom{4}{0} = 1, \quad \binom{4}{1} = 4, \quad \binom{4}{2} = 6, \quad \binom{4}{3} = 4, \quad \text{and} \quad \binom{4}{4} = 1,$$

Section A.3   Pascal's Triangle and the Binomial Theorem

and these are exactly the numbers that occur in the fourth line of Pascal's Triangle. We'll prove that the binomial coefficients always tell us the coefficients of $(a+b)^n$ in a moment, but first we need a lemma.

In Pascal's Triangle, each number is the sum of the two integers that are above it. The same property holds for binomial coefficients.

Proposition A.4. *Let m and r be integers, with $1 \le r \le m$. Then*
$$\binom{m}{r-1} + \binom{m}{r} = \binom{m+1}{r}.$$

Proof. Starting with the left-hand side, we have
$$\binom{m}{r-1} + \binom{m}{r}$$
$$= \frac{m!}{(r-1)!(m+1-r)!} + \frac{m!}{r!(m-r)!}$$
$$= \frac{m!r}{r(r-1)!(m+1-r)!} + \frac{m!(m+1-r)}{r!(m-r)!(m+1-r)}$$
$$= \frac{m!}{r!(m+1-r)!}\left((r)+(m+1-r)\right)$$
$$= \frac{m!(m+1)}{r!(m+1-r)!}$$
$$= \binom{m+1}{r}.$$

This proves the proposition. □

We can now write the first six rows of Pascal's Triangle as follows:

$$\binom{1}{0} \quad \binom{1}{1}$$
$$\binom{2}{0} \quad \binom{2}{1} \quad \binom{2}{2}$$
$$\binom{3}{0} \quad \binom{3}{1} \quad \binom{3}{2} \quad \binom{3}{3}$$
$$\binom{4}{0} \quad \binom{4}{1} \quad \binom{4}{2} \quad \binom{4}{3} \quad \binom{4}{4}$$
$$\binom{5}{0} \quad \binom{5}{1} \quad \binom{5}{2} \quad \binom{5}{3} \quad \binom{5}{4} \quad \binom{5}{5}$$
$$\binom{6}{0} \quad \binom{6}{1} \quad \binom{6}{2} \quad \binom{6}{3} \quad \binom{6}{4} \quad \binom{6}{5} \quad \binom{6}{6}.$$

The proposition says, for example, that $\binom{3}{1}+\binom{3}{2}=\binom{4}{2}$, which is exactly the same as saying that the binomial coefficient is the sum of the two above it. This implies that the binomial coefficients are the entries in Pascal's Triangle, since both can be generated in the same way.

We're now ready to prove our main result.

The Binomial Theorem. *Let n be a positive integer. Then*

$$(a+b)^n =$$
$$\binom{n}{0}a^n + \binom{n}{1}a^{n-1}b + \binom{n}{2}a^{n-2}b^2 + \cdots + \binom{n}{n-1}ab^{n-1} + \binom{n}{n}b^n.$$

Proof. We'll prove this using mathematical induction.

Step 1: When $n = 1$,

$$(a+b)^1 = a+b = \binom{1}{0}a + \binom{1}{1}b,$$

so our initial induction step is verified. (From the definition of the binomial coefficient, you can see that $\binom{n}{0} = \binom{n}{n} = 1$ for all positive integers n.)

Step 2: Assume that the binomial theorem is true for $n = k$, so

$$(a+b)^k =$$
$$\binom{k}{0}a^k + \binom{k}{1}a^{k-1}b + \binom{k}{2}a^{k-2}b^2 + \cdots + \binom{k}{k-1}ab^{k-1} + \binom{k}{k}b^k.$$

We now have to prove that the appropriate formula works for $n = k+1$. We begin by observing that

$$(a+b)^{k+1} = (a+b) \cdot (a+b)^k$$
$$= a(a+b)^k + b(a+b)^k.$$

Section A.3 Pascal's Triangle and the Binomial Theorem

We use the induction assumption to expand each $(a+b)^k$. This yields

$$a\left(\binom{k}{0}a^k + \binom{k}{1}a^{k-1}b + \binom{k}{2}a^{k-2}b^2 + \cdots + \binom{k}{k-1}ab^{k-1} + \binom{k}{k}b^k\right)$$

$$+ b\left(\binom{k}{0}a^k + \binom{k}{1}a^{k-1}b + \binom{k}{2}a^{k-2}b^2 + \cdots + \binom{k}{k-1}ab^{k-1} + \binom{k}{k}b^k\right)$$

$$= \binom{k}{0}a^{k+1} + \binom{k}{1}a^k b + \binom{k}{2}a^{k-1}b^2 + \cdots + \binom{k}{k-1}a^2 b^{k-1} + \binom{k}{k}ab^k$$

$$+ \binom{k}{0}a^k b + \binom{k}{1}a^{k-1}b^2 + \binom{k}{2}a^{k-2}b^3 + \cdots + \binom{k}{k-1}ab^k + \binom{k}{k}b^{k+1}$$

$$= \binom{k}{0}a^{k+1} + \left(\binom{k}{1} + \binom{k}{0}\right)a^k b + \left(\binom{k}{2} + \binom{k}{1}\right)a^{k-1}b^2$$

$$+ \cdots + \left(\binom{k}{k} + \binom{k}{k-1}\right)ab^k + \binom{k}{k}b^{k+1}.$$

Proposition A.4 allows us to replace the sums of binomial coefficients by a single binomial coefficient. For example,

$$\binom{k}{2} + \binom{k}{1} = \binom{k+1}{2}.$$

Also,

$$\binom{k}{0} = 1 = \binom{k+1}{0} \quad \text{and} \quad \binom{k}{k} = 1 = \binom{k+1}{k+1}.$$

Therefore, we obtain

$$\binom{k+1}{0}a^{k+1} + \binom{k+1}{1}a^k b + \cdots + \binom{k+1}{k}ab^k + \binom{k+1}{k+1}b^{k+1}.$$

We have proved that $(a+b)^{k+1}$ is given by the desired formula. Mathematical induction implies that the formula is true for all n. □

CHECK YOUR UNDERSTANDING

4. Compute $\binom{7}{3}$.

5. Find the coefficient of $a^2 b^6$ in the expansion of $(a+b)^8$.

A.4 Fibonacci Numbers

Fibonacci numbers occur in many situations in mathematics. In the following, we give a brief treatment of the properties that we need in this book.

Definition A.5. *The* **Fibonacci numbers** *are defined recursively by*

$$F_0 = 0, \ F_1 = 1, \ F_2 = 1, \ \text{and} \ F_n = F_{n-1} + F_{n-2} \ \text{for} \ n \geq 3.$$

As an example, we calculate the next four Fibonacci numbers:

$$F_3 = F_2 + F_1 = 1 + 1 = 2$$
$$F_4 = F_3 + F_2 = 2 + 1 = 3$$
$$F_5 = F_4 + F_3 = 3 + 2 = 5$$
$$F_6 = F_5 + F_4 = 5 = 3 = 8.$$

The notation F_n is also used for Fermat numbers (see Section 1.10). However, the context should always make it clear which type of number is being discussed.

There are many identities involving Fibonacci numbers. For example, $F_{n+3} + F_n = 2F_{n+2}$. To see why this is true, note that

$$\begin{aligned} F_{n+3} + F_n &= (F_{n+2} + F_{n+1}) + F_n \\ &= F_{n+2} + (F_{n+1} + F_n) \\ &= F_{n+2} + F_{n+2} = 2F_{n+2}. \end{aligned}$$

Some other identities are in the exercises.

The following result is known as **Binet's Formula**, in honor of Jacques

Section A.4 Fibonacci Numbers

Philippe Marie Binet (1786–1856), who, among other accomplishments, was the first to describe the rule for multiplying matrices. However, Binet's Formula was probably first discovered by de Moivre around 1730.

The **Golden Ratio** is defined to be

$$\phi = \frac{1+\sqrt{5}}{2} \approx 1.618034.$$

An easy calculation shows that

$$-\phi^{-1} = \frac{1-\sqrt{5}}{2} \approx -.618034.$$

Theorem A.6. *The nth Fibonacci number F_n is given by the expression*

$$F_n = \frac{1}{\sqrt{5}}\left(\phi^n - (-\phi)^{-n}\right). \qquad (A.4)$$

Proof. We will use strong induction to prove this.

Step 1: Verify that is true for $n = 1$ and $n = 2$.

$n = 1$:

$$\frac{1}{\sqrt{5}}(\phi^1 - (-\phi)^{-1}) = \frac{1}{\sqrt{5}}\left(\frac{1+\sqrt{5}}{2} - \frac{1-\sqrt{5}}{2}\right) = 1 = F_0.$$

$n = 2$:

$$\frac{1}{\sqrt{5}}(\phi^2 - (-\phi)^{-2}) = \frac{1}{\sqrt{5}}\left(\left(\frac{1+\sqrt{5}}{2}\right)^2 - \left(\frac{1-\sqrt{5}}{2}\right)^2\right)$$

$$= \frac{1}{\sqrt{5}}\left(\frac{6+2\sqrt{5}}{4} - \frac{6-2\sqrt{5}}{4}\right) = 1 = F_1.$$

Step 2: Assume that (A.4) is true for all $n \leq k$ and use this to prove it's true for $n = k+1$ This means that we assume that

$$F_m = \frac{1}{\sqrt{5}}(\phi^m - (-\phi)^{-m})$$

is true for $m = 1, 2, 3, \ldots, k$, and we want to prove that

$$F_{k+1} = \frac{1}{\sqrt{5}}(\phi^{k+1} - (-\phi)^{k+1}).$$

First, observe that
$$\phi^2 = \frac{6+2\sqrt{5}}{4} = \frac{3+\sqrt{5}}{2} = \phi + 1.$$

Multiply by ϕ^{k-1} to get
$$\phi^{k+1} = \phi^k + \phi^{k-1}. \tag{A.5}$$

Similarly,
$$(-\phi^{-1})^2 = \frac{6-2\sqrt{5}}{4} = \frac{3-\sqrt{5}}{2} = (-\phi^{-1}) + 1,$$

which tells us that
$$(-\phi)^{-(k+1)} = (-\phi)^{-k} + (-\phi)^{-(k-1)}. \tag{A.6}$$

Therefore,
$$\begin{aligned}
F_{k+1} &= F_k + F_{k-1} \quad \text{(from the definition of the Fibonacci sequence)} \\
&= \frac{1}{\sqrt{5}}\left((\phi^k - (-\phi)^{-k}) + (\phi^{k-1} - (-\phi)^{-(k-1)})\right) \\
&\quad \text{(from the induction hypothesis)} \\
&= \frac{1}{\sqrt{5}}\left(\phi^k + \phi^{k-1}\right) - \frac{1}{\sqrt{5}}\left((-\phi)^{-k} + (-\phi)^{-(k-1)}\right) \\
&= \frac{1}{\sqrt{5}}\left(\phi^{k+1} - (-\phi)^{-(k+1)}\right) \quad \text{(from (A.5) and (A.6))}.
\end{aligned}$$

This is (A.4) for $n = k+1$. By induction, the formula (A.4) is true for all positive integers. This completes the proof. □

If we look at the ratios
$$\frac{F_6}{F_5} \approx 1.600000, \quad \frac{F_{11}}{F_{10}} \approx 1.618182, \quad \frac{F_{16}}{F_{15}} \approx 1.618033,$$

we see that the ratios of successive Fibonacci numbers are approximately the Golden Ratio. This is explained by Binet's Formula:

Corollary A.7.
$$\lim_{n \to \infty} \frac{F_{n+1}}{F_n} = \phi.$$

Proof. From Binet's Formula, we have

$$\lim_{n\to\infty} \frac{F_{n+1}}{F_n} = \lim_{n\to\infty} \frac{\phi^{n+1} - (-\phi)^{-(n+1)}}{\phi^n - (-\phi)^{-(n)}}.$$

Dividing the numerator and denominator by ϕ^n yields

$$\lim_{n\to\infty} \frac{\phi - (-\phi)^{-(n+1)}\phi^{-n}}{1 - (-\phi)^{-n}\phi^{-n}}. \tag{A.7}$$

Since $|-\phi| > 1$, we see that $(-\phi)^{-(n+1)} \to 0$ as $n \to \infty$. Similarly, $\phi^{-n} \to 0$ and $(-\phi)^{-n} \to 0$ as $n \to \infty$. Therefore, the limit of the numerator in (A.7) is ϕ, and the limit of the denominator is 1. Putting everything together, we obtain

$$\lim_{n\to\infty} \frac{F_{n+1}}{F_n} = \phi,$$

as desired. □

If you know enough Fibonacci numbers, the corollary gives you a way to convert between miles and kilometers quickly. For example, 55 miles is approximately 88 kilometers (exactly 88.51392). This is because the successive Fibonacci numbers 55 and 88 have a ratio approximately $\phi \approx 1.618$, while one mile is 1.609344 kilometers, which is almost the same ratio. Similarly, if you know that 6765 and 10946 are successive Fibonacci numbers and you want to convert 110 kilometers to miles, you match the first digits and observe that 10946/6765 is approximately 110/68, so the answer is approximately 68 miles.

A.5 Matrices

A **matrix** is just a rectangular array of numbers. If you've ever used a spreadsheet where entries are specified by their row and column numbers, you've seen a matrix. All the matrices we use in this book have two rows and two columns and look like this:

$$\begin{pmatrix} a & b \\ c & d \end{pmatrix},$$

where the entries $a, b, c,$ and d are numbers. In this matrix, a is in the first row and first column, b is in the first row and second column, c is in the second row and first column, and d is in the second row and second column.

We also need to work with **vectors**. The vectors that we'll be using have 1 column and two rows:

$$\begin{pmatrix} e \\ f \end{pmatrix}.$$

For the Hill cipher in Chapter 7, we need to multiply a matrix times a vector and a matrix times a matrix.

To begin, here's the rule for multiplying a matrix by a vector:

$$\begin{pmatrix} a & b \\ c & d \end{pmatrix} \begin{pmatrix} e \\ f \end{pmatrix} = \begin{pmatrix} ae + bf \\ ce + df \end{pmatrix}$$

For example,

$$\begin{pmatrix} 3 & 2 \\ 5 & 8 \end{pmatrix} \begin{pmatrix} 4 \\ 7 \end{pmatrix} = \begin{pmatrix} 3 \cdot 4 + 2 \cdot 7 \\ 5 \cdot 4 + 8 \cdot 7 \end{pmatrix} = \begin{pmatrix} 26 \\ 76 \end{pmatrix}$$

Next, we'll use the method of multiplying a matrix times a vector to multiply a matrix times a matrix. Let's say you want to calculate MN, where

$$M = \begin{pmatrix} 7 & 10 \\ 3 & 4 \end{pmatrix} \quad \text{and} \quad N = \begin{pmatrix} 3 & 6 \\ 5 & 2 \end{pmatrix}.$$

Think of N as consisting of two vectors, so N is made up of the vectors

$$\begin{pmatrix} 3 \\ 5 \end{pmatrix} \quad \text{and} \quad \begin{pmatrix} 6 \\ 2 \end{pmatrix}.$$

For the product MN, we get the 2×2 matrix whose first column is

$$M \begin{pmatrix} 3 \\ 5 \end{pmatrix} = \begin{pmatrix} 7 & 10 \\ 3 & 4 \end{pmatrix} \begin{pmatrix} 3 \\ 5 \end{pmatrix} = \begin{pmatrix} 7 \cdot 3 + 10 \cdot 5 \\ 3 \cdot 3 + 4 \cdot 5 \end{pmatrix} = \begin{pmatrix} 71 \\ 29 \end{pmatrix}$$

and whose second column is

$$M \begin{pmatrix} 6 \\ 2 \end{pmatrix} = \begin{pmatrix} 7 & 10 \\ 3 & 4 \end{pmatrix} \begin{pmatrix} 6 \\ 2 \end{pmatrix} = \begin{pmatrix} 7 \cdot 6 + 10 \cdot 2 \\ 3 \cdot 6 + 4 \cdot 2 \end{pmatrix} = \begin{pmatrix} 62 \\ 26 \end{pmatrix}.$$

Section A.5   Matrices

Therefore,
$$MN = \begin{pmatrix} 7 & 10 \\ 3 & 4 \end{pmatrix} \begin{pmatrix} 3 & 6 \\ 5 & 2 \end{pmatrix} = \begin{pmatrix} 71 & 62 \\ 29 & 26 \end{pmatrix}.$$

In general, if
$$M = \begin{pmatrix} a & b \\ c & d \end{pmatrix} \quad \text{and} \quad N = \begin{pmatrix} e & f \\ g & h \end{pmatrix},$$
then
$$MN = \begin{pmatrix} ae+bg & af+bh \\ ce+dg & cf+dh \end{pmatrix}.$$

Notice that under this multiplication rule, if
$$M = \begin{pmatrix} a & b \\ c & d \end{pmatrix} \quad \text{and} \quad I = \begin{pmatrix} 1 & 0 \\ 0 & 1 \end{pmatrix},$$
then
$$MI = \begin{pmatrix} a\cdot 1 + b\cdot 0 & a\cdot 0 + b\cdot 1 \\ c\cdot 1 + d\cdot 0 & c\cdot 0 + d\cdot 1 \end{pmatrix} = \begin{pmatrix} a & b \\ c & d \end{pmatrix} = M.$$

Similarly, $IM = M$. Because of this, I is called the **identity matrix**. We now use the identity matrix to discuss how (and when) we can form the inverse of a matrix. Recall that if a is a nonzero number, then $a^{-1} \cdot a = 1$. Analogously, we want to compute the inverse of a matrix M. That is, given a matrix
$$M = \begin{pmatrix} a & b \\ c & d \end{pmatrix},$$
we want a matrix M^{-1} with the property that
$$M^{-1} M = I.$$

Fortunately, there is a formula: If $ad - bc \neq 0$, then
$$\begin{pmatrix} a & b \\ c & d \end{pmatrix}^{-1} = \frac{1}{ad-bc} \begin{pmatrix} d & -b \\ -c & a \end{pmatrix}.$$

Here, the notation
$$\frac{1}{ad-bc} \begin{pmatrix} d & -b \\ -c & a \end{pmatrix}$$

means that we divide every entry of the matrix by $ad - bc$.

It can be shown that if $ad - bc = 0$, then the matrix does not have an inverse.

Example.

$$\begin{pmatrix} 11 & 5 \\ 6 & 3 \end{pmatrix}^{-1} = \frac{1}{3}\begin{pmatrix} 3 & -5 \\ -6 & 11 \end{pmatrix} = \begin{pmatrix} 1 & -5/3 \\ -2 & 11/3 \end{pmatrix}.$$

We can check this:

$$\begin{pmatrix} 1 & -5/3 \\ -2 & 11/3 \end{pmatrix}\begin{pmatrix} 11 & 5 \\ 6 & 3 \end{pmatrix} = \begin{pmatrix} 1 & 0 \\ 0 & 1 \end{pmatrix}.$$

There's an issue that we need to mention. In the Hill cipher, the entries of the matrix are numbers mod n for some integer n. The number $ad-bc$ occurs in the denominator in our equation for the inverse of a matrix, and we need $ad - bc$ to have an inverse mod n in order for the inverse of the matrix to exist.

Finally, suppose v and w are vectors and M is a matrix and we have an equation

$$Mv = w,$$

where M is a matrix that has an inverse. Then

$$M^{-1}w = v.$$

For example,

$$\begin{pmatrix} 1 & 2 \\ 3 & 4 \end{pmatrix}\begin{pmatrix} 5 \\ 6 \end{pmatrix} = \begin{pmatrix} 17 \\ 39 \end{pmatrix}.$$

Since

$$\begin{pmatrix} 1 & 2 \\ 3 & 4 \end{pmatrix}^{-1} = \frac{-1}{2}\begin{pmatrix} 4 & -2 \\ -3 & 1 \end{pmatrix} = \begin{pmatrix} -2 & 1 \\ 3/2 & -1/2 \end{pmatrix},$$

we have

$$\begin{pmatrix} -2 & 1 \\ 3/2 & -1/2 \end{pmatrix}\begin{pmatrix} 17 \\ 39 \end{pmatrix} = \begin{pmatrix} 5 \\ 6 \end{pmatrix}.$$

CHECK YOUR UNDERSTANDING:

6. Compute $\begin{pmatrix} 1 & 2 \\ 3 & 4 \end{pmatrix}\begin{pmatrix} 4 & -2 \\ 1 & 5 \end{pmatrix}$.

7. Find the inverse of $\begin{pmatrix} 5 & 2 \\ 1 & 3 \end{pmatrix}$.

A.6 Problems

A.6.1 Exercises

Section A.1: Geometric Series

1. Evaluate $3 + \frac{3}{4} + \frac{3}{16} + \frac{3}{64} + \cdots + \frac{3}{4^n} + \cdots$.
2. Express $x = .91919191\cdots$ as a geometric series, and use the formula for the sum of a geometric series to write x as a rational number.
3. For what value of r does $1 + r + r^2 + r^3 + \cdots = 3$?
4. Why is there no r such that $1 + r + r^2 + r^3 + \cdots = 1/3$?

Section A.2: Mathematical Induction

5. Use induction to prove that $1 + 2 + 3 + \cdots + n = \frac{1}{2}n(n+1)$ for all $n \geq 1$.
6. Use induction to prove that $n + 2 \leq 3^n$ for all $n \geq 1$.
7. Use induction to prove that $1 + 2^2 + 3^2 + 4^2 + \cdots + n^2 = \frac{1}{6}n(n+1)(2n+1)$ for all $n \geq 1$.
8. Use induction to prove that $1 + 2^3 + 3^3 + 4^3 + \cdots + n^3 = \left(\frac{1}{2}n(n+1)\right)^2$ for all $n \geq 1$.
9. Here is a "proof" due to Pólya that all horses have the same color. We want to show that in every set of n horses, every horse has the same color. This is clearly true for $n = 1$. Assume that the statement is true for $n = k$. Now consider a set of $k + 1$ horses. Remove one horse, call it Ed, from the set. By the induction hypothesis, this set of k horses all have the same color. Now put Ed back into the set and remove another horse. Again, this set of k horses all have the same color, so Ed has the same color as the other horses. We conclude that all $k + 1$ horses have the same color. By induction, all horses have the same color. Where is the flaw in the reasoning?
10. The **Cantor Expansion** of a positive integer n is

$$n = a_m m! + a_{m-1}(m-1)! + \ldots + a_2 2! + a_1,$$

where $a_m \neq 0$ and $0 \leq a_k \leq k$ for each k. For example, the Cantor Expansion of 15 is $2 \cdot 3! + 1 \cdot 2! + 1$.
(a) Find the Cantor Expansion of 50, 73, and 533.
(b) Use strong induction and the fact that for every positive integer n there is an integer m with $m! \leq n < (m+1)!$ to show that every positive integer n has a unique Cantor Expansion.

Section A.3: Pascal's Triangle and the Binomial Theorem

11. (a) Compute $(a+b)^5$. What can you say about the divisibility of the coefficients by 5?
    (b) Compute $(a+b)^7$. What can you say about the divisibility of the coefficients by 7?
    (c) Compute $(a+b)^6$. Note that many of the coefficients are not divisible by 6.
    (d) Compute $(a+b)^{10}$. Note that many of the coefficients are not divisible by 10.

12. Show that $\binom{n}{k} = \binom{n}{n-k}$.

13. (a) Show that
$$\binom{n}{0} + \binom{n}{1} + \binom{n}{2} + \cdots + \binom{n}{n-1} + \binom{n}{n} = 2^n.$$
(*Hint:* Compute $(1+1)^n$.)
(b) Show that
$$\binom{n}{0} - \binom{n}{1} + \binom{n}{2} - \binom{n}{3} + \cdots + (-1)^n \binom{n}{n} = 0.$$
(*Hint:* Modify the hint for part (a).)

Section A.4: Fibonacci Numbers

14. Use induction to prove that $F_1 + F_2 + \cdots + F_n = F_{n+2} - 1$.

15. Prove that $F_1 + F_3 + \cdots + F_{2n-1} = F_{2n}$ for all $n \geq 1$
    (a) by induction.
    (b) by Binet's Formula plus Proposition A.1.

16. Cassini's identity says that $F_n^2 - F_{n+1}F_{n-1} = (-1)^{n-1}$ for all $n \geq 2$. Prove this
    (a) by induction.
    (b) by Binet's Formula.

Section A.6   Problems

17. (a) Prove that $\phi^2 - \phi^{-2} = \sqrt{5}$.
    (b) Use Binet's Formula to prove that $F_{2n} = F_{n+1}^2 - F_{n-1}^2$ for all $n \geq 2$.

18. The **Lucas numbers** 1, 3, 4, 7, 11, ... are defined by
    $$L_1 = 1, \quad L_2 = 3, \quad \text{and} \quad L_n = L_{n-1} + L_{n-2} \text{ for } n \geq 3.$$
    (a) Show that $L_n = \phi^n + (-\phi)^{-n}$.
    (b) Show that $\lim_{n \to \infty} L_{n+1}/L_n = \phi$.
    (c) Show that $F_{2n} = F_n L_n$.
    (d) Show that $F_n = L_{n-1} + L_{n+1}$.

19. (Fibonacci) Assume that each pair of rabbits produces a pair of rabbits every month, starting when the pair is 2 months old. Suppose you start with one pair of rabbits. After 1 month, you still have one pair. After 2 months, you have two pairs, the original pair and a pair that was just born. After 3 months, you have three pairs because a new pair is born. After 4 months, you have five pairs since the original pair and the second pair each produce a pair. How many pairs of rabbits are there after 12 months?

Section A.5: Matrices

20. Compute
$$\begin{pmatrix} 3 & 7 \\ 1 & 2 \end{pmatrix} \begin{pmatrix} 2 \\ 3 \end{pmatrix}.$$

21. Compute
$$\begin{pmatrix} 10 & 1 \\ 3 & 1 \end{pmatrix} \begin{pmatrix} 4 \\ 2 \end{pmatrix}.$$

22. Compute
$$\begin{pmatrix} 1 & 4 \\ 3 & 1 \end{pmatrix} \begin{pmatrix} 4 & 0 \\ 3 & 2 \end{pmatrix}.$$

23. Compute
$$\begin{pmatrix} -3 & 4 \\ 2 & 7 \end{pmatrix} \begin{pmatrix} 5 & 1 \\ 1 & 6 \end{pmatrix}.$$

24. Compute
$$\begin{pmatrix} 1 & 4 \\ 3 & 1 \end{pmatrix}^{-1}.$$

25. Compute
$$\begin{pmatrix} 7 & 3 \\ 3 & 1 \end{pmatrix}^{-1}.$$

A.6.2 Answers to "CHECK YOUR UNDERSTANDING"

1. This is a geometric series with $a = 1$ and $r = 2$. The formula says that the sum equals
$$\frac{2^n - 1}{2 - 1} = 2^n - 1.$$

2. The decimal can be rewritten as
$$.3 + .3(.1) + .3(.1)^2 + \ldots.$$
This is a geometric series with $a = .3$ and $r = .1$, so the sum is $a/(1-r) = .3/(1 - .1) = 1/3$.

3. The formula is true for $n = 1$, since $1 = 2^1 - 1$. Assume that it is true for $n = k$, so
$$1 + 2 + 4 + 8 + \cdots + 2^{k-1} = 2^k - 1.$$
Add 2^k to both sides of the equation:
$$1 + 2 + 4 + 8 + \cdots + 2^{k-1} + 2^k = 2^k - 1 + 2^k = 2^{k+1} - 1.$$
Therefore, the formula holds for $n = k+1$. By induction, the formula is true for all positive integers n.

4. Use the formula for binomial coefficients:
$$\binom{7}{3} = \frac{7!}{3!4!} = \frac{7 \cdot 6 \cdot 5 \cdot 4 \cdot 3 \cdot 2 \cdot 1}{3 \cdot 2 \cdot 1 \cdot 4 \cdot 3 \cdot 2 \cdot 1} = \frac{7 \cdot 6 \cdot 5}{3 \cdot 2 \cdot 1} = 35.$$

5. The coefficient is
$$\binom{8}{6} = \frac{8!}{6!2!} = \frac{8 \cdot 7}{2 \cdot 1} = 28,$$
since 6! cancels from the numerator and denominator.

6.
$$\begin{pmatrix} 1 & 2 \\ 3 & 4 \end{pmatrix} \begin{pmatrix} 4 & -2 \\ 1 & 5 \end{pmatrix} = \begin{pmatrix} 6 & 8 \\ 16 & 14 \end{pmatrix}.$$

7.
$$\begin{pmatrix} 5 & 2 \\ 1 & 3 \end{pmatrix}^{-1} = \frac{1}{13} \begin{pmatrix} 3 & -2 \\ -1 & 5 \end{pmatrix} = \begin{pmatrix} 3/13 & -2/13 \\ -1/13 & 5/13 \end{pmatrix}.$$

Appendix B

Answers and Hints for Odd-Numbered Exercises

Chapter 1

1. These use the definition of divisibility: $120 = 5 \cdot 24$, $165 = 11 \cdot 15$, $98 = 14 \cdot 7$.
3. (a) 1, 2, 4, 5, 10, 20
   (b) 1, 2, 4, 13, 26, 52
   (c) 1, 3, 5, 13, 15, 39, 65, 195
   (d) 1, 7, 29, 203
5. (a) True: If $10 \mid a$, then there exists k with $a = 10k$. Therefore, $ab = 10kb$, which says that $10 \mid ab$.
   (b) True: If $10 \mid a$ and $10 \mid b$, there exist j and k with $a = 10j$ and $b = 10k$. Therefore, $ab = 100(jk)$, which means that $100 \mid ab$.
   (c) False: Let $a = 1$, $b = 9$. Then $10 \nmid a$, $10 \nmid b$, but $10 \mid a + b$.
7. (a) True: If $c \mid a$, then there exists j with $a = cj$. Therefore, $ab = c(jb)$, which says that $c \mid ab$.
   (b) True: If $c \mid a$ and $c \mid b$, there exist j and k with $a = cj$ and $b = ck$. Therefore, $ab = c^2(jk)$, which means that $c^2 \mid ab$.
   (c) False: Let $c = 4$, $a = 1$, $b = 7$. Then $c \nmid a$, $c \nmid b$, but $c \mid a + b$.
9. Note that $n^2 - n = n(n-1)$. If this is a prime, we must have $n = \pm 1$ or $n - 1 = \pm 1$. These give $n^2 - n = 0, 2, 2, 0$. Therefore, $n = -1$ and $n = 2$ make $n^2 - n$ prime.
11. (a) $p = 5$, (b) $p = 3, 7$, (c) $p = 31$, (d) $|p - q| = 2$
13. (a) All odd primes, (b) None
15. $2 \cdot 3 \cdot 5 \cdot 7 \cdot 11 \cdot 13 + 1 = 59 \cdot 509$, $\quad 2 \cdot 3 \cdot 5 \cdot 7 \cdot 11 \cdot 13 \cdot 17 + 1 = 19 \cdot 97 \cdot 277$
17. For each prime p, there is an n with $n^2 < p < (n+1)^2$. Suppose there is a largest such n, call it N. Then all primes are less than $(N+1)^2$, which contradicts Euclid's theorem. Therefore, there are infinitely many such n.
19. (a) quotient = 1, remainder = 0, (b) quotient = 0, remainder = 9, (c) quotient = 24, remainder = 6, (d) quotient = 6, remainder = 0
21. (a) quotient = 4, remainder = 0, (b) quotient = 0, remainder = 45, (c) quotient = 1, remainder = 27, (d) quotient = 8, remainder = 10

23. $n = -3, -2, 0, 1$
25. (a) 2, (b) 1, (c) 5
27. (a) 12, (b) 3, (c) 1
29. Let $d = \gcd(2n - 1, 2n + 1)$. Then $d \mid (2n + 1) - (2n - 1)$.
31. Let $d = \gcd(b, c)$. Then $d \mid c$, so $d \mid a$. Also, $d \mid b$.
33. $(n + 1)(n! + 1) - ((n + 1)! + 1) = n$, so $\gcd(n! + 1, (n + 1)! + 1)$ divides n and $n! + 1$.
35. (a) $2 = 14 \cdot (-7) + 100$, (b) $6 = 6 \cdot 1 + 84 \cdot 0$, (c) $14 = 630 \cdot (-2) + 182 \cdot 7$, (d) $24 = 1848 \cdot 25 + 1776 \cdot (-26)$
37. Let $d = \gcd(n^2, n^2 + n + 1)$. Then $d \mid (n^2 + n + 1)_n^2$, so $d \mid n + 1$ and therefore $d \mid (n + 1)(n - 1) = n^2 - 1$.
39. (a) $13 = 2 \cdot 5 + 3, \quad 5 = 1 \cdot 3 + 2, \quad 3 = 1 \cdot 2 + 1, \quad 2 = 2 \cdot 1 + 0$
    (b) $1111111111111 = 100001000 \cdot 11111 + 111, \quad 11111 = 100 \cdot 111 + 11, \quad 111 = 10 \cdot 11 + 1, \quad 11 = 11 \cdot 1 + 0$
    (c) If $a = bq + r$ with $r > 0$, then $c = dQ + R$, where Q is q ones, each separated by $b - 1$ zeros, and the last 1 is followed by r zeros; and R is r ones.
41. (a) 194, (b) 21, (c) 133
43. Expand the final number in base 20: $a_0 + a_1 20 + a_2 20^2 + \cdots$. Then a_i is the number of people worth i billion.
45. If $n = rs$ with an odd number $r > 1$, then $a^n + 1$ has $a^s + 1$ as a factor.

Chapter 2

1. (a) $x = 2 + 4t, y = 1 - 3t$, (b) $x = 6 - 7t, y = 3 - 5t$,
   (c) $x = -5 + 23t, y = 2 - 9t$, (d) No solutions
3. (horses, oxen) $= (51, 9)$, or $(30, 40)$, or $(9, 71)$
5. Because $\gcd(a, b) = 1$, there is a solution x_0, y_0 with integers that are not necessarily positive. The general solution of $ax - by = c$ is $x = x_0 - bt, y = y_0 - at$. Let t be a large negative number.
7. Largest $= 38$, smallest $= 34$.
9. (a) (i) Use the solution for 10 and then add thirty 3-cent stamps
   (ii) Use the solution for 8 and then add eighty-four 3-cent stamps.
   (iii) Use the solution for 8 and then add ninety-eight 3-cent stamps.
   (b) Every number greater than $k + a - 1$ differs from a number on the list by a positive multiple of a.

Chapter 3

1. (a) $3^2 \cdot 5^4$, (b) 5625 is a square
3. Use Theorem 2.2 with $a = b$.
5. (a) Write $a = 2^{a_2} 3^{a_3} \cdots$ and $b = 2^{b_2} 3^{b_3} \cdots$. By Proposition 2.6, $na_p \leq nb_p$ for each p. Use Proposition 2.6 again to get $a \mid b$.
   (b) Write $a = 2^{a_2} 3^{a_3} \cdots$ and $b = 2^{b_2} 3^{b_3} \cdots$. Proposition 2.6 says that $ma_p \leq nb_p$ for each p. Use Proposition 2.6 again to get $a \mid b$.
   (c) Let $a = 4, b = 2, m = 1, n = 2$.

Appendix B 379

7. $d \mid c$, so $d \mid a+b$. Since $d \mid a$, we have $d \mid b$.
9. The answer is 3: If we have four consecutive integers, one of them is divisible by 4. An example of three consecutive squarefree integers is 1, 2, 3.
11. Let $n = 2^{n_2} 3^{n_3} \cdots$. Then $r = n_2$ and $m = 3^{n_3} \cdots$.
13. (a) p^3, (b) p

Chapter 4

1. (a) $2^3 \cdot 3^4 = 648$
   (b) $2^{15} \cdot 3^{10} \cdot 5^6$
3. We can write $n = ax + by$ with $x, y \geq 0$.
5. $n = 1, 2, 3, 6$
7. (a) If $x = a/b$ then $x^2 = a^2/b^2$.
   (b) If $\sqrt{2} + \sqrt{3}$ is rational, then $(\sqrt{2}+\sqrt{3})^2 = 5 + 2\sqrt{6}$ is rational. But $\sqrt{6}$ is irrational.
9. $x = 1/2$
11. $x = 5/3$
13. (a) Use $(\sqrt{2}+\sqrt{3})^2 = 5 + 2\sqrt{6}$ and $(\sqrt{2}+\sqrt{3})^4 = 49 + 20\sqrt{6}$.
    (b) The only possible rational roots of $x^4 - 10x^2 + 1$ are ± 1.
15. $(37, 684, 685)$
17. $(63, 16, 65), (33, 56, 65)$
19. $(24, 18, 30)$ and $(80, 18, 82)$
21. The case of odd n is given in the hint. If $n = 2^r k$, with $r \geq 2$, use the method preceding Theorem 3.5. If $n = 2k$ with k odd, construct a triple for k and double it.
23. Starting with an odd z: $n = (z+1)/2$, $m = (z-1)/2$.
    Starting with z that is a multiple of 4: $n = z/2$, $m = 1$.
25. $(x+y)(x-y) = p$ with $x > y > 0$ implies that $x + y = p$ and $x - y = 1$.
27. $(x, y) = (8, 7), (4, 1)$

Chapter 5

1. (a) 13, (b) 2, (c) 8, (d) 3, (e) 7, (f) 0
3. (a) $96 - 6 = 10 \cdot 9$, (b) $101 - (-9) = 10 \cdot 11$, (c) $-5 - 13 = 9(-2)$
5. $1, 3, 37, 111$
7. 9, because $12345 \equiv 9 \pmod{12}$
9. (a) Try $n \equiv 0 \pmod{2}$ and $n \equiv 1 \pmod{2}$. Both give $n(n+1) \equiv 0 \pmod{2}$.
   (b) Try $n \equiv 0, 1, 2, 3, 4, 5 \pmod{6}$.
   (c) $\binom{n+1}{2} = (n+1)n/2$ and $\binom{n+2}{3} = (n+2)(n+1)n/6$.
11. Try $n \equiv 0, 1, 2, 3, \ldots, 8 \pmod{9}$ and compute $n^3 + (n+1)^3 + (n+2)^3 \pmod{9}$. A more efficient way is to show that if $x \equiv 1 \pmod{3}$, then $x^3 \equiv 1 \pmod{9}$, and if $x \equiv -1 \pmod{3}$, then $x^3 \equiv -1 \pmod{9}$. Among any three consecutive integers, both of these occur and cancel each other. The remaining number is a multiple of 3; when it's cubed it's 0 mod 9.

13. 1 or 5

15. (a) True, (b) False, (c) True, (d) True, (e) True

17. Squares are 0 or 1 mod 3, so $x^2 + 3y^2$ cannot be 2 mod 3.

19. Let $q = p + 2$. (a) We must have $p \equiv 2 \pmod 3$ and $q \equiv 1 \pmod 3$.
(b) Look at the cases $p \equiv 1 \pmod 4$ and $p \equiv 3 \pmod 4$.
(c) Use Proposition 2.4.

21. Show that a cube is 0, 1, or -1 mod 9. The only ways to make a sum that's 0 mod 9 are $0 + 0 + 0$ and $0 + 1 + (-1)$ (in some order).

23. (a) Alice's strategy is to make the number of markers remaining be 0 mod $m + 1$ after she plays. If $n \equiv 0 \pmod{m+1}$, then Bob uses this strategy after Alice's first move and Bob wins.
(b) Alice's strategy is to make the number of markers remaining be 1 mod $m + 1$ after she plays.

25. Use the facts that $F_n + F_{n+1} = F_{n+2}$ and $F_{n+1} + F_{n+2} = F_{n+3}$.

27. (a) 6, (b) 1, (c) 6

29. (a) 4, (b) 1, (c) 4, (d) 1, (e) 1, (f) 3

31. $a = 6$

33. $a = 3$ and $b = 6$

35. (a) $x \equiv 8 \pmod{21}$, (b) $x \equiv 11 \pmod{19}$, (c) $x \equiv 12 \pmod{121}$

37. $n = 0, 5, 10, 15, 20, 25$

39. (a) $x \equiv 47 \pmod{100}$, (b) $x \equiv 94 \pmod{100}$

41. If $5 \nmid y$, divide by y to get $(x/y)^2 \equiv 2 \pmod 5$, which is impossible. If x, y are multiples of 5, then $x^2 - 2y^2$ is a multiple of 25.

43. $x \equiv 37 \pmod{231}$

45. 119

47. Let $p_1, \ldots, p_{100}$ be distinct primes. Solve $x \equiv -j \pmod{p_j^2}$ for $1 \le j \le 100$. Then $p_j^2 \mid x + j$ for $1 \le j \le 100$.

49. $x = -1$ solves the congruences. Add 60 to get $x = 59$.

51. Suppose p is a prime dividing $\gcd(b, n)$. Consider the case $p \mid m$ and the case $p \nmid m$, $p \mid n$.

Chapter 6

1. 1
3. 5
5. 4
7. Use $2222 \equiv 3 \pmod 7$ and $5555 \equiv 5 \pmod 6$ and Fermat's theorem to get $2222^{5555} \equiv 5 \pmod 7$. Use $5555 \equiv 4 \pmod 7$ and $2222 \equiv 2 \pmod 6$ and Fermat's theorem to get $5555^{2222} \equiv 4^2 \equiv 2 \pmod 7$.

9. If $p \mid a$, the statement is easy. Show that if $p \nmid a$, then $a^{(p-1)!} \equiv 1 \pmod p$.

11. Fermat's theorem says that $1^n + 2^n + 3^n + 4^n \pmod 5$ depends only on $n \bmod 4$. Check the cases $n = 1, 2, 3, 4$ to obtain the result.

13. 1

Appendix B 381

15. 18
17. 401
19. 2849
21. (a) 3, 4, 6 , (b) None, (c) 5, 8, 10, 12, (d) 11, 22, (e) None
23. (a) False, (b) True
25. (a) $\phi(N) = \phi(2)\phi(3)\phi(5)\cdots \geq 8$.
    (b) There is a prime p dividing N, and all primes are assumed to divide N.
    (c) Part (b) yields $\phi(N) = 1$, which is impossible by (a).
27. $(p-1)! \equiv (p-1) \cdot (p-2)! \equiv -((p-2)! \pmod{p}$.
29. Show that if $a + b \geq 2p - 1$ then $a \geq p$ or $b \geq p$.
31. (a) The right side is congruent to $(-1^2)(-2^2)(-3^2)\cdots(-((p-1)/2)^2)$ mod p.
    (b) If $p \equiv 1 \pmod{4}$, then $(p-1)/2$ is even.

Chapter 7

1. ciphertext = NCONDA; decryption function: $3x + 13$
3. plaintext = TODAY
5. plaintext = RAIN
7. encryption function: $3x + 1$, plaintext = DONE
9. A and N encrypt to B
11. (a) *DONE*, (b) *RIVER* and *ARENA*, (c) For example, *XY* decrypts to *HI* and to *NO*.
13. (a) *I NEED MONEY*, (b) *THIS ONE IS MORE DIFFICULT*
15. (a) *TO BE OR NOT TO BE THAT IS THE QUESTION*
    (b) *TO BE OR NOT TO BE THAT IS THE QUESTION*
17. *THE GOLD IS AT THE END OF THE RAINBOW*
19. $n = 47233, d = 28983$
21. (a) $c = 31, d = 29$, (b) $c = 84, d = 31$, (c) $c = 35, d = 31$, (d) $c = 70, d = 23$, (e) $c = 148, d = 37$
23. (a) $m = 6$, (b) $m = 80$, (c) $m = 63$
25. YES
27. (a) $c = 87694236463$, (b) $d = 53380876259$, (c) Check that $c^d \equiv 8051216 \pmod{n}$.
29. ATTHEMOVIE
31. TOBEORNOTTOBE
33. There exist x and y with $e_1 x + e_2 y = 1$. Then $c_1^x c_2^y \equiv m \pmod{n}$, where c_1, c_2 are the two ciphertexts.
35. (a) $r = pq - (p-1)(q-1) + 1$
    (b) The equation is $x^2 - (p+q)x + pq = (x-p)(x-q)$.
    (c) The quadratic equation is $x^2 - 978x + 218957$. The quadratic formula gives the roots 347 and 631.
37. She divides m_1 by 123 mod n. Note that $(123^e)^d \equiv 123 \pmod{n}$.

39. 01010101
41. 0, 0, 0, 1, 0, 0
43. $c_0 = 1$, $c_1 = 0$, $c_2 = 1$
45. $LADX$
47. $CASH$
49. $B_n = D(C_n) \oplus C_{n-1}$.
51. 5
53. $(2, 15)$ is the forgery, the secret is 18.

Chapter 8

1. 4
3. 2
5. 5
7. (a) Use Corollary 7.2, (b) 8
9. $\text{ord}_n(a) = 1$ means that $a^1 \equiv 1 \pmod{n}$.
11. (a) Since $3^{32} \equiv -1 \pmod{p}$, we have $3^{64} \equiv 1$. Theorem 7.1 says that $\text{ord}_p(3)$ divides 64 but does not divide 32.
    (b) Use Corollary 7.2.
13. Since $a^n \equiv 1 \pmod{a^n - 1}$ and no smaller exponent works (because a^n is too small), $\text{ord}_{a^n-1}(a) = n$. Now use Corollary 7.2.
15. Let $\ell = \text{lcm}(b, c)$ and $z = \text{ord}_{mn}(a)$. Then $a^\ell \equiv 1 \pmod{m}$ and $a^\ell \equiv 1 \pmod{n}$, so $a^\ell \equiv 1 \pmod{mn}$, so $z \mid \ell$. Since $a^z \equiv 1 \pmod{mn}$, we have $a^z \equiv 1 \pmod{m}$, so $b \mid z$. Similarly, $c \mid z$, so $\ell \mid z$.
17. $b^x \equiv b^y \pmod{n} \Leftrightarrow b^{x-y} \equiv 1 \pmod{n} \Leftrightarrow \text{ord}_n(b) \mid x - y$
19. (a) Show that $2^j \not\equiv 1 \pmod{13}$ for $1 \leq j \leq 11$. Alternatively, since $\text{ord}_{13}(2) \mid 12$, show that $2^j \not\equiv 1 \pmod{11}$ for $j = 1, 2, 3, 4, 6$.
    (b) 2, 6, 7, 11.
21. 3, 5, 6, 7, 10, 11, 12, 14
23. $(y^2)^{(p-1)/2} \equiv y^{p-1} \equiv 1 \pmod{p}$
25. If g is a primitive root, so is g^{-1}, by Exercise 12. Pair g with g^{-1}. Since $p > 3$, g does not pair with itself.
27. (a) The sum is $(p-2)(p-1)/2 = (p-1)/2 + (p-1)(p-1)/2$.
    (b) Use (a) and Proposition 7.6.
    (c) The numbers $g^0, g^1, g^2, \ldots, g^{p-2} \pmod{p}$ are a rearrangement of $1, 2, \ldots, p-1$.
29. (a) Use Proposition 7.6.
    (b) $x^2 \equiv y^2 + 2 + y^{-2} \equiv y^{-2}(y^4 + 1) + 2 \equiv 2 \pmod{p}$.
31. (a) If a is a square, then $x^2 \equiv (g^j)^2 \equiv a \equiv g^i \pmod{p}$. Conversely, if i is even, then $a \equiv (g^{i/2})^2$.
    (b) There are $(p-1)/2$ even numbers i and $(p-1)/2$ odd numbers i with $0 \leq i \leq p-2$.
33. (a) $.\overline{027}$, (b) 3

Appendix B   383

35. Here are three random examples:
$$1/47 = 0.\overline{0212765957446808510638297872340425531914893617}$$
$$1/19 = 0.\overline{052631578947368421}$$
$$1/11 = 0.\overline{09}$$

37. Use Proposition 7.11. Let $\ell = \text{lcm}(a,b)$ and $z = \text{ord}_{mn}(10)$. Then $10^\ell \equiv 1 \pmod{m}$ and $10^\ell \equiv 1 \pmod{n}$, so $10^\ell \equiv 1 \pmod{mn}$, so $z \mid \ell$. Since $10^z \equiv 1 \pmod{mn}$, we have $10^z \equiv 1 \pmod{m}$, so $a \mid z$. Similarly, $b \mid z$, so $\ell \mid z$.

39. 5

41. Raise both sides of $g^x \equiv h \pmod p$ to the power $(p-1)/2$ and obtain $(-1)^x \equiv h^{(p-1)/2}$.

Chapter 9

1. The key is 2.
3. The key is 18.
5. If Alice chooses $p-1$ as the secret exponent, then she sends $1 \equiv g^{p-1} \pmod p$ to Bob. When Bob raises this to a power, he still gets 1. Moreover, Eve sees Alice send 1 to Bob, so Eve knows that g^{ab} will still be 1.

   If Alice uses 1 as her secret exponent, then she sends g to Bob. Eve sees this and deduces that the exponent is 1. Therefore, when Bob sends $g^b \pmod p$ to Alice, Eve knows that this is g^{ab} and therefore obtains the key.

7. (a) $a=1$ and Alice computes that $g_2^n \equiv a \pmod p$.
   (b) $(g_i^n)^{(p-1)/d} = (g_i^{p-1})^{n/d} \equiv 1 \pmod p$ (note that n/d is an integer, so $1^{n/d} \pmod p$ is defined).
   (c) Let $a \equiv g^z \pmod p$. Then $a^{(p-1)/d} \equiv 1 \pmod p \Leftrightarrow z(p-1)/d \equiv 0 \pmod{p-1} \Leftrightarrow d \mid z$.
   (d) Bob calculates $g_2^{di} \pmod p$ for $0 \le i < (p-1)/d$ until he obtains a.

9. $(8, 19)$ is the forgery.
11. (a) $n = 33$, (b) $d = 3$, (c) $s = 29$
13. (a) $n = 51$, (b) $d = 13$, (c) $s = 5$
15. $m = 72$
17. $m_0 = 5$, $d = 23$, and $5^{23} \equiv 47 \pmod{57}$.

Chapter 10

1. $x \equiv 5$ or $6 \pmod{11}$
3. -1
5. Yes, it has solutions.
7. (a) No solutions because 2 is not a square mod 11.
   (b) Yes. There are solutions mod 11 and solutions mod 19. They can be combined via the Chinese Remainder Theorem.

9. (a) Note that $2^e \equiv 1 \pmod 3$ for every even integer e.
   (b) Since $p \equiv 1 \pmod 4$, $\left(\frac{3}{p}\right) = \left(\frac{p}{3}\right)$.
   (c) The order divides $p - 1 = 2^{2^m}$, so it must be a power of 2.
   (d) Use Euler's Criterion to show that
   $$-1 \equiv 3^{2^m - 1} \pmod p,$$
   so $\mathrm{ord}_p(3) \nmid 2^{2^m - 1}$.
   (e) Change 3 to a in (d).

11. Because $5 \equiv 1 \pmod 4$, we have
   $$\left(\frac{5}{p}\right) = \left(\frac{p}{5}\right) \quad \text{and} \quad \left(\frac{p}{5}\right) = \left(\frac{5}{q}\right).$$
   Now use $p \equiv q \pmod 5$.

13. Let g be a primitive root mod p. Write $b \equiv g^i \pmod p$ for some i. Show that
   $$b^{(p-1)/3} \equiv 1 \pmod p \Leftrightarrow i \equiv 0 \pmod 3,$$
   and $i \equiv 0 \pmod 3 \Leftrightarrow b$ is a cube mod p. (This proof needs $p \equiv 1 \pmod 3$ so that the exponent $(p-1)/3$ is an integer.)

15. $p \equiv 1, 3, 7, 9 \pmod{20}$.

17. (a) First suppose that $n \geq 4$. Then $q \equiv 1 \pmod 4$. Since $n \geq p$, we have $q \equiv 1 \pmod p$. Quadratic Reciprocity yields
   $$\left(\frac{p}{q}\right) = 1.$$
   If $n < 4$, then we must have $p = 3$ and $n = 3$.
   (b) The values of n are 11, 27, 37, 41, 73, 77, 116, 154. The values of q are $n! + 1$. For p, take any odd prime $p \leq n$. For example, we could always use $p = 3$ or $p = 11$.

19. (a) Note that $2^p \equiv 1 \pmod q$.
   (b) Use the Supplementary Law for 2, plus part (a).

21. Let $p \mid N$. Then 2 is a square mod p. Now use the Supplementary Law for 2 to get $p \equiv 7 \pmod 8$.

23. (a) If $n \equiv 0$ or $1 \pmod 2$, then $n^2 + n + 1 \equiv 1 \pmod 2$, so p must be odd.
   (b) Rewrite $4(n^2 + n + 1)$.
   (c) Part (b) says that -3 is a square mod p.

25. $y \equiv 5, 18 \pmod{23}$.

27. $y \equiv 6, 23 \pmod{29}$.

29. No.

31. $x \equiv 12, 18 \pmod{23}$.

33. -1

35. $+1$

37. (a) -1, (b) Squares make the Jacobi symbol equal $+1$, (c) Compute $\left(\frac{35}{11}\right) = -1$.

Appendix B 385

39. (a) +1, (b) No, it does not imply that 3 is a square mod 35, (c) Compute $\left(\frac{3}{5}\right) = -1$.
41. For $\left(\frac{3}{7}\right)$, the set of residues is $\{3, 6, 2\}$. One of them, namely 6, is above $p/2$, so $n = 1$ and $\left(\frac{3}{7}\right) = (-1)^1 = -1$.
    For $\left(\frac{3}{13}\right)$, the set of residues is $\{3, 6, 9, 12, 2, 5\}$. Two of them, namely 9 and 12, are above $p/2$, so $n = 2$ and $\left(\frac{3}{13}\right) = (-1)^2 = +1$.
43. Show that $n = (p-1)/2$.

Chapter 11

1. $2880 = 2^6 \cdot 3^2 \cdot 5$
3. $1152 = 2^7 \cdot 3^2$
5. $391 = 17 \cdot 23$
7. (a) $621 = 27 \cdot 23 = 3^3 \cdot 23$, (b) $621 = 3^3 \cdot 23$
9. Compute $2^{14} \equiv 4 \pmod{15}$
11. (a) $2^{11} - 1 = 2047$, (b) 1023 is a multiple of 11, (c) $2047 = 23 \cdot 89$ so 2047 is not prime. Now use the definition of strong pseudoprime.
13. Use Fermat's theorem for $p = 7, 11, 13, 41$ to show if $\gcd(a, 41041) = 1$, then $a^{41040} \equiv 1 \pmod{p}$ for each of these p.
15. Write $ed = 1 + k(p-1)(r-1)$. Use Fermat for p and the Carmichael property of r to show that $m^{ed} \equiv m \pmod{p}$ and $\pmod{r}$.
17. $\gcd(427 - 333, 893) = 47$ and $893 = 47 \cdot 19$
19. $(937 \cdot 1666)^2 \equiv (2 \cdot 3 \cdot 7)^2 \pmod{n}$ and $\gcd(937 \cdot 1666 - 2 \cdot 3 \cdot 7, 28321) = 223$. Therefore, $28321 = 127 \cdot 223$.
21. (a) Show that $4^{14} \equiv 1 \pmod{15}$, (b) Show that $4^7 \not\equiv \pm 1 \pmod{15}$, (c) Compute $\gcd(4^7 - 1, 15) = 3$ and obtain $15 = 3 \cdot 5$

Chapter 12

1. $41 = 5^2 + 4^2$
3. (a) Yes, (b) No, (c) No
5. $221 = 5^2 + 14^2$ or $221 = 10^2 + 11^2$
7. $1237 = 34^2 + 9^2$
9. $123 = 7^2 + 7^2 + 5^2 + 0^2 = 11^2 + 1^2 + 1^2 + 0^2 = 8^2 + 7^2 + 3^2 + 1^2$.
11. (a) Just multiply it out, (b) Again, multiply it out, (c) Take norms of everything in the formula in part (a).
13. Compute $x^3 \pmod 9$ for $0 \le x \le 8$. A sum of three cubes cannot be 4 or 5 mod 9.
15. 79

Chapter 13

1. (a) for 2: 1, 2;
   for 4: 1, 2, 4;
   for 15: 1, 3, 5, 15;

for 30: 1, 2, 3, 5, 6, 10, 15, 30;
for 60: 1, 2, 3, 4, 5, 6, 10, 12, 15, 20, 30, 60.
(b) $\sigma(2) = 3, \sigma(4) = 7, \sigma(15) = 24, \sigma(30) = 72, \sigma(60) = 168$
(c) $\sigma(4)\sigma(15) = 7 \cdot 24 = 168 = \sigma(60)$ but $\sigma(2)\sigma(30) = 3 \cdot 72 \neq 168 = \sigma(60)$.

3. $\sigma(20) = 42$
5. The case $n = 1$ is easy, so assume that $n \geq 2$. If $\phi(n) = n - 1$ then every positive integer less than n is relatively prime to n.
7. $\sigma(2^p - 1) = 1 + (2^p - 1) = 2^p$, so $\sigma(\sigma(2^p - 1)) = 1 + 2 + \cdots + 2^p = 2^{p+1} - 1$.
9. $\phi(5) = \phi(8) = 4$, $\phi(7) = \phi(9) = 6$, $\phi(5) = \phi(10) = 4$, $\phi(13) = \phi(21) = 12$.
11. Let $n = p^a q^b r^c \cdots$ be the prime factorization of n. If $\phi(n) = 14$, then there can be only one odd prime power factor (since each supplies $p - 1$, which is even). Therefore, $n = 2^a p^b$. We cannot have $2^a > 4$ nor $(p - 1)p^{b-1} > 14$. If $2^a > 2$, then there is no odd prime factor, so $n = 2^a$, which cannot yield $\phi(n) = 14$. Therefore, $n = 2^a p^b$ with $a = 0, 1$ and $(p - 1)p^{b-1} \leq 14$. Trying all cases yields no example with $\phi(n) = 14$.
13. Take the relation $2n = \sum_{d|n} d$ and divide by n. Note that if $d \mid n$, then n/d is a divisor of n.
15. The squares are open, the rest are closed.
17. $\sigma(pq) = (p+1)(q+1)$ and $\phi(pq) = (p-1)(q-1)$. Add them together.
19. (a) $\sigma(p^a) = 1 + p + p^2 + \cdots + p^a$ and each summand is odd.
(b) $\sigma(2^a) = 1 + 2 + 2^2 + \cdots + 2^a$, and each term except the first is even.
(c) Use (a) and (b), plus the multiplicativity of σ.
(d) If n is perfect, then $\sigma(n) = 2n$ is even.
21. (a) μ is multiplicative, so μ^2 is multiplicative.
(b) If $\gcd(m, n) = 1$, then the prime factors of m and n are distinct, so $\omega(mn) = \omega(m) + \omega(n)$.
(c) Prove the equation for prime powers (it becomes $1 = 1$) and then use multiplicativity.
23. There are $\binom{k}{k-j} = \binom{k}{j}$ divisors that have exactly j prime factors. Such d have $\mu(d) = (-1)^j$.
25. Because σ is multiplicative, if $\sigma(m)$ is prime then m is a prime power. Use the hint plus $\tau(p^a) = a + 1$.

Chapter 14

1. $p_0/q_0 = 1/1$, $p_1/q_1 = 5/4$, $p_2/q_2 = 16/13$, $p_3/q_3 = 21/17$, and $q_4 = 64$.
3. $a/b = 4/17$
5. $a/b = 6/11$
7. $43/10$
9. (a) $[2, \overline{1, 1, 1, 4}]$, (b) $(x, y) = (8, 3)$
11. (a) $[3, \overline{1, 6}]$, (b) $(x, y) = (4, 1)$
13. (a) $[n - 1, \overline{1, 2n - 2}]$, (b) $(x, y) = (n, 1)$
15. Obtain the equation $x^2 - 2y^2 = -1$ with $x = 2a + 1$ and $y = c$. The odd powers of $1 + \sqrt{2}$ yield $x_n + y_n\sqrt{2}$ that are solutions of $x_n^2 - 2y_n^2 = -1$. Note that x_n is always odd.

Appendix B 387

17. (a) 2, 3/2, 5/3
    (b) We have $p_k = p_{k-1} + p_{k-2}$ and $p_{-1} = 1 = F_1, p_0 = 1 = F_2$. So the p_k's satisfy the same recurrence relation as the Fibonacci numbers, and the two sequences start with indices shifted by 2. This implies that $F_k = F_{k-2}$ for all k. The proof for the q_k's is similar.

Appendix A

1. 4
3. $r = 2/3$
5. If it's true for $n = k$, then $1 + 2 + \cdots + k + (k+1) = \frac{1}{2}k(k+1) + (k+1) = \frac{1}{2}(k+1)(k+2)$.
7. If it's true for $n = k$, then $1^2 + 2^2 + \cdots + k^2 + (k+1)^2 = \frac{1}{6}k(k+1)(2k+1) + (k+1)^2 = \frac{1}{6}(k+1)(k+2)(2k+3)$.
9. The "proof" fails for the passage from $n = 1$ to $n = 2$.
11. (a) $(a+b)^5 = a^5 + 5a^4b + 10a^3b^2 + 10a^2b^3 + 5ab^4 + b^5$. All of the middle coefficients are multiples of 5.
    (b) $(a+b)^7 = a^7 + 7a^6b + 21a^5b^2 + 35a^4b^3 + 35a^3b^4 + 21a^2b^5 + 7ab^6 + b^7$. All of the middle coefficients are multiples of 7.
    (c) $(a+b)^6 = a^6 + 6a^5b + 15a^4b^2 + 20a^3b^3 + 15a^2b^4 + 6ab^5 + b^6$.
    (d) Coefficients are $1, 10, 45, 120, 210, 252, 210, 120, 45, 10, 1$
13. (a) $(1+1)^n = \sum_{j=0}^{n} \binom{n}{j} 1^j 1^{n-j}$.
    (b) $(1-1)^n = \sum_{j=0}^{n} \binom{n}{j} 1^j (-1)^{n-j}$.
15. (a) Use the identity $F_{2n} + F_{2n+1} = F_{2(n+1)}$.
    (b) $F_1 + F_3 + \cdots + F_{2n}$ can be written as
    $$(1/\sqrt{5})\left(\phi + \phi^3 + \cdots + \phi^{2n-1}\right)$$
    $$- (1/\sqrt{5})\left((-\phi)^{-1} + \cdots + (-\phi)^{-(2n-1)}\right).$$
    Sum the two geometric series, use the fact that $\phi^2 - 1 = \phi$, and simplify the result to get Binet's Formula for F_{2n}.
17. (a) $\phi^2 = (3+\sqrt{5})/2$ and $\phi^{-2} = (3-\sqrt{5})/2$, (b) Use Binet's Formula to rewrite $F_{n+1}^2 - F_{n-1}^2$ and then use (a) to simplify the resulting expression.
19. 233 pairs
21. $\binom{42}{14}$.
23. $\begin{pmatrix} -11 & 21 \\ 17 & 44 \end{pmatrix}$.
25. $\begin{pmatrix} -1/2 & 3/2 \\ 3/2 & -7/2 \end{pmatrix}$.

Index

≡, 95
$\lfloor x \rfloor$, 73
$\mu(n)$, 316, 319
ϕ-function, 138, 315, 318, 321
$\sigma(n)$, 311, 315, 318
$\sqrt{2}$, 79
$\tau(n)$, 315, 317, 320

abundant number, 311
Adleman, 7, 155, 170
AES, 154, 184
affine cipher, 157
al-Haitham, 128, 145
Al-Khodjandi, 3
Alford, 281
Alice, 153
amicable, 324
Apéry, 347
arithmetic progression, 93
Artin, 5, 208
asymmetric cryptosystem, 154

Babbage, 162
Bach, 284
Bachet, 342
bases, 37
Baudhāyana, 2
Bellaso, 162
Bernoulli number, 347
Bhāskara, 128, 145
billionaires, 49
Binet, 365
Binet's Formula, 364

binomial coefficient, 360
Binomial Theorem, 362
block cipher, 181
Bob, 153
Brahmagupta, 2
Brillhart, 288, 301
Brun, 341

Caesar, 157
Caesar cipher, 157
Cantor Expansion, 371
card shuffling, 214
cards, 233
Carmichael number, 281
Cassini's identity, 372
Catalan, 306
CBC, 194
Chebyshev, 5, 50
Chen, 6, 341
Chinese Remainder Theorem, 115, 129
cipher block chaining, 194
ciphertext, 153
Clay Mathematics Institute, 348
Cocks, 154
coin flipping, 231, 291
Collatz Conjecture, 50
compass and straightedge, 43
composite number, 19, 51
conclusion, 10
congruent, 95
continued fractions, 327

corollary, 14

de la Vallée-Poussin, 6, 348
de Moivre, 365
de Polignac, 342
decimal, 209, 211, 352
Dedekind, 4
deficient number, 311
DES, 154, 184
Diffie, 7, 155, 229
Diffie–Hellman Key Exchange, 229, 230
Diffie–Hellman Problem, 230, 243
digital signatures, 238
Diophantine equation, 55
Diophantus, 2, 55, 88, 342
Dirichlet, 3, 6, 344
discrete log, 216
discriminant, 344
Disquisitiones Arithmeticae, 97
divides, 17
divisibility, 104
Division Algorithm, 25, 67
divisor, 17

elliptic curve, 344
Ellis, 154
Eratosthenes, 23
Euclid, 1, 20, 21, 30, 40, 70, 74, 88, 312
Euclid's theorem, 19
Euclidean Algorithm, 30
Euler, 2, 4, 64, 201, 248, 297, 312, 341, 344, 345
Euler ϕ-function, 138, 315, 318, 321
Euler's criterion, 246
Euler's theorem, 139, 173
even, 10
eventually periodic, 211

exponentiation, 102
Extended Euclidean Algorithm, 32, 58, 67, 172

factorization, 275, 285, 286, 289
Fermat, 3, 4, 42, 125, 201, 297, 342
Fermat Factorization, 277, 294
Fermat number, 42, 135, 201
Fermat Primality Test, 279
Fermat's Last Theorem, 3, 150, 342, 343
Fermat's theorem, 134
Fibonacci, 128, 373
Fibonacci number, 48, 126, 364
Frey, 3, 344, 345
Fundamental Theorem of Arithmetic, 69

game, 126, 357
Gauss, 4, 43, 97, 208, 248, 306
Gauss's Lemma, 262
gcd, 27, 72
geometric series, 351
Germain, 3
Girard, 297
Goldbach, 6, 341
Goldbach's Conjecture, 6, 45, 341
Golden Ratio, 365
Granville, 281
Great Internet Mersenne Prime Search, 42, 285
greatest common divisor, 27
Green, 93
Gupta, 209

Hadamard, 6, 348
Hardy, 7
hash function, 243
Heath-Brown, 209

Index   391

Helfgott, 6, 341
Hellman, 7, 155, 229
Hermes, 43
Hermite, 301
Hilbert, 4, 306, 348
Hill, 182
Hill cipher, 182
Hippasus, 1
Hooley, 209
horses, 371
hotline, 178
Huygens, 327
hypothesis, 10

induction, 353
infinitude of primes, 20, 21, 149
inverse, 112, 182, 369
irrational, 15, 79

Jacobi, 305
Jacobi symbol, 256

Kasiski, 162
key, 153
key exchange, 229
kilometers, 367
Kummer, 3, 4, 344

Lagarias, 348
Lagrange, 2, 145, 218, 301
Lamé, 3
Langlands, 3
lcm, 76
least common multiple, 76
Legendre, 3, 248, 344
Legendre symbol, 246
Lehmer, 288
lemma, 14
Lenstra, H., 323
LFSR, 178

limerick, 77
linear congruences, 108
linear feedback shift registers, 178
Littlewood, 50
Lucas, 41, 202
Lucas number, 373
Lucas-Lehmer Test, 285

Möbius function, 316, 319
Möbius inversion, 319
matrix, 182, 367
Maynard, 342
Mazur, 3
Mersenne, 40, 297
Mersenne number, 40, 203, 284
Midy's theorem, 212
Mihăilescu, 307
miles, 367
Millennium Prize problems, 348
Miller, 284
Miller-Rabin Primality Test, 284
Mirimanoff prime, 150
mod, 95
modular exponentiation, 102
modularity, 345
modulus, 95
Morrison, 288
multiple, 17
multiplicative function, 312, 315
multiplicative inverse, 112
Murty, 209

Newton–Raphson Method, 82

odd, 12
one-time pad, 177
order, 199

Pólya, 371
Pascal, 359

Pascal's Triangle, 359
Pell, 2
Pell's equation, 334
perfect number, 311
period, 210
pi, 328
plaintext, 153
Plimpton, 1
poker, 233, 243
Pomerance, 281, 284
Postage Stamp Problem, 60
Powers, 288
primality test, 285
primality testing, 279
prime number, 19, 68
prime number race, 50
Prime Number Theorem, 6, 279
primitive root, 203
private key, 154
proof by contradiction, 15
proposition, 14
pseudoprime, 279, 289
public key, 154
public key cryptography, 154
public key cryptosystem, 170
Pythagorean triple, 83, 125
Pythagoreans, 1

quadratic equations, 255
quadratic formula, 255
quadratic nonresidue, 245
Quadratic Reciprocity, 247
quadratic residue, 245
queens, 121

rabbits, 373
Rabin, 284
rational, 15, 79
Rational Root Theorem, 81
Regius, 40

relatively prime, 28, 51
Ribet, 3, 345
Riemann, 6, 348
Riemann Hypothesis, 345, 348
Riemann zeta function, 345
Rivest, 7, 155, 170
Rivoal, 348
Robin, 348
RSA, 7, 155, 169, 290
RSA signatures, 239
Rubinstein, 50

Sarnak, 50
Scientific American, 166
secret sharing, 184
Selfridge, 284
Serre, 3
Serret, 301
Shamir, 7, 155, 170
shift cipher, 156, 188
Shimura, 3
short message attack, 175
sieve, 341
sieve of Eratosthenes, 23
Smith, 301
square roots, 253
squarefree, 73, 129, 130
squaring, 103
stream cipher, 176
Strong Fermat Primality Test, 282
strong induction, 356
strong pseudoprime, 283
Suetonius, 157
sum of four squares, 301
sum of two squares, 297
Sun Tzu, 4, 115
supplementary laws, 248
symmetric cryptosystem, 154

Index 393

Takagi, 5
Taniyama, 3
Tao, 93, 342
Taylor, 3, 345
theorem, 14
Thue, 297
Thue's lemma, 298, 302
transposition cipher, 166
Tunnell, 3
twin prime problem, 341
twin primes, 125

vector, 368
Vigenère, 162
Vigenère cipher, 160
Vinogradov, 6, 341
voting, 26

Wagstaff, 284
Wantzel, 43
Waring, 145, 306
Weber, 5
Weil, 3
Well-Ordering Principle, 36, 356
Wieferich prime, 150
Wiles, 3, 14, 345
Wilson, 145
Wilson's theorem, 145, 225

Zagier, 5
zeta function, 345
Zhang, 342
Zsigmondy's theorem, 151
Zudilin, 348